RELATIVISTIC QUANTUM METAPHYSICS

A FIRST PRINCIPLES BASIS FOR THE STANDARD MODEL OF ELEMENTARY PARTICLES

STEPHEN BLAHA, PH.D.

Pingree-Hill Publishing

Pingree-Hill Publishing
P. O. Box 368
Auburn, NH 03032

ISBN: 978-0-9819049-7-9

This book is printed on acid free paper.

Cover Credits
Cover by Stephen Blaha © 2009.

rev. 00/00/03 February 8, 2010

To John, Natasha & Stephen
Who Give Meaning to My Life

Logic is the basis of Ultimate Reality

Some Other Books by Stephen Blaha

The Algebra of Thought & Reality: The Mathematical Basis for Plato's Theory of Ideas, and Reality Extended to Include A Priori Observers and Space-Time; Second Edition (ISBN: 9780981904931, Pingree-Hill Publishing, Auburn, NH, 2009)

To Far Stars and Galaxies: Second Edition of Bright Stars, Bright Universe (ISBN: 9780981904962, Pingree-Hill Publishing, Auburn, NH, 2009)

The Metatheory of Physics Theories, and the Theory of Everything as a Quantum Computer Language (ISBN: 097469584X, Pingree-Hill Publishing, Auburn, NH, 2005)

A Complete Derivation of the Form of the Standard Model With a New Method to Generate Particle Masses SECOND EDITION (ISBN: 9780981904900, Pingree-Hill Publishing, Auburn, NH, 2008)

The Origin of the Standard Model: The Genesis of Four Quark and Lepton Species, Parity Violation, the ElectroWeak Sector, Color SU(3), Three Visible Generations of Fermions, and One Generation of Dark Matter with Dark Energy (ISBN: 0974695882, Pingree-Hill Publishing, Auburn, NH, 2007)

Physics Beyond the Light Barrier: The Source of Parity Violation, Tachyons, and A Derivation of Standard Model Features (ISBN: 0974695874, Pingree-Hill Publishing, Auburn, NH, 2007)

A Derivation of ElectroWeak Theory based on an Extension of Special Relativity; Black Hole Tachyons; & Tachyons of Any Spin (ISBN: 0974695866, Pingree-Hill Publishing, Auburn, NH, 2006)

Quantum Theory of the Third Kind: A New Type of Divergence-free Quantum Field Theory Supporting a Unified Standard Model of Elementary Particles and Quantum Gravity based on a New Method in the Calculus of Variations (ISBN: 0974695831, Pingree-Hill Publishing, Auburn, NH, 2005)

Quantum Big Bang Cosmology: Complex Space-time General Relativity, Quantum Coordinates™ Dodecahedral Universe, Inflation, and New Spin 0, ½, 1 & 2 Tachyons & Imagyons™ (ISBN: 0974695815, Pingree-Hill Publishing, Auburn, NH, 2004)

A Unified Quantitative Theory Of Civilizations and Societies: 9600 BC - 2100 AD (ISBN: 0974685858, Pingree-Hill Publishing, Auburn, NH, 2006)

Available on bn.com, Amazon.com, and other web sites as well as at better bookstores (through Ingram Distributors).

PREFACE

A reading of a goodly number of the many books on the various branches of Philosophy and Metaphysics leads this author, a physicist by way of education and an investigator in many areas of science, economics and history, to conclude that Philosophy and Metaphysics have yet to come to terms with modern scientific research in physics, mathematics, logic, biology, computer theory, and cognition. Part of the problem – particularly as regards studies of Reality – is a lack of understanding, or incomplete understanding, of what has become the language of these areas of endeavor – mathematics. Mathematics is a requirement to appreciate what we have learned of Reality not only in physics but increasingly in biological areas related to understanding the human Mind and its relation to Reality. We see this, for example, in computer networks developed to simulate neural processing, learning and other aspects of processes taking place in the human mind.

Failing to understand mathematics and not completely realizing the implications of modern science particularly quantum theory and the theory of relativity, philosophers and metaphysicians revert to verbal descriptions that answer some questions but do not truly give a correct view of Reality as it is known today from scientific investigations. These efforts make sense for everyday phenomena but at the quantum level and at high velocities these efforts fail to come to grips with Reality. *And so the claims of universality of existing philosophy and metaphysics are simply not true.* The discussions of Reality are simplistic (although sometimes subtle) in view of our current knowledge of the universe, use terminology that is not well-defined and often vague, and often engage in verbal gymnastics that do not grasp the essence of aspects of Reality. Despite these shortcomings there are genuine grains of wisdom in philosophical and metaphysical studies.

Our plan in this book is to examine the true core of philosophy and metaphysics, taking account of quantum and relativity theory as it applies to physical Reality, and to develop a line of reasoning that ultimately leads us to Reality as it is currently understood at the most fundamental level – the Standard Model of Elementary Particles and General Relativity. *We will derive the form of The Standard Model in detail showing the origin of parity violation, the Strong interaction and the origin of its peculiar symmetry.*

This book develops new formalisms for Logic that are of interest in themselves and also provide the Platonic bridge to Reality. The bridge to Reality will then be explored in detail. We anticipate that the current "fundamental" level of physical Reality may be based on a still lower level and/or may have additional aspects remaining to be found. However the effects of certain core features such as quantum theory and relativity theory will persist even if a lower level of Reality is found, and these core features suggest the form of a new natural Philosophy and a new Metaphysics of physical Reality. We have coined the phrase "Relativistic Quantum Metaphysics" for this new metaphysics of physical Reality. One might also describe our generalization of the encompassing Philosophy as Relativistic Quantum Philosophy.

It will be objected that vast areas of Philosophy and Metaphysics exist outside of the domain of physical Reality. To that objection we answer that most of those areas – the materialistic areas – emerge ultimately from physical Reality. As science evolves it is becoming increasingly clear that all events and phenomena in our universe ultimately follow from fundamental physical Reality. Some phenomena, such as the processing of abstract concepts in the human mind, are very distantly related to fundamental physical Reality, and it may be difficult (and perhaps presently impossible) to derive a complex phenomena from fundamental physical Reality. Nevertheless the connection exists despite our inability to derive it. So we regard all things (except the spiritual) as emanating from fundamental physical Reality. Our present inability to derive connections is a matter of happenstance; but the principle of a unitary Reality based on fundamental physical Reality has the force of historical trends behind it as the developments in the twentieth century clearly show.

This book starts by describing aspects of Philosophy and Metaphysics relevant to the study of current physical Reality introducing a new natural Philosophy and a new Metaphysics of physical Reality based on quantum theory: Quantum Metaphysics. Part of this development are new Logics, Operator Logic, Probabilistic Operator Logic, and Quantum Operator Logic, developed in earlier books by this author. Using them we are led to develop a connection to the beginnings of The Standard Model of Elementary Particles. Then we derive the form of The Standard Model in detail based on the new Operator Logics. This derivation is an expanded version of Blaha (2008) and (2009). Metaphysical implications of The Standard Model are then considered – Relativistic Quantum Metaphysics and a Generalized Metaphysics for all possible types of universes. (The Operator Logics described here are revisions and expansions of the Operator Logic formalisms described in earlier books by this author. They are the same as Blaha (2010a) and the first part of this book through chapter 13 is Blaha (2010a) reproduced for the sake of completeness.) While mathematics is essential in the latter stages of the book we have tried to present it with sufficient text discussion to make what it is doing understandable to the non-mathematical reader. Generally we will avoid using the jargon of Philosophy, Logic and Physics as much as possible. But the second part of this book is necessarily mathematical. In it we describe the basis of the form of The Standard Model in a more fundamental set of principles.

CONTENTS

1. EPISTEMOLOGY 1

1.1 EPISTEMOLOGY 1

1.2 ORIGIN OF HUMAN KNOWLEDGE OF PHYSICAL REALITY 3

1.3 SCOPE OF HUMAN KNOWLEDGE OF PHYSICAL REALITY 3

1.4 NATURE OF HUMAN KNOWLEDGE OF PHYSICAL REALITY 3

1.5 LIMITATIONS ON HUMAN KNOWLEDGE OF PHYSICAL REALITY 4

1.6 MIND-BODY "PROBLEM", AND REALITY 5

2. WHAT IS METAPHYSICS? 8

2.1 DEFINITIONS 8

2.2 A SUMMARY OF CLASSICAL ("EVERYDAY") METAPHYSICS 10

2.3 ARE THERE ANY TRUTHS IN CLASSICAL METAPHYSICS? 11

3. RELATIVISTIC QUANTUM METAPHYSICS 14

3.1 REDUCTION OF REALITY TO THE PRESENTLY KNOWN ULTIMATE REALITY 14

3.2 A METAPHYSICAL VIEW OF ELEMENTARY PARTICLES 15

3.2.1 Particles and Substances 15

3.2.2 Forms and Properties 17

3.2.3 Change, Interactions and Causation 18

3.2.4 Modality 21

3.2.5 From Generalized Metaphysics to The Standard Model 22

4. LOGIC AND PHYSICS 23

4.1 LOGIC IS THE KEY TO METAPHYSICS 23

4.2 A LOGIC VIEW OF PHYSICAL EXPERIMENTS 24

4.3 FILTRATION STAGES IN A QUANTUM EXPERIMENT AND THEIR MAPPING TO THE FORM OF A PROPOSITION 24

5. OPERATOR LOGIC 27

5.1 LOGIC AND ITS PARADOXES 27

5.2 PREDICATE CALCULUS 28
5.3 DEDUCTIVE SYSTEMS AND THEIR CALCULI 29
5.4 UNIVERSES OF DISCOURSE 30
5.5 OPERATOR REPRESENTATION OF A STATEMENT 32
5.5.1 Dirac Notation vs. (x, y) Notation for Inner Products 33
5.5.2 Gödel Numbers 34
5.5.3 Statements Expressed in Gödel Numbers 36

6. LINEAR VECTOR SPACES FOR UNIVERSES OF DISCOURSE 37

6.1 THE SEMANTIC UNIVERSE OF DISCOURSE 37
6.2 THE CALCULUS UNIVERSE OF DISCOURSE 41
6.3 VARIABLES IN OPERATOR LOGIC 41
6.4 LINKING UNIVERSES OF DISCOURSES LINEAR VECTOR SPACES 41
6.5 PREDICATE CALCULUS 43
6.6 GENERAL APPROACH TO SUBJECTS AND DOMAINS OF PREDICATES 44
6.7 FORMALISM FOR OPERATOR LOGIC 45
6.8 A SIMPLIFIED EXAMPLE 46
6.9 THE PRIMITIVE TERMS OF UNIVERSES OF DISCOURSE 48
6.10 PREDICATE DOMAINS AND PARADOX AVOIDANCE 49
Liar Paradox 50
Grelling Paradox 50
Barber Pseudoparadox 50
Berry Paradox 51
Russell's Paradox 51
Cantor Paradox 52
Burali-Forti Paradox 52
Richard Paradox 53
Gödel's Undecidability Theorem 54
Conclusion 56

7. SENTENTIAL CALCULUS IN OPERATOR LOGIC 57

7.1 TRUTH VALUES OF SIMPLE SENTENCES AND CLAUSES 58
7.2 TRUTH VALUE OF COMPOUND SENTENCES AND CLAUSES 61
7.3 TRUTH VALUE OF THE NORMAL FORM 63

8. WHY OPERATOR LOGIC? 64

8.1 OPERATOR LOGIC IS THE CORRECT FORMALISM FOR LOGIC 64
8.2 ADVANTAGES OF OPERATOR LOGIC...64
8.3 OPERATOR LOGIC RAISING AND LOWERING OPERATORS......................................65
8.3.1 Eigenvalue Operators that are not Functions of Variables 65
8.3.2 Eigenvalue Operators that are Functions of Variables 68
8.3.3 Effect of Raising and Lowering Operators on Universes of Discourse........69
8.3.4 Multiple Lowering Operator Case 71

9. HIERARCHIES OF UNIVERSES OF DISCOURSE 73

9.1 HIERARCHIES OF SUBUNIVERSES..73
9.2 SUBUNIVERSES OF A SIMPLE ARITHMETICAL UNIVERSE OF DISCOURSE 74
9.2.1 Additive Universe of Discourse..75
9.2.2 Multiplicative Universe of Discourse...75
9.3 SUBUNIVERSES OF EUCLIDEAN GEOMETRY 75
9.4 A SUBUNIVERSE OF ELECTROMAGNETISM...76
9.5 PRODUCTS OF UNIVERSES 78

10. MATRIX FORM OF OPERATOR LOGIC 80

10.1 SPINOR REPRESENTATION OF OPERATOR LOGIC 80
10.2 A SYMMETRY OF THE SPINOR MATRIX REPRESENTATION 82
10.3 DIRECT SUM OF SPINOR REPRESENTATIONS OF UNIVERSES – DIRAC MATRICES 83
10.3.1 Independent Universes...84
10.3.2 Mixing Between the Universes..86
10.4 VECTOR OPERATOR LOGIC 88
10.4.1 Vector Statements...90
10.4.2 S_3 Symmetry of Vector Universe of Discourse 91
10.4.3 Raising and Lowering Operators to Define Vector Subuniverses..............91

11. PROBABILISTIC AND QUANTUM OPERATOR LOGICS 94

11.1 QUANTUM OPERATOR LOGIC IS NOT FUZZY LOGIC 95
11.2 PROPOSITIONS ARE ANALOGOUS TO EXPERIMENTS...95
11.3 OVERVIEW OF THE RELATION BETWEEN OPERATOR LOGIC, PROBABILISTIC OPERATOR LOGIC & QUANTUM OPERATOR LOGIC...96

11.3.1 Operator Logic Formulation *96*
11.3.2 Probabilistic Operator Logic Formulation *96*
11.3.3 Quantum Operator Logic Formulation *97*
11.4 PROBABILISTIC OPERATOR LOGIC FORMALISM 97
11.4.1 Probabilistic Example "if ... then ..." Modus Ponens *101*
11.4.2 Probabilistic Operator Logic Example *102*
11.4.3 The Probability of Undecidability *103*
11.5 QUANTUM OPERATOR LOGIC 104
11.5.1 Quantum Operator Logic Formalism *105*
11.5.2 Operator Term Space *109*
11.5.3 The Term Space of Quantum Operator Logic *112*
11.5.4 Quantum Operator Logic Projection Operators *117*
11.6 QUANTUM OPERATOR LOGIC SUBUNIVERSES GENERATED BY LOWERING OPERATORS 118

12. FROM IDEAS TO REALITY: FIRST PRINCIPLES 120

12.1 SIX GENERAL PRINCIPLES 120
12.2 LOGIC AND METAPHYSICS 121
12.3 PLATONIC CONCEPTION OF THE RELATION OF IDEAS TO REALITY 123
12.4 OPERATOR LOGIC AND QUANTUM OPERATOR LOGIC EXIST INDEPENDENT OF OUR KNOWLEDGE OF THEM 124
12.5 THE REALM OF IDEAS 124
12.6 ROLE OF THE OBSERVER IN THE REALM OF REALITY 125
12.6.1 The Observer in Quantum Theory *125*
12.6.2 The Observer in Operator Logic & Quantum Operator Logic *127*
12.7 SPACE AND TIME 128
12.7.1 The Necessity of Time, and an Arrow of Time, in Operator Logic and Quantum Operator Logic *128*
12.7.2 Why add Space to the Logics? *129*
12.8 BEING AND EXISTENCE OF THE MATERIAL WORLD 130
12.8.1 The Substance of Particles *131*
12.8.2 The Form of Particles *134*
12.8.3 Being as Form *134*
12.8.4 Origin of Being *135*
12.9 THE ORIGIN AND MECHANISM OF PHYSICAL LAWS 135

13. PRÉCIS OF THE TRANSITION FROM IDEAS TO REALITY.................. 137

13.1 THE ASPECTS OF REALITY 137
13.2 MATTER 137
13.3 MATTER IS INSUBSTANTIAL 138
13.4 THE KNOWLEDGE BASE OF REALITY 139
13.5 PARTICLES AS MONADS 142
13.6 DIRAC-LIKE EQUATIONS FOR SPIN ½ MATTER 142

14. THE NATURE OF THE PHYSICAL DERIVATION/CONSTRUCTION OF A FUNDAMENTAL THEORY OF PHYSICS 147

14.1 WHAT IS THE FORM OF THE FUNDAMENTAL THEORY OF PHYSICS? 147
14.2 A THEORY FOR ITS TIME 148
14.3 DERIVATION VS. CONSTRUCTION OF A PHYSICAL THEORY 149
14.4 WHAT PURPOSE DOES A CONSTRUCTION OR DERIVATION SERVE? 151
14.5 THE RIGOR OF A DERIVATION 151
14.6 CONSISTENCY AND COMPLETENESS OF A SET OF POSTULATES 152
14.7 THE DIFFERENCE BETWEEN A MATHEMATICAL-DEDUCTIVE SYSTEM AND A FUNDAMENTAL SCIENTIFIC THEORY 153
14.8 MODALITY AND PHYSICAL THEORIES 154

15. TIME, SPACE, AND THE COMPLEX LORENTZ GROUP........................ 156

15.1 ARE TIME AND SPACE CONTINGENT OR NECESSARY? 156
15.2 TRANSFORMATIONS BETWEEN COORDINATE SYSTEMS 157
15.3 THE LORENTZ GROUP 159
15.4 THE NATURE OF $\Lambda(\Omega, \text{U})$ FOR COMPLEX Ω 162
15.5 COMPLEX LORENTZ GROUP 163
15.6 FASTER-THAN-LIGHT TRANSFORMATIONS 163
15.7 LEFT-HANDED SUPERLUMINAL TRANSFORMATIONS 164
15.7.1 Cosh-Sinh Representation of Left-Handed Superluminal Boosts............ 165
15.7.2 General Velocity Transformation Law – Left-Handed Superluminal Boosts 167
15.7.3 Left-Handed Transformations Multiplication Rules 167
15.7.4 Inverse of Left-Handed Transformations 169
15.8 RIGHT-HANDED SUPERLUMINAL TRANSFORMATIONS 170
15.9 INHOMOGENEOUS LEFT-HANDED LORENTZ GROUP TRANSFORMATIONS 172

15.10 Inhomogeneous Right-Handed Extended Lorentz Group 175
15.11 General Forms of Superluminal Boosts 175
15.11.1 "Lepton-like" Left-Handed Boosts *176*
15.11.2 "Lepton-like" Right-Handed Boosts *177*
15.11.3 "Quark-like" Left-Handed Boosts *177*
15.11.4 "Quark-like" Right-Handed Boosts *178*
15.11.5 "Quark-like" Boosts *178*
15.11.6 "Dirac-like" Boosts *179*

16. FROM A 16-DIMENSIONAL COMPLEX GENERAL RELATIVITY TO THE COMPLEX LORENTZ GROUP 180

16.1 Complex, Sixteen Dimensional General Relativity (Optional) 180
16.2 Flat Complex 16-Dimensional Space-time; GL(16), U(16), and L(16) (Optional) 184
16.2.1 Relation of GL(16) to U(16) *184*
16.2.2 The Complex, 16-Dimensional Lorentz Group *185*
16.3 GL(4)⊗GL(4) Subgroup (Required) 187
16.4 The Complex, 4-Dimensional Lorentz Group, L(C) 188
16.5 New Notation for GL(4) Generators Bringing Out the Subgroups: L(C), U(3) and U(1) 189
16.5.1 L(C) Lorentz Group Generators *191*
16.5.2 U(1) Generator *192*
16.5.3 U(3) Generators *192*
16.5.4 Gell-Mann SU(3) Generators *192*
16.5.5 Additional U(1) Generator needed for U(3) *192*
16.6 The GL(4) Subgroup Transformations have Complex Coefficients ... 193
16.7 How GL(4) and GL'(4) Combine to Yield Four Fermion Species and Fermion Generations 194

17. FERMION PARTICLE SPECIES 196

17.1 Four Fermion Species 196
17.2 Deriving the Free Dynamical Equations of Motion for Fermions 196
17.2.1 Dirac Equation – Free Charged Lepton Dynamical Equation *198*
17.2.2 L(C) Complex Boost for a Dirac Spinor *200*
17.2.3 Tachyon Dirac Equation – Free Neutral Lepton Dynamical Equation ... *201*

17.2.3.1 Form Leading to a Left-Handed Neutrino Theory......201
17.2.3.2: Form Leading to a Right-handed Neutrino Theory......202
17.2.4 Quark Dirac Equation – Free Up-type Quark Dynamical Equation......*203*
17.2.5 Left-handed Quark Dirac Equation – Free Down-type Quark Dynamical Equation......*205*
17.3 FREE LAGRANGIANS OF THE FOUR FERMION SPECIES......206
17.3.1 Free Charged Lepton Lagrangian......*206*
17.3.2 Free Neutral Lepton Lagrangian......*207*
17.3.3 Free Up-type Quark Lagrangian......*207*
17.3.4 Free Down-type Quark Lagrangian......*207*
17.4 CANONICAL COMMUTATION RELATIONS OF THE FOUR FERMION SPECIES......208
17.4.1 Free Charged Lepton Commutation Relations......*208*
17.4.2 Free Neutral Lepton Commutation Relations......*208*
17.4.3 Free Up-type Quark Commutation Relations......*210*
17.4.4 Free Down-type Quark Commutation Relations......*211*
17.5 FOURIER EXPANSIONS OF FREE FERMION FIELDS......212
17.5.1 Free Charged Lepton Field Fourier Expansion......*213*
17.5.2 Free Neutral Lepton Field Fourier Expansion......*213*
17.5.3 Free Up-Type Complexon Quark Field Fourier Expansion......*214*
17.5.4 Free Down-Type Complexon Quark Field Fourier Expansion......*216*
17.5.4.1 Left-Handed Down-Type Complexon Quark Field Fourier Expansion......216
17.5.4.2 Right-Handed Down-Type Complexon Quark Field Fourier Expansion......217
17.6 A GLOBAL SU(3) SYMMETRY OF COMPLEXON QUARKS......218
17.6.1 Global SU(3) Generators......*221*

18. ELECTROWEAK SECTOR OF THE STANDARD MODEL......223

18.1 L(C) TRANSFORMATIONS OF DIRAC AND TACHYON EQUATIONS......223
18.2 EXTENSION OF DIRAC & TACHYON EQUATIONS TO A DOUBLET FORMAT......224
18.3 NON-INVARIANCE OF THE EXTENDED EQUATION'S ACTION UNDER AN L(C) TRANSFORMATION......227
18.3.1 The Lepton Doublet......*229*
18.4 L(C) COVARIANCE OF COMPLEXON (QUARK) DOUBLET DYNAMICAL EQUATIONS......230
18.4.1 Doublet Dynamical Equation for Complexon Quarks......*232*

18.4.2 Non-Invariance of the Generalized Free Quark Complexon Action under an L(C) Boost ... *234*
18.5 POSTULATES OF THE DERIVATION OF THE LEPTONIC ELECTROWEAK SECTOR 234
18.6 DERIVATION OF THE LEPTONIC ELECTROWEAK SECTOR OF THE STANDARD MODEL ... 235
18.7 POSTULATES OF THE QUARK ELECTROWEAK SECTOR ... 239
18.8 DERIVATION OF THE QUARK ELECTROWEAK SECTOR OF THE STANDARD MODEL ... 240
18.9 GEOMETRICAL BASIS OF PARITY VIOLATION AND ELECTROWEAK THEORY ... 245

19. COLOR SU(3) ... 246

19.1 TWO POSSIBLE APPROACHES TO COLOR SU(3) ... 246
19.2 COMPLEXON BASIS OF COLOR SU(3) AND THE STRONG INTERACTIONS ... 248
19.3 THE ONE GENERATION STANDARD STANDARD MODEL ... 252
19.4 THE ONE GENERATION COMPLEXON STANDARD MODEL ... 254
19.4.1 Renormalization and Divergence Issues ... *256*
19.5 TWO-TIER FORMULATION OF THE ONE GENERATION COMPLEXON STANDARD MODEL ... 257
19.5.1 Simple Two-Tier Formalism ... *257*
19.5.2 Two-Tier L(C)-based Standard Model Theory ... *258*
19.5.3 Two-Tier Feynman Propagators for Real Space-time ... *260*
19.5.4 Complexon Feynman Propagators ... *262*
19.6 REASON FOR PARITY VIOLATING FORM OF THE STANDARD MODEL ... 263

APPENDIX 19A. COMPLEXON GAUGE FIELD DYNAMICS ... 264

19A.1 COLOR SU(3) ... 264
19A.2 PURE COMPLEXON GAUGE GROUPS ... 265
19A.3 PURE GAUGE COMPLEXON PATH INTEGRAL FORMULATION AND FADDEEV-POPOV METHOD ... 266
19A.3.1 Faddeev-Popov Application to the Complexon Condition ... *268*
19A.4 COMPLEXON QUARK-GLUON PERTURBATION THEORY ... 269
19A.5 COMPLEXON QUARK ELECTROWEAK INTERACTIONS ... 271

20. ORIGIN OF THE GENERATIONS ... 273

20.1 INTRODUCTION ... 273
20.2 THE FERMION GENERATIONS ... 274

20.2.1 Generation Field index ... *275*
20.3 SPECULATION: FOURTH GENERATION = DARK MATTER? ... 275

21. A POSSIBLE ORIGIN FOR PARTICLE MASSES—THE DIMENSIONAL MASS MECHANISM ... 277

21.1 THE ORIGIN OF MASS ... 277
21.2 THE ORIGIN OF K (WITH DIMENSION $[\text{MASS}]^2$) IN THE ROBERTSON-WALKER METRIC ... 278
21.3 ORIGIN OF THE FERMION MASSES AND MASS MIXING MATRICES ... 279
21.4 ORIGIN OF THE MASSIVE VECTOR BOSON MASSES ... 282

22. THE "STANDARD" STANDARD MODEL ... 284

22.1 POSTULATES ... 284
22.2 THE STANDARD STANDARD MODEL LAGRANGIAN ... 289
22.3 RENORMALIZATION ... 291

23. THE COMPLEXON STANDARD MODEL ... 292

23.1 POSTULATES ... 292
23.2 THE COMPLEXON STANDARD MODEL LAGRANGIAN ... 297
23.3 RENORMALIZATION ... 299

24. RESOLVED AND UNRESOLVED ISSUES ... 300

24.1 RESOLVED ISSUES IN THE STANDARD STANDARD MODEL AND THE COMPLEXON STANDARD MODEL ... 300
24.2 UNRESOLVED ISSUES IN THE STANDARD STANDARD MODEL AND THE COMPLEXON STANDARD MODEL ... 300
24.3 THE BREAKDOWN MECHANISM ... 301
24.4 DETERMINATION OF THE VALUES OF THE COUPLING CONSTANTS AND MASS MATRICES ... 301

25. EXPERIMENTAL TESTS ... 303

25.1 EXPERIMENTAL TESTS TO NAIL DOWN THE FORM OF THE STANDARD MODEL 303
25.2 TACHYONIC NEUTRINOS AND DOWN-TYPE QUARKS ... 303
25.3 QUARKS AND GLUONS: COMPLEXONS OR "NORMAL" ... 303
25.4 A 4TH GENERATION OF FERMIONS? ... 304
25.5 HIGGS MECHANISM OR DIMENSIONAL MASS MECHANISM ... 305

26. SUMMARY OF THE EIGHT POSSIBLE VARIATIONS OF THE DERIVED STANDARD MODEL DUE TO EXPERIMENTAL UNCERTAINTIES 306

27. THE LAYER BELOW THE STANDARD MODEL LAGRANGIAN 307

28. BEYOND THE STANDARD MODEL 309

28.1 APPROACHES TO A DEEPER LEVEL OF REALITY 309
28.2 EXTREMUM RULE FOR THE LAGRANGIAN OF PHYSICAL REALITY? 310
28.3 AXIOMATIC APPROACH TO THE KNOWLEDGE BASE OF REALITY 314

29. RELATIVISTIC QUANTUM METAPHYSICS 318

29.1 METAPHYSICAL TRUTHS IN THE STANDARD MODEL 318
29.2 "WHAT ARE" QUESTIONS 319
29.3 WHAT WE HAVE LEARNED OF METAPHYSICAL IMPORT 319
29.3.1 Substance and Properties 319
29.3.2 Time and Space 321
29.3.3 Causation 322
29.3.4 Origin of Physical Laws – The Knowledge Base 322

30. GENERALIZED METAPHYSICS 323

30.1 GOALS OF GENERALIZED METAPHYSICS 323
30.2 LOGIC IS PAN-UNIVERSAL 324
30.3 ALL NONTRIVIAL POSTULATES IN ALL UNIVERSES ARE CONTINGENT 324
30.4 TOWARDS A GENERALIZED METAPHYSICS OF UNIVERSES 325

31. LOGIC, LANGUAGE AND THE UNIVERSE 327

31.1 LOGICAL EQUIVALENCE OF LANGUAGES 327
31.2 OPERATOR LANGUAGES 328
31.3 QUANTUM LANGUAGES, GRAMMAR, TURING MACHINES, COMPUTERS, AND COMPUTER PROGRAMS 329

REFERENCES 331

ABOUT THE AUTHOR 335

TABLES & FIGURES

Figure 1.1. Why do theories created by the human mind so closely mirror Reality? The author believes it is due to Reality being "derived" from Logic, the mechanism of thought. This topic will be discussed in detail later. 6

Figure 2.1. The three sources of metaphysics and their interrelations with "Everyday" Metaphysics. 9

Figure 2.2. A visualization of the domain of applicability of Classical Metaphysics.... 13

Figure 3.1. A Feynman Diagram for "Causation" and "Change" in The Standard Model. It depicts the decay of a heavy quark particle into three particles: a lighter quark q', an electron e and an electron-type neutrino ν_e. It illustrates one form of Change.. 19

Figure 3.2. Another Feynman Diagram for "Causation" and "Change" in The Standard Model. It depicts the interaction (force exchange) between two particles which can lead them to change direction or to change the type of particles that they are. There are two simple diagrams shown. Actually these are the simplest diagrams of an infinite set of diagrams that contribute to the change in the two initial particles to the two final particles.. 19

Figure 3.3. Yet another Feynman Diagram for "Causation" and "Change" in The Standard Model. It depicts the gravitational interaction (force exchange) between two clumps of matter via graviton particles on particles within the clumps, which lead the clumps to change direction. This illustrates the force of gravity at the quantum level. As yet the understanding of quantum gravity is subject to dispute. But most physicists would agree that gravitons exist and are the fundamental "carriers" of the force of gravity. Again these are the simplest diagrams of an infinite set of diagrams that contribute to the process of change. Clump1 and Clump2 each consist of a large number of particles in lumps of matter. The thick "dots" represent lots of particles and lots of gravitons being exchanged between the particles in the clumps. The result of the extremely large number of gravitons exchanged is the clumps change direction and approach each other since gravity is a strictly attractive force according to the General Theory of Relativity. 20

Figure 3.4. A visualization of the steps leading from a Generalized Metaphysics to Relativistic Quantum Metaphysics and thence to The Standard Model. 22

Figure 4.1. A two filter experiment that selects particles with the exact velocity of 100 miles per second from a stream of incoming particles with a variety of velocities. 24

Figure 6.1 A table of the syntactical and semantic operators associated with a statement. 43

Figure 9.1. A hierarchy of universes with the lowering operators indicated for the transition to each subuniverse. 73

Figure 12.1. A three stage experiment beginning with a system in the state with eigenvalue x of observable X, having an intermediate state 2 with the eigenvalue y of observable Y, and ending in state 3 with eigenvalue z of observable Z. Time proceeds from left to right in the diagram. 125

Table 12.1. The number of spinor components for various numbers of space-time dimensions. 129

Figure 12.2. Diagrams showing some ways in which particles interact with each other and transform to other types of particles. In diagram A a particle "splits" into two particles. In B two particles combine to produce an "outgoing" particle. In C two particles interact in some complex way represented by the filled circle to produce two outgoing particles which may or may not be the same particles. In D a particle transforms (decays) into three particles. 132

Figure 15.1. Depiction of two coordinate systems. The "primed" coordinate system is moving with velocity **v** in the positive x direction with respect to the unprimed coordinate system. 157

Figure 15.2. Three reference frames used to establish transformation multiplication rules. 168

Figure 16.1. The hierarchy of groups that appear in the breakdown from GL(16) and lead to the known 3 generation fermion spectrum. 195

Figure 19.1. Generation of the ElectroWeak sector and color SU(3) from L(C). This diagram is symbolic. For example General Relativity actually originates in 16-dimensional space-time which then breaks down to the real 4-dimensional space-time we experience. 247

Figure 19A.1. Quark self-energy diagram in which a quark emits and subsequently absorbs a gluon. The quark momentum is $p = (p^0, \mathbf{p_r} + i\mathbf{p_i})$ and the gluon momentum is $q = (q^0, \mathbf{q_r} + i\mathbf{q_i})$ with conservation of energy, and separate conservation of real spatial momentum and imaginary spatial momentum at each vertex. 270

Figure 19A.2. Quark self-energy diagram in which a quark emits and subsequently absorbs a photon. The quark momentum is $p = (p^0, \mathbf{p_r} + i\mathbf{p_i})$ and the photon momentum is $q = (q^0, \mathbf{q_r})$ with conservation of energy, and separate conservation of real spatial momentum and imaginary spatial momentum at each vertex. 271

Figure 19A.3. Photon self-energy diagram with a quark loop. The photon momentum is $q = (q^0, \mathbf{q_r})$ and quark loop momentum is $p = (p^0, \mathbf{p_r} + i\mathbf{p_i})$. 272

Figure 20.1. Group hierarchy of extended "space-time" leading to the Standard Model. Nature takes L(C) from one GL(4) subgroup to define the space-time structure, and U(3) from GL'(4) to define 3 generations and their mixing. (Note: the ' serves only to differentiate between the GL(4)'s.) 274

Figure 26.1. A Table of the possible variants of the "basic" Standard Model due to experimental uncertainties. 306

1. Epistemology

If one wishes to begin a trip, or a study, one should begin at the beginning. This chapter is the beginning of a study that leads to an understanding of physical Reality at its currently most fundamental level – the Standard Model of Elementary Particles combined with the General Theory of Relativity.[1] Some primary issues that we will examine in the next few chapters with a view towards our goals are Epistemology – the nature, scope and limitations of human knowledge; Metaphysics – the nature of ultimate Reality; and Logic – the true relationships between the parts of ultimate Reality. The author believes that the impact of Quantum Theory and the Special Theory of Relativity on these areas of study has yet to be fully appreciated. One of our goals is to provoke a thorough reanalysis of these subjects from an Operator Logic viewpoint.

1.1 Epistemology

Epistemology is the study of human knowledge. Since our goal is to understand one specific area of knowledge – physical Reality – our approach to epistemology will be limited to the nature, scope, and limitations of human knowledge of physical Reality. Physical Reality can be viewed as consisting of three parts: cosmological physical Reality (the universe in the large), everyday earthly (planetary) physical Reality, and quantum physical Reality (Reality at the quantum level on the very smallest distance scales). These parts of Reality conveniently break up into distance scales: very large distances – cosmological distances; the distances and sizes of everyday experience – ordinary distances; and very small distances – quantum scale distances. Each of these areas of human knowledge has a different epistemology. Almost all work in the

[1] It is clear that the current "fundamental" level is either an interim level preliminary to a deeper level or is only part of a broader fundamental level. The major evidence for this view is Dark Matter and Dark Energy – phenomena that have appeared in Cosmological attempts to account for observed astronomical data. Nevertheless the Standard Model of Elementary Particles has repeatedly been proved to be an accurate description of known earthly experimental data and most cosmological data.

epistemology of knowledge of physical Reality has been at the level of everyday earthly Reality.

Before turning to a study of the epistemology of very large-scale and very small-scale knowledge we should understand the makeup of epistemological theory. One way to begin is to consider an isolated human mind with absolutely no sensory inputs, and then to consider what physical senses are available and how their capabilities and limitations affect the ability of the human mind to acquire knowledge and process it.

So let us first consider an awake human mind (or any intelligent mind) in *total* isolation with no prior, or current, sensory input and ask what sort of knowledge it can obtain. First it is clear that any knowledge that it might have would be the result of imagination.[2] Conceivably the mind could form some concepts. Perhaps the simplest concept would be number. But not having anything to count why should it conceive of number.[3] Similarly all other concepts derive from sensory perceptions and so it is difficult to believe that any concepts would occur to the isolated mind except possibly concepts/ideas of a genetic origin – if such exist.

But there is a difficulty in the mind itself. Modern research has established that sensory inputs affect the wiring of the human mind as the mind responds by processing the inputs. So a totally isolated mind could be viewed as an "unwired" mind. The thought processes that would take place in such a mind would appear to be haphazard. Thus we are led to conclude that an isolated mind without sensory input and without inherent genetic information would in general have no knowledge and be incapable of thought. The young Helen Keller, who lacked hearing and sight but had a sense of touch, is an (imperfect) example of an isolated mind. Ms. Keller was not rational until after much schooling.

If we now consider the human mind with its usual physical senses: sight, hearing, smell and touch, then a totally different view emerges. For the senses, by providing input to the mind, gives it items to

---

[2] The possibility of inherent genetic "knowledge" present in the mind (such as Jungian archetypes) is not excluded.

[3] Even for people who have sensory input the concept of number is not easy. For example members of some tribes cannot count beyond 2.

consider, behavior to analyze, and motives (self-preservation and the acquisition of wealth and power) for survival.

1.2 Origin of Human Knowledge of Physical Reality

The nature of human knowledge of physical Reality at the everyday level of experience has been a subject of discussion for thousands of years. We will not enter to that still ongoing discussion other than to say that it is largely based on direct observation using the human senses of physical Reality.

Rather we will examine the nature of human knowledge at the quantum and cosmological levels. In both cases the nature of human knowledge is observation of physical Reality through instrumentation of varying degrees of complexity, which is interpreted using a theoretical analysis that can be quite complex and often uses probabilistic interpretations. As a result the nature of human knowledge in these areas of Reality is indirect. Theoretical analyses stand between our knowledge of Reality and experimental results. A change in the analysis of an experiment usually leads to a different "knowledge" of the corresponding part of Reality.

1.3 Scope of Human Knowledge of Physical Reality

Thus the scope of knowledge of Reality is in a sense limited by our ability to analyze what we are observing. And so quantum and cosmological knowledge of Reality has a layer of theory between us and observational data. In some areas of science such as elementary particle physics the theoretical analysis of the data is extremely complex and problematic. As our ability to do experiments to determine features of physical Reality becomes greater the theoretical analysis will correspondingly become more critical.

1.4 Nature of Human Knowledge of Physical Reality

In the three areas of knowledge of physical Reality we see a pattern or structure of knowledge. First one has a set of facts relating to some aspects of physical Reality. Then these facts are usually organized into a theory based on principles or axioms that presumably are "deeper" then the amalgamation of facts and from which the facts may be derived. Lastly, in an increasingly larger number of theories of various aspects of physical Reality, the theory is put into a mathematical form. Euclidean geometry is considered the example par excellance of a mathematicized theory. Many theoretical areas of physical Reality espouse the goal of becoming a mathematical-deductive theory – a theory with a set of fundamental axioms from which theorems can be derived that describe all phenomena in the area.

Presumably all aspects of physical Reality follow from a basic Theory of Everything but the connections between the various specialized theories is often hard to derive. The Theory of Everything is expected by many scientists to follow from a unified theory of fundamental particles and gravitation. But the derivation of psychological phenomena, for example, from the Theory of Everything is a long and tortuous process that has many gaps currently – primarily at the level of the understanding of the chemistry of the human brain and its relation to psychological states.

1.5 Limitations on Human Knowledge of Physical Reality

The sources of human knowledge are observation, direct or indirect, and rational analysis of facts acquired through observation. To obtain factual knowledge about Reality we must be able to information by direct observation using our senses or through instruments that play the role of extending the senses. For example, we cannot see ultraviolet light with our eyes but we can devise an instrument that can "see" ultraviolet light.

When everyday phenomena were the stuff of knowledge then it appeared that we could have a total knowledge of physical reality. However the situation changed dramatically in the twentieth century. First quantum theory showed us that certain combinations of facts could not simultaneously be ascertained at the quantum level (small distances). For example the position and momentum of an electron cannot be simultaneously exactly determined due to the Uncertainty Principle.

In addition, more recently, it has become clear that the amount of matter and the amount of energy in the cosmos are different from that determined by astronomical observations. There is an "unseen" part of the universe. All of our senses and instruments for observation are based ultimately on the four known forces of nature: gravity, electromagnetism, the weak interactions and the strong interaction. The latter two forces are only detectable at high energies (short "quantum-scale" distances). The unseen matter in the universe, Dark Matter (about eighty-five per cent the total matter in the universe) is only known from its gravitational effects at the present time. It seems to be either undetectable through the stronger forces of nature or , perhaps, only very weakly detectable through the stronger interactions. That being the case, Dark Matter can "interpenetrate" (occupy the same space) as the ordinary matter that we see, and form its own unseen part of the universe. Similarly we see the effects of Dark Energy[4] in the cosmos – particularly in the rate of expansion of the universe. Here again we see the effects but do not see the entity.

We conclude that our senses, even our extended senses through instrumentation, cannot currently acquire significant knowledge of most of the universe.

Everyday epistemology and metaphysics did not anticipate and has not explained the unseen, and unseeable, aspects of the universe. Therefore any claims that metaphysics explains ultimate Reality are incorrect. The mission of this book is to begin the extension of metaphysics to the current empirical views of Reality.

1.6 Mind-Body "Problem", and Reality

A topic of interest for many years has been the "Mind-Body" problem. Many people see a discrepancy between the intellectual processes taking place in the mind and the qualities of the human body—particularly the human brain. However, recent research on the human brain using new techniques such as MRI have revealed that specific parts of the brain handle input from the senses, emotions, and even certain thought processes such as number processing (mathematics). We are therefore developing an understanding of the relation between the mind or consciousness and the physical-chemical structure of the human brain.[5] In a very real sense the human mind is now being seen as a very complex computer, much beyond our bigest and fastest electronic computers, yet still a computing apparatus. Mathematical networks have been developed that simulate aspects of learning, thinking and memory.[6]

---

[4] Energy of unknown origin constituting about seventy-five per cent of the energy in the universe.

[5] Scientists using fMRI (functional Magnetic Resonance Imaging) have recently found signatures of consciousness. They have detected coordinated activity across the entire brain. They have also detected conscious processing of images in the mind. See A. Schurger et al, Science, DOI: 10.1126/Science .1180029 (November, 2009).

[6] Cf. the papers and books of Professor Steven Grossberg of Boston University and collaborators and colleagues in the field.

If all the aspects of the mind can be reduced to mechanisms of the human brain then the major remaining issue is the nature of consciousness. Consciousness is a conglomeration of the features of the mind that give us a coherent "picture" from which we can observe, process, and respond to external and internal events. Consciousness also generally has a continuity of transitions from one state to another. Perhaps the best simple description of consciousness is a self-controlled computer or television set that undergoes transitions based on internal processing of internal states and external sensory input.

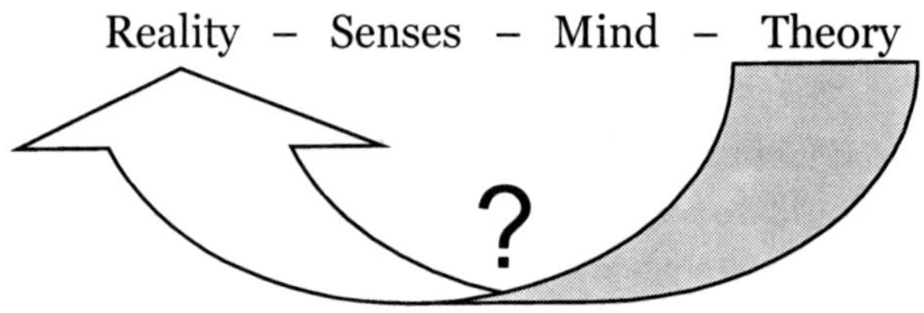

Figure 1.1. Why do theories created by the human mind so closely mirror Reality? The author believes it is due to Reality being "derived" from Logic, the mechanism of thought. This topic will be discussed in detail later.

We now have a strong beginning to understanding the role of the various parts of the brain in processing sensory input, making the body perform mechanical activity, memory, creative thought and abstract thinking in general. Abstract thought takes one of two forms: coordinated brain activity and connectivity of thinking through logical progression. Creative thought arises from the association of ideas, the logical progression of ideas, changed views of concepts (a familiar word or idea that is evoked but viewed differently from its previous view in the mind), and serendipitous "mistakes."[7]

Based on this line of reasoning the phrase "Mind-Body Problem" is a misnomer at best. It would better described as the Mind-Reality

[7] An interesting area that requires investigation is the role of writing as a memory aid. Before writing was developed, and up to the past few centuries, the human mind carried the burden of remembering a train of thought. Since writing became common there has been a remarkable increase in theoretical knowledge – particularly in the sciences – due to the interaction between thoughts in the mind and thoughts written on paper. A poignant case in point is that of Professor Stephen Hawking who has remarked that he regrets the impact on his work of not being able to work out thoughts on paper or a blackboard.

Problem. We create theories with our minds that describe natural phenomena. Mathematical theories of physics have been extremely accurate in describing natural phenomena – physical Reality. On a number of occasions physicists have raised the question: how can our somewhat strange mathematical theories accurately describe (or perhaps better said – accurately mirror) physical Reality. Does Reality do the computations in real wall clock time (actually more seemingly instantaneously) to make events happen according to our calculations? That is a fundamental question which metaphysics should address.

Epistemology frames our view of knowledge of Reality and addresses the question of the theories that organize our knowledge of Reality. Thus it provides the backdrop for the study of metaphysics.[8]

---

[8] The human mind is capable of simulating sensory input. Hypnotists do this when they cause a subject to think they see or feel something illusory. An individual can do this as well. It commonly happens in dreams. It can also happen through a form of self-hypnosis. We are familiar with this phenomena at the level of the senses. But it is possible that a mind can create a mental image of another universe and cause this universe to evolve according to some set of physical laws concocted by the mind. Then we would have a mental implementation of a form of the "many worlds" hypothesis that appears in metaphysics and modal logic. We might call this type of thought process creating "Universes of the Mind."

2. What is Metaphysics?

Turning and turning in the widening gyre
The falcon cannot hear the falconer;
Things fall apart; the centre cannot hold;
W. B. Yeats –*The Second Coming*

2.1 Definitions

Metaphysics means different things to different people. Some say it is the "Theory of Ultimate Reality."[9] Others say it is the "Theory of the Nature of Abstract Entities."[10] Yet others express alternate views.

We shall take Metaphysics to be the contemporary theory of ultimate Reality realizing that our views of Reality have changed drastically in the past hundred odd years and are likely to change again in the future. Rather than wait for "ultimate Reality" to surface[11] we will construct the Metaphysics of presently known Reality. In doing so we shall also explore the abstract entities that make up the world[12] as we know it focusing particularly on quantum Reality. Quantum Reality is quite different from the world of everyday Reality that is the primary focus of most metaphysical discussions. The theory of Relativity also alters our view of Reality in a dramatic way.

The combination of the mathematics of quantum theory and relativity open a new world of metaphysics that we call *Relativistic Quantum Metaphysics*. This new metaphysics is simpler in some ways than traditional discussions of the metaphysics at the level of everyday experience. It is also different from everyday metaphysics in many significant ways. Since relativity and quantum theory are at a deeper level than everyday phenomena we can say with assurance that

---

[9] van Inwagen (2009).
[10] Lowe (2002).
[11] If perhaps it has not as yet surfaced.
[12] The word world represents our universe or the set of all universes if there is more than one.

relativistic quantum metaphysics is closer (actually as close as we can currently get) to ultimate Reality.

We will thus view Metaphysics as composed of three parts (See Fig. 2.1.): the Metaphysics of everyday experience,[13] Quantum Metaphysics,[14] and Cosmic Metaphysics.[15] Because physics has shown us that the union of Relativity Theory and Quantum Theory is essential to the understanding of natural phenomena in general, we consider a combined metaphysics based on the combination of these physical theories.

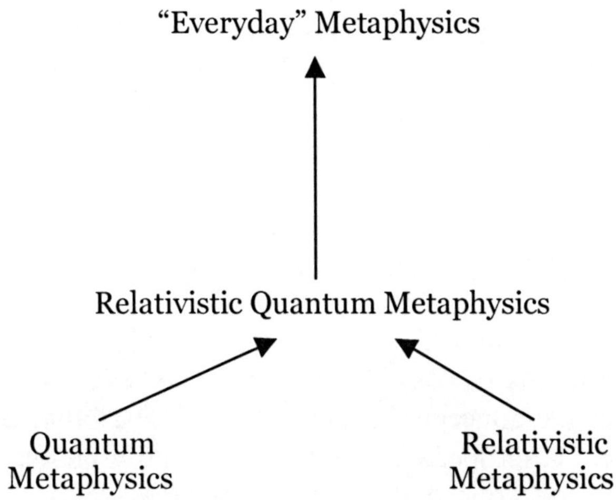

Figure 2.1. The three sources of metaphysics and their interrelations with "Everyday" Metaphysics.

"Everyday" Metaphysics has traditionally been applied to many areas of experience including physical phenomena, thought, the mind, ethics, and Theology. Excepting Theology we will view all aspects of "Everyday" Metaphysics as ultimately based on Relativistic Quantum Metaphysics. The chain connecting Relativistic Quantum Metaphysics to

---

[13] This is the metaphysics of the past 2500 years.

[14] The metaphysics of the ultra-small and of high energies whose empirical basis was only discovered in the twentieth century. Its nature is very different from "everyday" metaphysics as the following chapters will show.

[15] The metaphysics of the cosmos based on the Special and General Theories of Relativity. This metaphysics also is vastly different from "everyday" metaphysics.

aspects of "Everyday" Metaphysics may be quite long and involved—the links of the chain in many cases are not as yet forged. Yet we will take a solidly empiricist position[16] that this chain is fact and that progress on all fronts is being made. "Everyday" metaphysics is derivative from the much deeper Reality of Relativistic Quantum Metaphysics.

2.2 A Summary of Classical ("Everyday") Metaphysics

Classical Metaphysics[17] is the metaphysics hitherto developed based on our experience of everyday phenomena. Therefore it addresses issues relevant at that level of Reality.

Classical Reality consists of many different substances with many different properties. As a result Classical Metaphysics is in part concerned with categorizing the many substances and exploring the interrelationships between them and their properties. It considers substratum and bundle theories relating groups of properties and substances or entities. It also considers the nature and role of propositions (essentially statements), states of affairs (situations), events, and facts.

Classical Metaphysics has found it necessary to introduce the concept of a plurality of possible worlds.[18] The "existence" of this set of worlds enables the concepts of necessity and possibility of Modality to have meaning. Is a property necessary to an entity or sometimes part of an entity? – A modal question. With many worlds we can (in a simplified way) consider whether an entity in the many worlds always has a property (necessary) or sometimes has a property (contingent).

Another topic of interest in Classical Metaphysics is Causation: the study of cause and effect, their connection, and whether causes necessitate their effects. The study of classical causation is interesting because it attempts to describe these features in general terms that to this author does not reflect what actually happens in ultimate Reality according to scientific empirical data in the quantum and relativistic regime.

---

[16] In this choice we see our view as similar to that of Hume and other empiricists.

[17] It seems appropriate to call "everyday" metaphysics classical based on an analogy with the nomenclature of classical physics and quantum physics.

[18] The multiple worlds concept has no relation to many universes theories in physics since the multiple worlds exist at the conceptual level and are used to clarify the notions of modality: does an entity have a necessary property (absolutely required) or a contingent property (a property that it might have in one world but not in another world).

Actions and events are viewed as types of entities in Classical Metaphysics. Their nature is analyzed in a qualitative, abstract manner in contrast to what we encounter at the deeper levels of physical Reality.

A similar statement can be made about studies of the Nature of Time and Space, as well as some studies of space-time that seek to bring in a measure of the Reality of the Special Theory of Relativity. Attempts are made to abstract the nature of space and time that, in this author's view, do not correspond to Reality. As a statement becomes more abstract it loses content. And the studies of space and time simply do not elucidate their nature.

Change takes place in time. The Classical Metaphysics of change are perhaps meaningful at the level of everyday change but do not corresponds to what we have learned of change at the quantum level – the ultimate source of change.

We conclude – knowing that we have only briefly discussed Classical Metaphysics – that it is limited to the subject area from which it originated – everyday objects, properties, and change. It does not address the profound issues that Quantum Theory and the Theory of Relativity have raised. Therefore Classical Metaphysics does not address ultimate physical Reality. In fact, it tends to mask it with a veneer of depth and generality that it truly does not have. Classical Metaphysics cannot be extended to quantum and relativistic entities.

2.3 Are There Any Truths in Classical Metaphysics?

Van Inwagen (2009) raised two questions that support the somewhat negative tone of the previous section:[19] "Why is there no such thing as metaphysical information? Why has the study of metaphysics yielded no established facts?" To be fair there are certain principles of metaphysics that were proposed by Leibniz[20] that can be viewed as metaphysical information. Leibniz proposed three basic principles:[21]

1. The Principle of Sufficient Reason
2. The Principle of Identity
3. The Principle of Perfection

---

[19] van Inwagen (2009) p. 11.
[20] Rescher (1967).
[21] Rescher (1967) chapter 2.

These principles, and their consequences, are to be understood in Leibniz's view as axioms within the framework of generalized subject-predicate logic. Leibniz viewed propositions (statements) about substances or entities (existents) as having a subject-predicate form[22] or as reducible to a combination of sub-statements having subject-predicate form. He used a form of analysis (*Leibnizean analysis*) of propositions that first determined the set of properties of a subject (substance) and then scanned the list to see if the predicate (a property) appeared in the list. Thus a proposition is true if the predicate is in the list of properties of a subject and false otherwise.

The Principle of Sufficient Reason is every true proposition is analytic. Thus it asserts every true proposition can be reduced to subject-predicate form or a combination thereof and each predicate can be shown to be in the set of the subject's properties using Leibnitzean analysis.

The Principle of Sufficient Reason is illustrative of the abstract nature of the above three principles as well as Leibniz's four subsidiary principles:[23]

1. The Identity of Indiscernibles
2. The Principle of Plenitude
3. The Law of Continuity
4. The Principle of Harmony

The reader is directed to Rescher (1967) and other sources for information on these principles since they will not be germane to Relativistic Quantum Metaphysics. They are too general to be of value in the study of ultimate physical Reality. However Leibniz's focus on subject-predicate Logic as the underlying framework of his metaphysics will be seen to be a precursor to part of our approach to Relativistic Quantum Metaphysics.

In addition to his principles Leibniz suggested that there are many possible worlds in a physical, tangible sense – not as a conceptual construct. And he gave a criterion for the selection of our world as the world of Reality:

---

[22] Example: "The house is green." has house as the subject and "is green" as the predicate.

[23] See Rescher (1967) chapter 4.

> The world which is most perfect … is the simplest in its axioms and the richest in phenomena.[24]

This principle is an extremal principle in the sense of extremal principles in mathematics and physics. Leibniz is famous for developing calculus (at the same time as Newton) and for his extremal principles which have been of great value in physics.

We will discuss the possibility of a selection principle for physical Reality with our universe as the solution (or a possible solution) at a subsequent point in our presentation.

We conclude our brief view of Classical Metaphysics with a figure, Fig. 2.2, that indicates where it applies in the range of Reality in this universe (world).

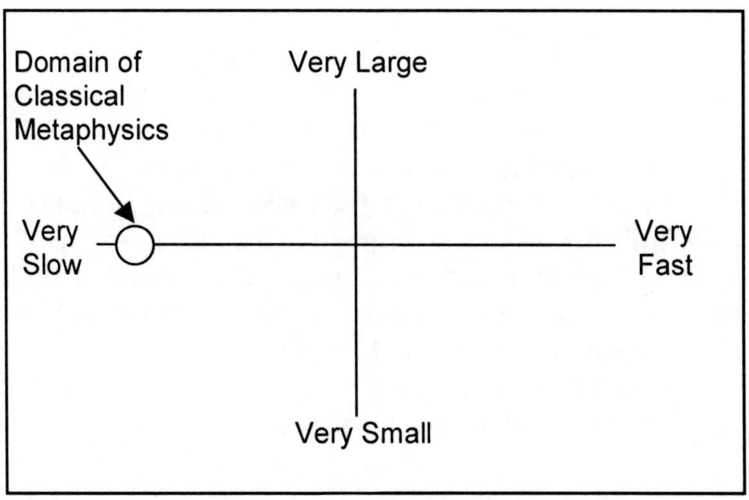

Figure 2.2. A visualization of the domain of applicability of Classical Metaphysics.

[24] Quoted in Rescher (1967) p. 19.

3. Relativistic Quantum Metaphysics

3.1 Reduction of Reality to the Presently Known Ultimate Reality

Classical Metaphysics was developed over the millenia based on concepts that arose in everyday experience with a small admixture from early science up to at most the nineteenth century. Since 1900 there has been an explosion of knowledge – primarily scientific knowledge – that calls for a new metaphysics of physical Reality.

First, we have come to the realization that everyday phenomena which appear complex in nature and involve a staggering variety of different substances, properties, and events at the physical level, the mental level and the social level are all ultimately derivative[25] from a vastly simpler underlying physics that is embodied in the Standard Model of Elementary Particles and the General Theory of Relativity.[26]

We will begin with a description of the overall nature of this "all encompassing" physics and then proceed to develop Relativistic Quantum Metaphysics. Fig. 2.1 shows our view of Classical Metaphysics in relation to Relativistic Quantum Metaphysics.

So we consider the progressive depth of view of physical Reality (omitting gravity for a while in our discussion):

All Everyday →	Molecular →	Atomic →	Elementary Particles
Phenomena	Level	Level	Level

[25] We will not consider Theological questions and the nature of God because we view this area as not part of physical Reality although the spiritual and the physical may interact with each other.

[26] We say that realizing that aspects of gravity and the cosmos (dark matter and dark energy for example) as well as features of elementary particles (including the issue of whether there is a deeper theory of elementary particles such as a Superstring theory) remain to be understood. These unknowns are not relevant to "everyday" phenomena and thus our development of Relativistic Quantum Metaphysics does provide a complete underpinning for "everyday" phenomena and thus the domain of Classical Metaphysics.

All substances at the everyday level with all their properties are consequences of the nature and behavior of elementary particles. The set of elementary particles, their nature and their behavior are extremely well described by The Standard Model of Elementary Particles. We will examine features of The Standard Model in some detail in subsequent chapters.

3.2 A Metaphysical View of Elementary Particles

At this point we will consider general features of elementary particles and The Standard Model of interest to metaphysics.[27] We also will consider the theories of Special and General Relativity from a metaphysical perspective.

3.2.1 Particles and Substances

In everyday Reality we see a multitude of substances. In the case of elementary particles it appears that there is only one substance[28] but the substance can take different forms with different properties. Each form is a type of elementary particle. Since the late 1960's it has become clear that there is a basic set of particles: eighteen types of quarks and six types of leptons[29] that combine to make protons, neutrons, atoms, and other less familiar bound states (combinations). The forces of nature are gravity, electromagnetism, the Weak Interactions and the Strong Interactions. Quarks and leptons exert these forces on each other. Each force has an associated particle or set of particles that embody the force: gravity has gravitons, electromagnetism has photons, the Weak Interactions have vector bosons, and the Strong Interactions have eight gluons. In total, there are thirteen "force particles." Thus the number of known elementary particles is thirty-seven. Many physicists believe that

---

[27] Again we note that The Standard Model, although it describes elementary particles and their behavior very well, is either an incomplete theory (Dark Energy and Dark Matter are not in The Standard Model.) or it is based on a deeper theory of particles such as Superstring theory. In later chapters we provide a line of reasoning (a derivation with some steps remaining to be justified) of the major features of The Standard Model.

[28] The nature of Dark Matter and Dark Radiation is an open question. They may constitute another substance or possibly two substances. However experimental results reported in the past month (October, 2009) suggest that Dark Matter can transform or decay into normal matter and energy. If these results are correct then Dark Matter is of the same substance as normal matter.

[29] An electron is a lepton. Neutrinos are also leptons.

there are more particles such as Higgs bosons remaining to be discovered. The known particles comprise The Standard Model.

If there are thirty-seven types of elementary particles how do we know they are composed of the same substance? The reason is quite simple. These particles can transform into each other through the decay of a particle into other particles, and can transform into each other when they interact (or exert a force) on each other. Although they differ in properties their ability to undergo transfers into each other strongly supports the idea that there is one substance that can take thirty-seven (or perhaps more) different forms with different properties. These properties include the spin of a particle and numbers associated with their form that we call internal quantum numbers.

Can we take the alternate view that there are thirty-seven different substances? Yes, but it does not change the fact of transformations between the particles, and Ockham's Razor – the simplest solution is usually the correct solution – would support the view that we have one substance capable of assuming thirty-seven different forms. This discussion shows that Relativity and Quantum Theory upon which this Reality rests gives us a far simpler metaphysics than Classical Metaphysics.

Having reduced the number of substances to one substance, the question naturally arises what is that substance? The answer is that it is unknowable. The various forms it can assume are known but the substance itself is an unknowable. We can only specify the forms it takes and their properties. Taking a note from Logic we simply call it a *primitive term* – an undefinable term which we can use to build a mathematical theory. To the reader who might object to this answer we point to Euclid's geometry, which has several primitive terms in its axioms. Thus primitive terms are to be expected in a theory. Some day if a more fundamental theory than The Standard Model appears, then it also can be expected to have primitive terms as well.

Lastly, the reader might ask what of the vacuum? Is it nothing or a substance? The Standard Model, which is a Relativistic Quantum Field theory, has a vacuum which is composed of an infinite number of all the different particles. But these particles are uniformly distributed throughout space so the forces that they might have exerted on any individual particle of matter cancel each other[30] and thus we think of the vacuum as empty. The vacuum is thus also composed of the substance of particles in infinite, uniform, quantity.

Sometimes the vacuum undergoes a quantum fluctuation creating a particle-antiparticle pair for an instant. Thus the vacuum has a dynamic aspect. But the extremely short existence of the particle-antiparticle pair usually makes vacuum fluctuations unobservable. Yet we know that they happen because they exert effects in the observed properties of matter.

Our discussion has led us to the conclusion that there is only one kind of substance in the deepest known physical theory. And so we have our first Relativistic Quantum Metaphysics principle:[31]

I. There is only one substance in the world (universe). It can assume a variety of forms as particles, which we call fundamental elementary particles.

Currently there are thirty-seven known fundamental particles and thus a corresponding number of forms.

3.2.2 Forms and Properties

Classical Metaphysics is much concerned with properties. It considers them individually for entities, in bundles of properties that entities might possess, and as forming an entity in the form of a substratum.

We have learned from the study of the physical nature of matter that the thirty-seven known elementary particles (and any new ones that might be found) have bundles of properties consisting of energy, momentum, spin, and internal quantum numbers. We will call the bundle of properties of a particle its *form*. Thus we come to a second Relativistic Quantum Metaphysics principle:

II. Each particle's form consists of a finite number of properties: its energy, momentum, spin, and internal quantum numbers.

---

[30] This is a simplification in that a charged particle, for example, can distort the vacuum attracting particles of opposite charge to it which "mask" (renormalize) part of the charge.

[31] SuperString Theories, one set of possible theories of ultimate Reality, assume all elementary particles in the universe are made of extremely small mathematical strings. Thus these theories conform to principle I.

This extreme simplicity makes the theoretical developments in the study of properties in Classical Metaphysics not relevant to the ultimate Reality of the world (universe) that we inhabit.

However, Classical Metaphysical studies may be relevant to the study of possible worlds if they are extended to a wider framework than the picture of ultimate Reality just discussed. In that wider framework Relativity and Quantum theory, may be supplemented, or replaced, by other theories. Thus a wider metaphysics would be required.

3.2.3 Change, Interactions and Causation

Causation and change have been much studied in the abstract in Classical Metaphysics. The abstractness of the work in these areas and the variety of everyday phenomena have led to a plethora of ideas. But this plethora is little related to the provisional ultimate Reality embodied in Relativistic Quantum Field theory in general and The Standard Model of Elementary Particles in particular – the deepest verified theory of physical Reality – supported by a large amount of positive experimental data with no significant contradictory experimental data.[32] (The Standard Model is a specific Quantum Field Theory.)

In the Standard Model causation and change result from the interactions (forces) between elementary particles. The interactions can be at very small distances or large distances between individual elementary particles, or when particles combine (through forces such as electromagnetism or the strong interaction) the interactions can accumulate to cause tangible effects at everyday distances between everyday objects.

At the most fundamental level the decay of an elementary particle and the interaction between two elementary particles take place through the thirteen fundamental interaction particles mentioned earlier: gravitons, photons, three vector bosons, and eight gluons. They can be visualized using Feynman diagrams. These diagrams, developed by Richard Feynman, associate a straight line with each of the eighteen "matter type" particles and a "wiggly" line with each of the thirteen

[32] We state again that it is known that there are areas that are not part of the Standard Model and there may be a deeper theory waiting in the wings. These difficulties are not roadblocks to viewing the Standard Model provisionally as ultimate physical Reality since The Standard Model will still hold for current experimental data within a larger theory or a deeper theory. In these cases it becomes a good approximation for its domain of experimental data.

"interaction particles." They are viewed from left to right. (Time can be considered as increasing from left to right in these diagrams although that is not strictly correct.) See Figs. 3.1 – 3.3 for simple Feynman diagrams illustrating decays and interactions.

Thus we again see that the complex issues and ensuing discussions in Classical Metaphysics are reduced in Relativistic Quantum Metaphysics to interactions between particles or combinations of particles – much simpler, less abstract, and not subject to dispute. Causation and Change are reduced to single or multiple particle interactions.

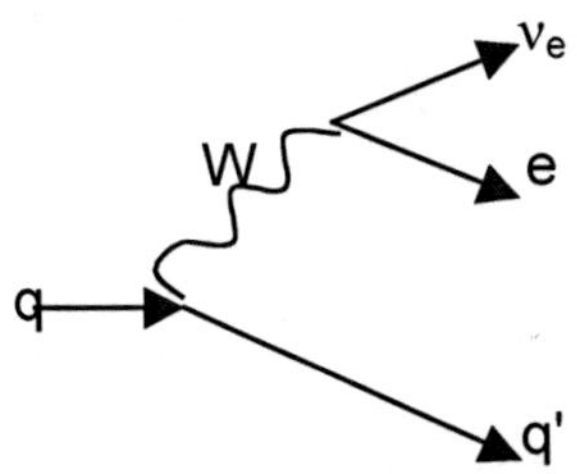

Figure 3.1. A Feynman Diagram for "Causation" and "Change" in The Standard Model. It depicts the decay of a heavy quark particle into three particles: a lighter quark q', an electron e and an electron-type neutrino ν_e. It illustrates one form of Change.

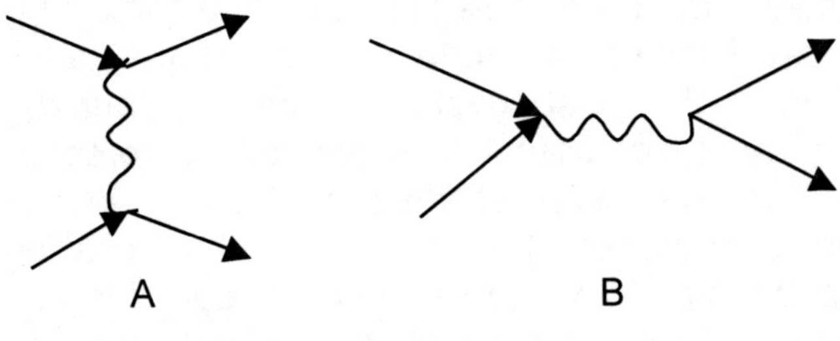

Figure 3.2. Another Feynman Diagram for "Causation" and "Change" in The Standard Model. It depicts the interaction (force exchange) between two particles which can lead them to change direction or to change the type of particles that they are. There are two simple diagrams shown. Actually these are the simplest diagrams of an infinite set of diagrams that contribute to the change in the two initial particles to the two final particles.

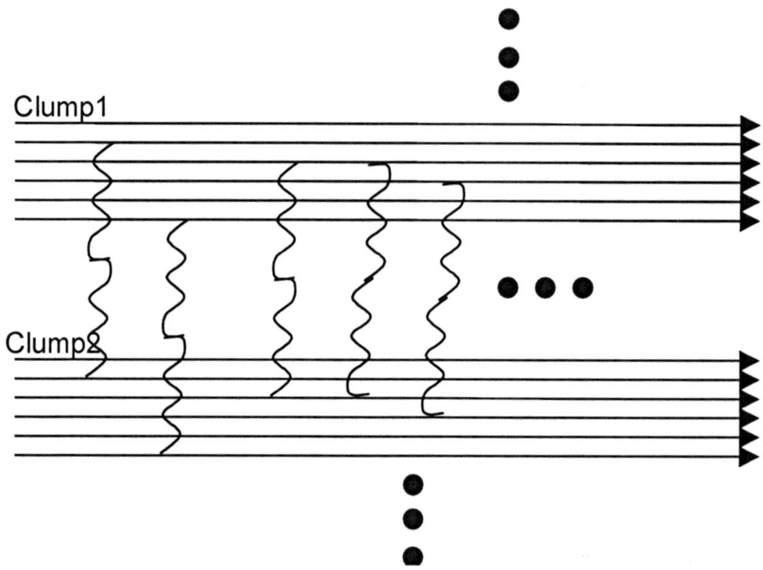

Figure 3.3. Yet another Feynman Diagram for "Causation" and "Change" in The Standard Model. It depicts the gravitational interaction (force exchange) between two clumps of matter via graviton particles on particles within the clumps, which lead the clumps to change direction. This illustrates the force of gravity at the quantum level. As yet the understanding of quantum gravity is subject to dispute. But most physicists would agree that gravitons exist and are the fundamental "carriers" of the force of gravity. Again these are the simplest diagrams of an infinite set of diagrams that contribute to the process of change. Clump1 and Clump2 each consist of a large number of particles in lumps of matter. The thick "dots" represent lots of particles and lots of gravitons being exchanged between the particles in the clumps. The result of the extremely large number of gravitons exchanged is the clumps change direction and approach each other since gravity is a strictly attractive force according to the General Theory of Relativity.

The general, abstract approach of Classical Metaphysics to Causation and Change is defective in not properly dealing with quantum phenomena.[33] However if one wishes to study all possible universes then Classical Metaphysics suitably generalized to include Quantum and Cosmological Reality may support the study of alternate universes. This would open a new sphere of activity for generalized classical metaphysics.

Considerations of the sort illustrated by by Figs. 3.1 – 3.3 lead us to the following additional principle for Relativistic Quantum Metaphysics:

III. At the level of ultimate Reality, Causation and Change take place through interactions (forces) mediated (conveyed) by particles.

3.2.4 Modality

Modality is concerned with questions of necessity and contigency. Is a property necessary to an entity or is a property merely a possible part of an entity? Modality appears in Classical Metaphysics in relation to questions about the nature of entities.

Part of the framework of Modality is the existence of multiple worlds in at least a conceptual sense. Thus if a certain entity existed in all worlds and always had a certain property then that would imply that the property was *necessary* to the entity. If, on the other hand, a certain entity existed in all worlds and had the property in some worlds but not in others then the property would be a *contingent* part of the entity.

Questions of Modality will appear when we consider Relativistic Quantum Metaphysics. We will face situations where a property may be necessary or contingently part of an entity.

It is perhaps appropriate to raise a question of Modality with respect to Relativistic Quantum Metaphysics. We were led to consider a generalization of Classical Metaphysics due to the Theory of Relativity and Quantum Theory. However, it is not clear that we should not consider an even more general metaphysics where Relativity and Quantum Theory are contingent "properties." This *Generalized*

---

[33] For example Leibniz and others suggest that Change must be continuous. It is a known fact that electrons make jumps ("quantum leaps") between orbits in atoms. They are never between atomic levels. Quantum Theory shows the inadaquacy of Classical Metaphysics view of Causation and Change.

Metaphysics would consider worlds where Relativity and/or Quantum Theory did not necessarily hold and where alternatives to Relativity and Quantum Theory might hold. Generalized Metaphysics would then have the goal of determining why our world, which embodies ultimate physical Reality is based on the Theory of Relativity (Special and General) and on Quantum Theory.

3.2.5 From Generalized Metaphysics to The Standard Model

Generalized Metaphysics as we conceived it in the previous subsection assumes the three fundamental principles stated earlier but does not assume the Theory of Special or General Relativity, or Quantum Theory. It assumes an infinity of possible worlds (universes) could exist. It then proceeds to develop selection criteria that lead to the ultimate Reality with which we are familiar. Thus the space-time of Special and General Relativity must be selected.[34] Secondly, the selection criteria must choose Quantum Theory as we know it. Finally the detailed theory of elementary particles and their interactions must follow from the selection criteria.[35] The result would be a true ultimate theory of Reality. Fig. 3.4 symbolizes the steps just outlined.

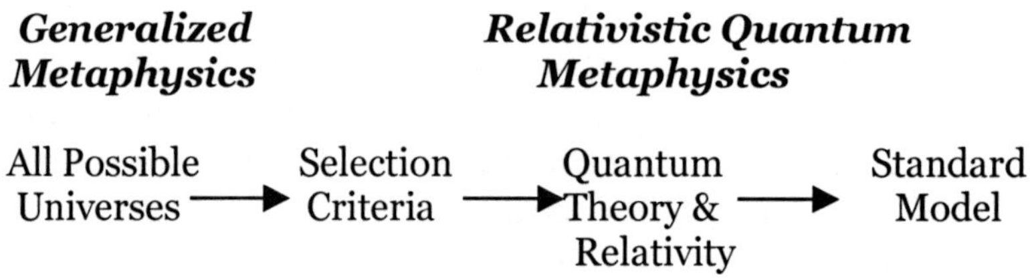

Figure 3.4. A visualization of the steps leading from a Generalized Metaphysics to Relativistic Quantum Metaphysics and thence to The Standard Model.

[34] There is an ongoing discussion of various possible forms of General Relativity. Einstein's formulation or something like it is the most favored form at present. Consequently the selection ctriteria, if found, could be of help in choosing the correct form.

[35] It is of course possible that Quantum Theory and the General Theory of Relativity emerged through mere chance or some arbitrary fluke. In these cases there would be no selection criteria. Ultimate Reality would then be an "accident."

4. Logic and Physics

An experiment is a proposition.

4.1 Logic is the Key to Metaphysics

Classical Metaphysics is largely composed of attempts to define fundamental principles, to clarify concepts and to categorize entities. The key basis of these attempts is Logic. This was pointed out many years ago by such philosophers as Leibniz.[36] Logic enables philosophers to use fundamental principles to derive new principles. Logic enables philosophers to clarify concepts and their differences. Logic is used to categorize and differentiate between entities.

Thus Logic is the engine of Philosophy and Metaphysics as Reality is the meat of Metaphysics.

In considering Metaphysics Leibniz felt that propositions of the subject-predicate form or reducible to a combination of clauses of the subject-predicate form sufficed to meet the needs of metaphysical discussion and analysis. This view has been disputed by some metaphysicians but from the viewpoint of physical Reality it appears the view of Leibniz is correct.

Consequently we will consider (physical?) Logic as applying to axioms and propositions that have subject-predicate form or are reducible to subject-predicate form.

We will relate Logical propositions to experiments on physical Reality. We will then show that defects in Logic that lead to paradoxes can be remedied by a new form of logic called *Operator Logic* in chapter 5. Subsequently we will use Operator Logic to directly construct a basis for physical Reality in a manner reminiscent of Plato's Theory of Ideas and Reality, and their connection.

[36] Rescher (1967) pp. 22-23.

4.2 A Logic View of Physical Experiments

At first glance experiments on physical Reality have no direct connection to Logic. However as Blaha (2009), and his earlier work, showed a quantum experiment (specifying intermediate stages of the experiment and a specific result) can be stated as a proposition in subject-predicate format or a proposition consisting of a set of clauses in subject-predicate format. Thus quantum experiments can be mapped in a one-to-one fashion with propositions. We describe the form of the map in the next section.

4.3 Filtration Stages in a Quantum Experiment and Their Mapping to the Form of a Proposition

If we consider an example of a proposition such as

The car is a small, green, Ford.

we see that the proposition is equivalent to

The car is small, and the car is green, and the car is a Ford.

Now if we consider the set of all cars then we see that the first clause restricts the set of all cars to the subset of small cars. The second clause restricts the subset of small cars to green cars, and the clause restricts the subset to the smaller subset of Ford cars. Each restriction to a yet smaller subset we can call a *filtration* in the sense that each restricting clause filters a set to a smaller subset.

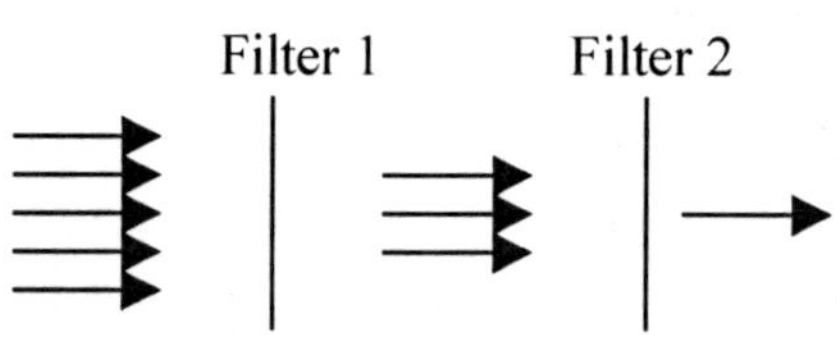

Figure 4.1. A two filter experiment that selects particles with the exact velocity of 100 miles per second from a stream of incoming particles with a variety of velocities.

Now let us consider a (marginally) quantum experiment[37] that progressively filters a stream of particles that have a range of velocities so that only particles with a velocity of 100 miles per second emerge. Let us assume two filters: filter 1 only lets particles with velocities greater than or equal to 100 miles per second pass through, filter 2 only lets particles less than or equal to 100 miles per second pass through. As a result only particles with a velocity exactly equal to 100 miles per second pass through the two filters. Fig. 4.1 illustrates this simple experiment.

This experiment can be expressed as a proposition:

A "stream of particles" "passes through a filter that allows only particles with velocity greater than or equal to 100 miles per second through" and then "passes through a filter that only allows particles of velocity less than or equal to 100 miles per second through" so that only "particles with a velocity that is exactly 100 miles per second" "emerge."

Or, in short,

"A stream of particles with a range of velocities" "is reduced to a stream of particles with a velocity of 100 miles per second."

We use quotes "…" to indicate subjects and predicates.

Clearly there is a conceptual correspondence between logical propositions and experimental propositions. In both cases the predicates can be described as filters that perform a selection upon the set specified by the subject or by the implied set that the subject is generally a member of. In the above logic proposition, the subject begins as a member of the set of all cars and is then filtered to a very specific subset of the set of all cars.

Therefore it is clear that we can view subject-predicate logic propositions and their generalizations to subject-predicate clauses as a series of filtrations similar to the filtrations that take place in quantum physics experiments. We define this feature as the fourth principle:

IV. Subject-Predicate propositions in Logic can be viewed as filtrations analogous to filtrations that take place in quantum physics experiments.

---

[37] The filters are assumed quantum in nature.

With this concept as a beginning point we will proceed to examine Logic, point out some paradoxes,[38] and develop a new form of Logic that we have called Operator Logic. This new logic resolves the issue of paradoxes including Gödel's Undecidability Theorem. It also furnishes a starting point for a transition from the world of Ideas (Logic propositions) to the world of Reality. Pursuing this line of development we eventually reach the form of The Standard Model of Elementary Particles.

[38] There are an infinite number of them.

5. Operator Logic

In this chapter we will establish a new formalism for Logic that we call Operator Logic. We will use Operator Logic as the basis for Relativistic Quantum Metaphysics in following chapters.

5.1 Logic and its Paradoxes

Modern Logic is a very well developed field of studies with many notable achievements. A cloud overhanging Logic is the presence of a number of paradoxes. The number of paradoxes described in the literature is of the order of ten. But they represent an infinity of paradoxical propositions so the problem they represent is significant. Among the better known paradoxes is Gödel's Undecidability Theorem. This theorem asserts that for certain mathematical-deductive systems (and there are many) there exist propositions that can be proven to be true or proven to be false – thus the undecidability.

A number of philosopher-logicians[39] have suggested a mechanism to avoid paradoxes. This mechanism applies to propositions having a subject-predicate form or reducible to a set of clauses – each of which has a subject-predicate form. The mechanism is to assume that each predicate has a set of allowed subjects – a *domain* of allowed subjects. Subjects that are not in the set are excluded from a subject-predicate clause for that predicate. By limiting a domain of subjects, paradoxical propositions can be avoided. As Blaha (2009) points out the Gödel Undecidability Theorem is also evaded.

The procedure of specifying a domain of allowed subjects for each possible predicate is a "manual" procedure in contemporary Logic. However it is possible to create a mathematical formalism that implements domains for predicates and enables propositions to be evaluated as inner products in a linear vector space. This formalism is called Operator Logic because it associates projection operators for

[39] Leibniz may have been the first to suggest this mechanism but without thought of resolving paradoxes.

subjects and predicates that lead to zero-valued inner products for undecidable or paradoxical propositions.[40]

While Operator Logic is of interest in its own right as a means of defining paradox free propositions and mathematical-deductive formalisms, it is also the starting point of the development of a metaphysics that leads to our current understanding of ultimate physical Reality. We shall pursue that possibility later in this book.[41]

5.2 Predicate Calculus

We start by considering the internal logical form of statements[42] and develop a *first order predicate calculus*. The first order predicate calculus deals with simple statements of the form (in English):

subject predicate

as well as more complex forms. Predicates typically contain a verb and often express a property. For example, a simple predicate is "is mortal". A statement generated from this predicate would have the form:

X is mortal.

where X is a subject such as Socrates. In more complex cases there can be multiple subjects and multiple clauses with differing predicates in a sentence.

A more general predicate calculus can be defined that enables more complex statements to be reduced to the form of subject-predicate clauses that are joined by connectives such as "and," "or," and "not." For example the complex statement

The sky and my car are the same color blue.

can be restated as

The sky is blue and and my car is the same color blue.

---

[40] Blaha (2009) and earlier work.

[41] It is very similar to the development in Blaha (2009).

[42] We use statement and proposition interchangeably.

Each clause before and after "and" has a subject-predicate form. Statements about substances in Metaphysics appear to all fall within the framework of general subject-predicate statements.

5.3 Deductive Systems and Their Calculi

Theories usually consist of two types of statements: axioms (assumptions) and theorems. There are many theories of various aspects of mathematics, physics, and other sciences. Some examples include Euclidean geometry, quantum field theory, elementary number theory and so on. All or parts of these theories can be formalized with a set of axioms expressed in terms of primitive concepts, from which theorems are derived. The axioms plus the set of derivable theorems constitutes an *axiomatic system*.[43]

Axiomatic systems can be viewed as consisting of two types: deductive systems[44] and calculi.[45] A *deductive system* has *semantic*[46] characteristics—it defines the primitive terms of the axioms (usually in terms of "real" things such as real lines, real physical particles and their properties, real properties of integers, and so on). Thus the implications of the axioms are "real" as well and describe the reality associated with the deductive system. In physics a deductive system describes an aspect of nature. Examples include the Landau-Ginzberg theory of superconductivity, and Newton's mechanics. In mathematics examples are Euclid's geometry and Gödel's theory of integer numbers. Statements derived in a deductive system are true – a semantic property.

A *calculus* is an axiomatic system without a definition of primitive terms. Theorems are derived following specified rules of derivation with no reliance on semantics. Thus they have syntax without semantics. Most calculi have one or more corresponding deductive systems. It is possible to have a calculus without any known corresponding deductive system. Despite that, a calculus is best viewed as the skeleton of a deductive system. The deductive system constitutes the flesh.

---

[43] Kleene (1967) pp. 198 – 201.

[44] Another name for some deductive systems is mathematical-deductive system.

[45] Other names used for calculi (singular calculus) are formal system, formalism, or logistic system.

[46] We use semantic in its usual sense as the "meaning" of something.

The discussion of a *calculus*, its definition and its consequences, is the *metatheory* or *metalanguage* or *syntax language* of the calculus. The investigation of a calculus is done within the metalanguage and called *metamathematics* or *proof theory*. The calculus itself is often called the *object theory* or *object language*.

While there is much more to be said on this topic, the preceding discussion is all that we need to develop the features of Operator Logic. The reader is referred to the References at the end of this book, and other books on Logic, for more details.

5.4 Universes of Discourse

A generalization of the constituent properties of a statement is needed to conclusively eliminate paradoxical statements from Logic. This generalization has an analogue in the measurement theory of Quantum Mechanics.[47] The generalization is based on the axioms:

0. Every statement appears in a universe of discourse. A *universe of discourse* is a set of statements and phrases. We denote a universe of discourse state in the form $|\Omega_u>$ and treat it as a state in a linear vector space using the notation and concept of quantum state introduced by Dirac (described later).
1. Every statement, and every part of a statement, in a universe of discourse has an associated logical status operator[48] S, which is a projection. These operators when applied to the universe of discourse state $|\Omega_u>$ have two possible types of results. They can yield a zero value or a non-zero state. (A statement yielding a zero value does not have a truth value. A statement yielding a non-zero value does have a truth value – it is either true or false.) Thus $S|\Omega_u> = 0$ indicates undecidable, while the expectation value[49] (inner product) $<\Omega_u|S|\Omega_u> \neq 0$ indicates decidable – either true or false.
2. The status operator S of a statement is the product of the status operators of the phrases (terms) in the statement. A phrase can be

---

[47] See Blaha (2009). The reader who may be apprehensive of the use of linear vector spaces, which is implicit in this introductory discussion, may be reassured somewhat by noting that linear vector spaces together with their mathematical rules can be viewed as a language on a par with human languages and symbolic languages for use in Logic.

[48] Also called a truth value operator.

[49] Inner products are often denoted in mathematics with parentheses such as (x, Ay) where x and y are vectors and A is an operator. Dirac's equivalent notation is $<x|A|y>$ if A is Hermitean.

one word or a finite set of words comprising a concept (an idea) or object. So symbolically we can write

$$S = S_1\ S_2\ S_3\ S_4\ S_5\ \ldots \tag{5.1}$$

where S_1, S_2, S_3, S_4, S_5 and so on represent the status operators of the phrases in the statement S.

3. The status operator of a phrase S_i is determined by its use in the universe of discourse. It is a projection operator satisfying $S_i^2 = S_i$.
4. Projection operators are not necessarily commutative. (This will be discussed in more detail subsequently.)

If we now consider the paradoxical statement and form a trivial universe of discourse with

This sentence is true. (5.2)

and some related terms, we find a simple universe of discourse consists of the subject "This sentence" and the possible predicates "is true" and "is false." We now assign status operators:

Item	Status Operator
"This sentence"	P
"is true"	P
"is false"	1 – P

(5.3a)

with the result

Statement	Status Operator Product
"This sentence is true"	PP = P
"This sentence is false"	P(1 – P) ≡ 0

(5.3b)

where P is a projection operator. Note $P(1 - P) = (P - P) \equiv 0$. Consequently we find "This sentence is false." does not have a truth value while eq. 5.2 is does have a truth value. A paradoxical statement becomes a non-paradox by the introduction of a new formalism for statements, subjects, and predicates: namely status operators.

Another simple example is the Grelling paradox:

"Heterological is heterological."

If we define a simple universe of discourse with the status operators:

Item	Status Operator
"Heterological"	P
"is heterological"	1 – P

(5.4a)

then

Statement	Status Operator Product
"Heterological is heterological."	$P(1 - P) \equiv 0$

(5.4b)

we see the paradox is resolved: the statement eq. 5.4b does not have a truth value. A predicate containing a word, which is also a subject, may have a different status operator.

In the case of more complex universes of discourse, one introduces as many status operators as needed: P_1, P_2, P_3, and so on in such a way as to exclude paradoxical statements. This can be systematically implemented by establishing all combinations of subjects and predicates that are paradoxical and defining projection operators so as to exclude paradoxical statements from the set of valid statements.

The discussion in this section is meant to give the flavor of Operator Logic without obscuring it with numerous technical details.

5.5 Operator Representation of a Statement

In the previous section we introduced status operators, which are projection operators. We can extend the operator formalism to include interpreting statements as the "eigenvalues" of sequences of eigenvalue operators where the "eigenvalues" are words, phrases, or sequences of logical symbols.[50] We will call these strings of characters "eigenvalues"

[50] Thus terms can be primitive terms, defined terms, functions, variables, and logical and mathematical symbols.

as well, although eigenvalues are normally numeric.[51] A *term* is one or more words or symbols. The terms that compose a statement can be viewed as occurring in consecutive "stages" (although the exact ordering is language dependent.)

We can develop an operator formalism that will yield statements as a sequence of eigenvalues of operators.

5.5.1 Dirac Notation vs. (x, y) Notation for Inner Products

A simple example that illustrates the use of Dirac notation is to define a state vector with two "eigenvalues" such as

$$|\Omega_l\rangle = |\text{"The car", "is red"}\rangle \tag{5.5}$$

and two operators, Subject and Predicate, satisfying the "eigenvalue" equations:

$$\text{Subject}|\Omega_u\rangle = \text{"The car"}\ |\Omega_u\rangle \tag{5.6}$$
$$\text{Predicate}|\Omega_u\rangle = \text{"is red"}\ |\Omega_u\rangle \tag{5.7}$$

with the resulting inner product[52]

$$\langle\Omega_u|\text{Subject} + \text{Predicate}|\Omega_u\rangle \equiv (\Omega_u,(\text{Subject} + \text{Predicate})\Omega_u) = \text{"The car is red"} \tag{5.8}$$

where a + sign is used as a separator between the operators.

In the present example the "eigenvalues" are strings of characters. Mathematically, we can relate each string of characters (symbols) in a one-to-one manner[53] to a real numeric value by mapping strings into Gödel numbers, denoted gn(i). By mapping strings into Gödel numbers[54] we can establish numbers as eigenvalues for our

---

[51] We convert terms to numeric values – Gödel numbers – at a later point in the discussion. Thus treating terms as eigenvalues is legitimatized by a mapping from terms to numbers.

[52] The Operator Logic eigenvalue operators and the projection operators are all hermitean in this book. Thus $(\alpha, O\beta) = (O\alpha, \beta)$ due to O's hermiticity and we can use Dirac notation $\langle\alpha|O|\beta\rangle$ without ambiguity due to O's hermiticity.

[53] Due to the "Fundamental Theorem of Arithmetic", namely that every integer number greater than one, which is not a prime itself, has a unique representation in terms of prime number factors.

[54] It should be noted that Leibniz developed a procedure to map strings to integers. See Rescher (1967) p. 132 and cited references to Leibniz.

eigenvalue operators. The set of operators can then be defined within the context of a linear vector space formalism. Thus we reduce the various eigenvalue operators of interest to self-adjoint linear vector space operators.

5.5.2 Gödel Numbers

Gödel numbers are real positive numbers. They can be viewed as the eigenvalues of self-adjoint operators A_i corresponding to Subject and Predicate in the above:

$$A_{Subject}|\Omega_u> = gn(\text{“The car”})|\Omega_u> \qquad (5.9)$$
$$A_{Predicate}|\Omega_u> = gn(\text{“is red”})|\Omega_u>$$

and $|\Omega_u>$ can be viewed as the direct product of eigenstates of $A_{Subject}$ and $A_{Predicate}$

$$|\Omega_u> = |\ gn(\text{“The car”})>|\ gn(\text{“is red”})> \qquad (5.10)$$

Using plus signs as separators we express the statement as

$$(A_{Subject} + A_{Predicate})|\Omega_u> = \text{“The car is red”}|\Omega_u> \qquad (5.11)$$

A string expression such as

AyyBx

is assigned a Gödel number by combining the assigned numerical values of each symbol in the string. The numerical values become exponents of a product of prime numbers greater than one. For example the symbols in the preceding string expression could have the symbol numbers: A ≡ 3, y ≡ 9, B ≡ 5 and x ≡ 7. Then its <u>G</u>ödel <u>n</u>umber is

$$gn(\text{“AyyBx”}) = 2^3 3^9 5^9 7^5 11^7 \qquad (5.12)$$

where the numbers being exponentiated are the ordered set of prime numbers.

In general, if E is an expression (string) of symbols, β_1, β_2, β_3, β_4, … β_n and v_1, v_2, v_3, v_4, … v_n is the sequence of integers corresponding to these symbols, then the Gödel number of E is the integer

$$gn(E) = \prod_{m=1}^{n} Pr\,(m)^{v_m} \tag{5.13}$$

where $Pr\,(m)^{v_m}$ is the mth prime number raised to the power v_m with the 1st prime number taken to be 2. If E is an empty string (containing no symbols) then gn(E) = 1.

The definition of Gödel numbers, and their consequent uniqueness, results in the following conclusions and definitions:

1. If g = gn(E), then the inverse relation is defined to be E = gi(g).

2. If E_1 and E_2 are expressions (terms) such that $gn(E_1) = gn(E_2)$, then $E_1 = E_2$.

3. If $E_1, E_2, E_3, \ldots, E_n$ are a finite sequence of expressions, then the Gödel number of the sequence is

$$gn(E_1, E_2, E_3, \ldots, E_n) = \prod_{m=1}^{n} Pr\,(m)^{gn(E_m)} \tag{5.14}$$

4. The Gödel number of an expression is never equal to the Gödel number of a sequence.

5. If $E_1, E_2, E_3, \ldots, E_n$ and $E'_1, E'_2, E'_3, \ldots, E'_n$ are both finite sequences of expressions and

$$gn(E_1, E_2, E_3, \ldots, E_n) = gn(E'_1, E'_2, E'_3, \ldots, E'_m) \tag{5.15}$$

then n = m, and for k = 1, 2, …, n

$$E_k = E'_k$$

and the strings to which they correspond are identical.

5.5.3 Statements Expressed in Gödel Numbers

Statements can be expressed in terms of Gödel numbers. In this subsection we describe the general interpretation of expressions involving eigenvalue operators.

If eigenvalue operators are added as in eq. 5.11 then the sum of eigenvalue operators can be viewed as a symbolic list of Gödel numbers that is not arithmetically added or multiplied. As we read a statement from word to word the statement "unfolds" and becomes particularized, more precise and more specific. In this sense the symbolic sum of eigenvalue operators specifies the statement when applied to the state that represents the system. Placing eigenvalue operators adjacent to each other[55] is interpreted as equivalent to the logical relation "or." We add an additional term "is green" with operator $A_{\text{is green}}$ to the previous example and use the state

$$|\Omega_2> = |\text{ gn("The car")}>|\text{ gn("is red")}>|\text{ gn("is green")}>$$

Then the statement

$$(A_{\text{Subject}} + A_{\text{is red}}A_{\text{is green}})|\Omega_2> \equiv \text{"The car is red or is green."}|\Omega_2>$$

exemplifies the "or" case. Alternately, we could define "or" as a term with a corresponding Gödel number.

$$(A_{\text{Subject}} + A_{\text{is red}} + A_{\text{or}} + A_{\text{is green}})|\Omega_3> \equiv \text{"The car is red or is green."}|\Omega_3>$$

where

$$|\Omega_3> = |\text{ gn("The car")}>|\text{ gn("is red")}>|\text{ gn("is green")}>|\text{ gn("or")}>$$

---

[55] This corresponds with the "or" notation of Hilbert (1950) p. 12 where XY = X v Y.

6. Linear Vector Spaces for Universes of Discourse

A statement is more than a statement.

We will define two types of universes of discourse: a semantic universe of discourse and a calculus universe of discourse. A *calculus universe of discourse* consists of undefined primitive terms, axioms, and the theorems derived from these axioms. Statements in a calculus are provable, not provable or undecidable.

A *semantic universe of discourse* consists of primitive terms defined in some fashion, axioms, and the theorems derived from the axioms. Statements in a semantic universe are true, false, or undecidable.

A calculus universe of discourse can be viewed as a semantic universe of discourse stripped of its definition of primitive terms and reduced to a "skeleton of algebra" with truth reduced to provability.

6.1 The Semantic Universe of Discourse

A *semantic universe of discourse* consists of a set of primitive and defined terms, related to some Reality,[56] that are used to formulate the axioms and theorems of a deductive system. The terms are mapped to Gödel numbers. Each term E has a unique Gödel number gn(E). Each Gödel number is assumed to be an eigenvalue of a unique, self-adjoint, eigenvalue operator that we denote A_E. Each A_E has two eigenvalues E = gn(" … "), the Gödel number of a non-empty string, or E = gn("") = 0. We denote an eigenstate as |E>. The eigenstate |0> corresponds to the empty string. If there are n primitive terms in a universe of discourse, then the *universe of discourse state* $|\Omega_u>$ is a direct product of all term eigenstates with the form:

$$|\Omega_u> = |E_1, E_2, E_3, \dots , E_n> = \prod_{m=1}^{n} |E_m> \qquad (6.1)$$

[56] Related to Reality in the sense that each primitive term is "defined" in terms of real world features.

where each E_i can be a non-zero Gödel number, or zero. Note: since n is the total number of terms in the semantic universe of discourse it must be finite since the number of words in the English language, or any language expressing semantic meanings, is finite. Each eigenvalue operator A_{E_k} satisfies the equation[57]

$$A_{E_k}|E_1, E_2, E_3, \ldots , E_n> = gn(E_k)|E_1, E_2, E_3, \ldots , E_n> \qquad (6.2)$$

for k ε [1, n]. *All eigenvalue operators in a universe of discourse in this chapter commute and are thus compatible.*[58]

A semantic statement in this notation has the symbolic form

$$\sum_{n \in \{m\}} A_{E_n} \qquad (6.3)$$

for some subset {m} of the set of terms in the universe of discourse. If applied to the eigenstate eq. 6.1 we obtain the *symbolic* sum

$$\sum_{n \in \{m\}} A_{E_n}|E_1, E_2, E_3, \ldots , E_n> = \sum_{n \in \{m\}} gn(E_n)|E_1, E_2, E_3, \ldots , E_n> \qquad (6.4)$$

The Gödel numbers in eq. 6.4 are not literally added together. Rather the terms in the symbolic summation on the right side are individually rendered in a language[59] using the inverse relation of Gödel numbers gi(g), (which is property 1 in section 5.5.2) on each term of the symbolic sum. Additional words such as "a" and "the" are added to the statement implicitly to make it grammatically correct.

---

[57] **Each eigenvalue operator and its corresponding projection operators operate on the corresponding eigenstate in the product of eigenstates comprising $|\Omega_u>$. Thus, although we do not make it explicit to avoid complicated expressions, one can consider each eigenvalue operator and its projections to be a direct product with the implicit factors being identity operators for the spaces spanned by the other eigenvectors in $|\Omega_u>$ of the other eigenvalue operators.** This also applies in the discussions in the following chapters.

[58] In Quantum Operator Logic, described in a later chapter, eigenvalue operators do not necessarily commute.

[59] Spaces and other required grammatical constructs such as "a" and "the" are assumed to be automatically inserted when translated into a language.

Thus we have a linear vector space formalism for the semantics of a deductive system. We thus "turn the tables" and use the very well known mathematics of linear vector spaces as the "language" of our semantic universe—rather than English or an artificial symbolic language, which both contain the possibilities of paradox and inconsistency.

One immediate benefit of a linear vector space formulation is that each self-adjoint eigenvalue operator A_E has a corresponding projection operator[60] P_E that projects a composite state

$$|E_{tot}> = a|E_1> + b|E_2> \tag{6.5}$$

into a specific eigenvalue state $|E_1>$ or $|E_2>$: [61]

$$P_{E_1}|E_{tot}> = a|E_1> \tag{6.6}$$

In addition the operator $1 - P_E$ projects a state $|E>$ to 0. Thus

$$(1 - P_{E_1})|E_1> = 0 \tag{6.7}$$

The projection operators for eigenvalue states are combined with semantic status projection operators P'_E that are used to determine whether a statement has a truth value:

$$P_{Etot} = P'_E P_E \tag{6.8}$$

The set of these projection operators $\{P_{Etotk}, (1 - P_{Etotk})\}$, becomes the status operators discussed in section 5.4. *All projection operators in a universe of discourse in this chapter commute.* The general form of the projection operator expression corresponding to eq. 6.4 that determines whether the statement has a truth value is

$$F(P'_{E_1}, P'_{E_2}, \ldots) = \prod_{n \in \{m\}} P'_{E_n} \tag{6.9}$$

---

[60] Mackey (1963) pp. 64-65.

[61] The state $|0>$ corresponds to the empty string.

If

$$<\Omega_u|F(P'_{E_1}, P'_{E_2}, \ldots)|\Omega_u> = 0 \quad (6.10)$$

then the corresponding statement has no truth value. If

$$<\Omega_u|F(P'_{E_1}, P'_{E_2}, \ldots)|\Omega_u> \neq 0 \quad (6.11)$$

then the corresponding statement has a truth value that may be true or false.

The product of the projections P_{E_n} will be non-zero for the universe state (eq. 6.1), and zero if any of the eigenvalues of the corresponding eigenvalue operator A_{E_n} for some $n \in \{m\}$ is zero.[62]

The form of $F(P'_{E_1}, P'_{E_2}, \ldots)$ given above is for a simple subject-predicate statement. The form of $F(P'_{E_1}, P'_{E_2}, \ldots)$ is more complex for more complex statements reducible to a set of subject-predicate clauses (that depend on the logical constructs "and", "or", and so on). It will be discussed later.

We can view a projection operator as projecting or "filtering" a state and producing a new state in general.[63] Each projection operator in a series of projection operators filters a state step by step to produce a final state. The most general form of projections for a simple subject-predicate statement is:

$$F(P'_{E_1}, P'_{E_2}, \ldots) \prod_{n \in \{m\}} P_{E_n}|E_1, E_2, E_3, \ldots, E_n> \quad (6.12)$$

It determines whether a statement has a truth value (a non-zero value) or not (a zero value) when one takes its inner product with the initial universe state:

$$<E_1, E_2, E_3, \ldots, E_n|F(P'_{E_1}, P'_{E_2}, \ldots) \prod_{n \in \{m\}} P_{E_n}|E_1, E_2, E_3, \ldots, E_n> \quad (6.13)$$

---

[62] Universe states containing some zero eigenvalues occur in the case of subuniverse states, which are described later.

[63] This is similar to the role projection operators play in the analysis of quantum physics experiments. See Blaha (2009) for a more detailed discussion.

yields either zero (no truth value) or non-zero (has a truth value).

6.2 The Calculus Universe of Discourse

A *calculus universe of discourse* consists of a set of primitive terms, and terms defined in terms of them. They are unrelated to any reality and have only a symbolic significance. They are used to formulate the axioms and theorems of a deductive system. A calculus may correspond to one or more semantic deductive systems. The formulation of a calculus universe of discourse is analogous to that of a semantic universe of discourse.

The Operator Logic for a calculus universe of discourse is identical to that of the semantic universe of discourse as described in section 6.1.

6.3 Variables in Operator Logic

If a primitive term is in fact a function of one or more variables x_1, x_2, …, then the corresponding eigenvalue operator and projection operator are also functions of these variables: $A_E(x_1, x_2, \ldots)$ and $P_{Etot}(x_1, x_2, \ldots)$ respectively for semantic and calculus universes of discourse.

If a free variable v appears in a statement, then it has an eigenvalue operator $A_v(v)$, and a projection operator, $P_v(v)$. A function f(x) has an eigenvalue operator $A_f(x)$ with projection operators $P_f(x)$. The generalization to functions of several values is direct.

6.4 Linking Universes of Discourses Linear Vector Spaces

The calculus and semantic Linear vector space statements are linked using a generalization of Tarski's method of combining a calculus sentence with its semantic meaning through eq. 6.14:

$$\text{SM is true if, and only if, S} \qquad (6.14)$$

as a starting point where S is a statement and SM is its semantic equivalent. If S is true, then SM is true. If S is false, then SM is false. If S is Gödel undecidable, then SM can be true or false.

Clearly, statement by statement, we can map semantic statements in a deductive system to syntactic statements in its corresponding calculus. We can always define a *perfect map* on the operator level in which eigenvalues and their corresponding eigenvalue operators have a one-to-one map as do the projection operators.

For example, consider a semantic statement with the form of eq. 6.4. Its corresponding product of projection operators is

$$P_S = F(P'_{E_1}, P'_{E_2}, \ldots) \prod_{n \in \{m\}} P_{E_n} \tag{6.15}$$

The product of projection operators for the corresponding calculus statement is

$$Q_S = F(Q'_{E_1}, Q'_{E_2}, \ldots) \prod_{n \in \{m\}} Q_{E_n} \tag{6.16}$$

where Q_{E_n} is the calculus projection operator corresponding to P_{E_n}, and Q'_{E_n} is the status operator corresponding to P'_{E_n}. The operator equivalents of eq. 6.14 and its associated comments are:

1. Q_S is non-zero if, and only if, P_S is non-zero (indicating that the statement can be proven true or false, or provable or not provable, in both universes of discourse.)
2. Q_S is zero if, and only if, P_S is zero (indicating that the statement is Gödel undecidable in both universes of discourse – it cannot be proved and cannot be disproved.) However, the statement may be shown true or false (or provable or not provable) by other means – for example, from experiments.

where it is assumed that these projection operator expressions are evaluated between universe states.

The reader will note that these points differ from the Tarski formulation (eq. 6.14) in several ways. Tarski's formulation assumes that the statements, S and SM, are either true or false – contrary to the implications of Gödel's Undecidability Theorem. Points 1 and 2 above are consistent with the Undecidability Theorem. Item 2 provides a calculational approach to determine "decidability."

English Language Statement	Semantic System Statement	Calculus Statement
Eigenvalue Operator Sum	Semantic Eigenvalue Operator Sum	Syntax Eigenvalue Operator Sum
Status Projection Operator Product	Semantic Projection Operator Product	Syntax Projection Operator Product

Figure 6.1 A table of the syntactical and semantic operators associated with a statement.

Earlier we considered a simple example of a two statement semantic system, which is also its calculus:

(1) This sentence is false.
(2) This sentence is true.

Statement (1) is undecidable – being neither true nor false. Statement (2) is true. We do not differentiate between the semantic system and the calculus since they are essentially identical.

The general case of the operator representation of an English (or other) language statement in a semantic system and its corresponding calculus statement is displayed in Fig. 3.1.

6.5 Predicate Calculus

Returning to an earlier example (subsections 5.5.1 – 5.5.3) we see that the predicate calculus can be implemented in our linear vector space representations. First we note the statement generated from the Subject and Predicate operators is

$$((\text{Subject}) + (\text{Predicate}))|\Omega_u> = \text{“The car is red”}\ |\Omega_u> \qquad (6.17)$$

which again we note is implemented via corresponding self-adjoint operators with Gödel number eigenvalues.

One can define projection operators:

$$P_{\text{Subject(“The car”)}} = P_1$$

and (6.18)

$$P_{\text{Predicate(“is red”)}} = P_2$$

The simplicity of the example allows us to treat semantic and calculus operators similarly. The product of the projection operators for the statement, when evaluated for the universe state is not identically zero, and so the statement has a non-zero truth value of true or false.

This simple example can be directly extended to more complex statements with multiple subjects and clauses.

6.6 General Approach to Subjects and Domains of Predicates

In a given universe of discourse we can introduce a projection operators P_i for each predicate, or for primitive terms within a predicate as the case may be. Then, for each predicate P_i, we determine the class of acceptable subjects (subjects, which when combined with the predicate to form a statement, form a valid statement that is either true or false). We can assign a projection operator P_i to each member of the class of allowed subjects.

Unacceptable subjects (subjects, which, when combined with P_i to form a statement, form an invalid statement that has no truth value) are assigned the projection operator $1 - P_i$.

Thus, in a universe of subjects and predicates, each predicate has a class of acceptable subjects (its *domain* D of allowed subjects) and a complementary class of unacceptable subjects.

We implement domains of predicates assigning a set of projections to each predicate—one truth value projection for each subject in the domain of subjects for the predicate. For predicate p we will denote its *predicate projection operator* as

$$P_{ptot} = P'_p P_p \tag{6.19}$$

using the projection operators defined in the discussion of eqs. 6.5 – 6.8. The *state projection operator* P_p when applied to an eigenstate $|p\rangle$ of the predicate satisfies $P_p|p\rangle = |p\rangle$ if p is not a predicate consisting of an empty string (which has Gödel number zero). ($P_p|0\rangle = 0$ for an empty string.) The *predicate truth value projection operator* P'_p is defined by[64]

$$P'_p = \prod_{i \varepsilon D} P'_i \prod_{j \notin D} (1 - P'_j) \quad (6.20)$$

where the combined set of projections denoted above as P'_i and P'_j are the predicate truth value projection operators for all subjects in the universe of discourse. P'_p satisfies

$$P'_p P'_k = P'_k P'_p = P'_p \quad (6.21)$$

for predicate p with subject k contained in D, and

$$P'_p P'_k = P'_k P'_p = 0 \quad (6.22)$$

for predicate p with subject k *not* contained in D.

For each subject s we associate a *subject projection operator:*

$$P_{stot} = P'_s P_s \quad (6.23)$$

where P_s is the *subject state projection operator.* When P_s is applied to a subject eigenstate $|s\rangle$ it satisfies $P_s|s\rangle = |s\rangle$ if the subject is not an empty string (which has Gödel number zero). ($P_s|0\rangle = 0$ for an empty string.) P'_s is the *subject truth value projection operator* of the subject s.

6.7 Formalism for Operator Logic

In this section we will look at the operator formalism for a semantic universe of discourse. A *statement* is represented by an inner product[65] containing a sum of the eigenvalue operators of the statement:

$$\text{statement} = \langle\Omega_u|A_1 + A_2 + A_3 + \ldots |\Omega_u\rangle \quad (6.24)$$

---

[64] We have called these operators, and corresponding subject operators, status operators in earlier chapters.

[65] Expressions of the form $\langle\alpha|O|\beta\rangle$ are inner products that are called *expectation values* in quantum mechanics and are often denoted as $(\alpha, O\beta)$ in mathematics texts where O is a hermitean operator. The Operator Logic eigenvalue operators and the projection operators are all hermitean. Thus $(\alpha, O\beta) = (O\alpha, \beta)$ due to O's hermiticity and we can use Dirac notation $\langle\alpha|O|\beta\rangle$ without ambiguity due to O's hermiticity.

where A_i is the ith eigenvalue operator, and its truth value[66] is given by

$$TV = \text{Truth Value} = \langle\Omega_u|\ F(P'_{E_1}, P'_{E_2}, \ldots) \prod P_{E_n} |\Omega_u\rangle \qquad (6.25)$$

TV = 0 indicates the statement has no truth value. TV ≠ 0 indicates the statement has a truth value: true or false. The projections appearing in F and the product are those corresponding to the eigenvalue operators in the statement.

We will define the *universe state* $|\Omega_u\rangle$ for a universe with q primitive terms as

$$|\Omega_u\rangle = |E_1, E_2, E_3, \ldots, E_q\rangle \qquad (6.26)$$

where each eigenvalue E_i is a non-zero Gödel number and the set of integers 1, 2, …, q is denoted {m} and $\langle\Omega_u|\Omega_u\rangle = 1$. The hermitean conjugate of $|\Omega_u\rangle$ is denoted $\langle\Omega_u|$. Our equations use Dirac's bra-ket notation for states and inner products.

A statement is composed of adjoining (a symbolic sum of) Gödel numbers: gn(i). Using the inverse function gi(gn(i)) on each adjoining Gödel number (separated by symbolic plus signs) the statement can be "translated" into English or some other human or symbolic language. The plus signs act as separators and are subsequently discarded. Grammatical constructs such as "a" and "the" can be added as needed to make a grammatically correct statement.

6.8 A Simplified Example

One simple example of a semantic universe has the primitive terms (and corresponding eigenvalue operators):

Primitive Term	Eigenvalue Operator	Truth Value Projection Operator
"This sentence"	A_1	P_1
"Some Sentence"	A_2	P_2
"is true"	A_3	$P_3 = P_1$
"is false"	A_4	$P_4 = 1 - P_1$

[66] In the case of a calculus universe of discourse truth value becomes provability.

where we assign projection operators to each term using an operator P defined below. The primitive terms consist of two predicates, "is true" and "is false". The other primitive term is the subject: "This sentence". The operator $P_2 = P$ projects an eigenstate as follows

$P_1|\Omega_1\rangle$ = |"This sentence", "Some sentence", "is true", "is false">

while

$$(1 - P_1)|\Omega_1\rangle = 0$$

and

$(1 - P_1)|$ "", "Some sentence", "is true", "is false"> =
= | "", "Some sentence", "is true", "is false">

We can construct four simple statements from these primitive terms:

English Statement	Eigenvalue operator Sum	Projection Operator Inner Product[67] $\langle\Omega_1|\text{product}|\Omega_1\rangle$
"This sentence is true"	$A_1 + A_3$	$P_1P_3 = P_1 \neq 0$
"This sentence is false"	$A_1 + A_4$	$P_1P_4 = P_1(1 - P_1) \equiv 0$
"Some sentence is true"	$A_2 + A_3$	$P_2P_3 = P_2P_1 \neq 0$
"Some sentence is false"	$A_2 + A_4$	$P_2P_4 = P_2(1 - P_1) \neq 0$

where the zero and non-zero values in the third column result from taking the inner product $\langle\Omega_1|\text{product}|\Omega_1\rangle$ using the universe state $|\Omega_1\rangle$.

From the preceding table we see that the statements represented by $A_1 + A_3$, $A_2 + A_3$ and $A_2 + A_4$ have truth values (either true or false). The statement represented by $A_1 + A_4$ does not have a truth value.

The statements represented by $A_1 + A_3$, $A_2 + A_3$ and $A_2 + A_4$ have the truth value true from semantic considerations. This result is not due to the projection operator results. Projection operators only determine whether a statement has a truth value, or does not have a truth value, *at this point in our discussion.*[68]

[67] The universe states are not displayed in this column to save space.

[68] At a later point in the development they will be used to determine the truth or falsity of such statements.

For each of the predicates A_3 and A_4 we identified a domain of subjects that lead to statements that have a truth value. For A_3 we find A_1 and A_2 are subjects that leads to statements with a truth value and thus constitute the domain of A_3. In the universe specified above we find A_4 has a domain consisting of A_2.

It is clear that the identification of the domain of subject terms for each predicate that leads to statements with a truth value is the key to excluding paradoxical statements. We will see this in more detail after discussing primitive terms in general in our formulation. This discussion can be directly generalized to more complex statements.

6.9 The Primitive Terms of Universes of Discourse

The set of primitive terms of a universe of discourse consists of "string eigenvalues" that are mapped to the Gödel numbers of the corresponding eigenvalue operators of the universe. Each primitive (and defined) term maps to one eigenvalue operator. Then all statements in the universe of discourse can be expressed as symbolic sums of the corresponding eigenvalue operators.

The primitive terms are the key to the entire universe of discourse. We now consider primitive terms in human and symbolic languages.

Statements can be expressed in primitive terms in human languages such as English or in artificial symbolic languages. A problem that arises—particularly in human languages—is that there are different ways of expressing the same statement and different ways of specifying the same primitive terms. A particularly good example of this issue appears when one compares a primitive term in German, which can create words by combining other words, and English where the equivalent of a German primitive term may be a phrase.

In this section we point out that one can define a map between the primitive terms of two semantic universes of discourse that have the same calculus universe of discourse. Thus no problem can arise if we choose to use differing primitive terms in two semantic universes of discourse with the same calculus universe of discourse. A famous example of this situation is the calculus of Euclidean geometry *without* the fifth postulate. Depending on the definition of "point", "straight line" and "congruence" one obtains Euclidean geometry or a non-Eucliean geometry.[69]

Thus the issue of primitive terms is important in the consideration of semantic universes of discourse. We will now consider general features of primitive terms.

A universe of discourse consists of

1. Primitive terms
2. Axioms expressed in terms of the primitive terms
3. Terms defined as combinations of primitive terms
4. A set of theorems derivable from the axioms

The primitive terms in a calculus universe of discourse are undefined. The primitive terms in a semantic universe of discourse are given a meaning external to the universe of discourse. Frequently the meaning is based on physical or mathematical entities.

If two semantic universes of discourse have the same calculus universe of discourse then the primitive terms in each semantic universe must map to the other on a one-to-one basis. (It is possible that a primitive term in one universe of discourse maps to a pair of (or several) primitive terms in the other universe of discourse. However the pair of (or several) terms always appear together in the other universe of discourse with the result that they are effectively one term.)

6.10 Predicate Domains and Paradox Avoidance

From the discussion of previous sections it is clear that each predicate has a domain of allowed subjects.[70] The subjects that are not in the domain of allowed subjects can be used to form statements. But these statements would have no truth value.[71] They would be neither true nor

---

[69] Weyl (1950) p. 81.

[70] In this respect our view is contrary to Leibniz's view of 1686, and other views, that the predicate is contained in the concept of the subject in each true statement. Our view does agree with Frege's use of a functional notation wherein the predicate is the function and a subject is the variable. The set of subjects constitutes the domain of the function. Further, it agrees with Hilbert (1950), "To each predicate corresponds a certain 'class' of objects, consisting of all objects for which the predicate holds." p. 46.

[71] Truth value becomes provability in a calculus universe of discourse.

false and thus would properly be called undecidable or paradoxical statements.[72]

We will now discuss the resolution of most of the well-known paradoxes using our new formalism.

Liar Paradox

The statement of the Liar Paradox has a subject such as Subject = "The statement that I am now making" and predicate "is a lie." If the subject is reflexive as it appears implicitly in the predicate, which might be better phrased as "is my lie", then a paradox is evident. The elimination of this paradox is accomplished by limiting the domain of the predicate to non-reflexive subjects. In the Operator Logic formalism the domain of the predicate operator is limited to non-reflexive subject operators. One implements this by assigning a projection P to the predicate and the projection (1 – P) to all reflexive subject operators. Thus paradoxical statements of this type have no truth value and the Operator Logic formalism excludes such paradoxes.

Grelling Paradox

Some adjectives have the property that they apply to themselves. Such adjectives can be called autological. Other adjectives do not apply to themselves. These adjectives can be called heterological. The Grelling paradox is:

"Heterological is heterological."

If we identify the primitive terms as the predicate "is heterological" and the subject "Heterological", then assigning eigenvalue operators to each term, and the projection operator P as the subject projection operator, and (1 – P) as the predicate projection operator gives the Grelling statement no truth value, thus eliminating the paradox.

Barber Pseudoparadox

The Town mayor issues an order: "The one town barber must shave those men in the village who do not shave themselves." In the simplest

[72] Herbrand, Schmidt, Wang, Hintikka, Hailperin, Lightstone & Robinson, Gilmore, and Quine have considered approaches within the framework of earlier formalisms based on restricting the subject domain.

Operator Logic form the subject is "The one town barber" and the predicate is "must shave those men in the village who do not shave themselves." Assigning eigenvalue operators to each term and the subject projection operator P with the predicate projection operator (1 – P) results in the statement not having a truth value. Thus the paradox is avoided.

Berry Paradox

The number of positive integers that can be named in English in less than a fixed number of syllables is finite. Thus there must be a least integer that cannot be so named. However, " the least integer that cannot be named in English in less than fifty syllables" is an English name of less than fifty syllables. Thus the least integer has a name contrary to the assumption and thus a paradox.

Resolution:

This superficial paradox is based on an ambiguity in the use of the word "named." If it means "numerically named" (in words representing a numerical value) there is no paradox. If it means "any type of name" then an apparent paradox appears until one realizes that it is then a tautology "The least integer that cannot be named in English in less than fifty syllables is the least integer that cannot be named in English in less than fifty syllables."

Russell's Paradox

From experience we know we can consider classes of things; classes of integer numbers, classes of cars, and so on. We can consider classes of classes such as the class of all classes of cars in the various big cities of America. There are two interesting varieties of classes: proper classes and improper classes. *Proper classes* are classes, which are not members of themselves. For example, the class of all cars in China is proper. *Improper classes* are classes, which are members of themselves. For example, the class of all classes is a member of itself and thus improper.

Let us define the Russell class R as the class of all proper classes. Paradox: if R is a proper class, it is a member of itself, and is thus by definition an improper class. If R is an improper class, it is not a member of itself and therefore, by definition, it is a proper class. Thus there is no resolution of this paradox according to conventional logic.

Resolution:

The primitive terms of this paradox are the subject, "the class of all proper classes" and the predicate, "is a proper class." After associating an eigenvalue operator and projection operator with each term we see that the statement "the class of all proper classes is a proper class" has no truth value if we specify a projection operator P for the subject and (1 – P) for the predicate. This is again a case where the paradox is avoided by the predicate having a non-reflexive domain.

Cantor Paradox

According to the theory of cardinal (infinite) numbers the set of all subsets of a set C has a higher cardinal number than C. If C is the set of all sets, then the preceding statement is a contradiction.

Resolution:

The Cantor paradox can be changed to the following by simple substitution for C:

The set of all subsets of the set of all sets has a higher cardinal number than the set of all sets.

The subject can be simplified to give:

The set of all sets has a higher cardinal number than the set of all sets.

which is a manifestly verbal paradox but not an Operator Logic paradox.

If we identify the subject term as "The set of all sets" and the predicate term as "has a higher cardinal number than the set of all sets", and define corresponding eigenvalue and projection operators, then the statement can be shown to have no truth value eliminating the paradox. We choose the projection operator for the subject as P and the projection operator for the predicate as (1 – P) with the result that the statement has no truth value within the framework of Operator Logic. Again the paradox is avoided by the predicate having a non-reflexive domain.

Burali-Forti Paradox

The Burali-Forti paradox is based on the theory of transfinite (infinite) numbers. The theory proves a) every well-ordered set has a unique ordinal number; b) any set of ordinals, that is placed in a natural order such that each element contains all its predecessors, has an ordinal number which is greater than any preceding element in the set; and c) the set A of all ordinals placed in natural order is well-ordered. Then by theorems a and c, A has an ordinal number n. Since n is in A we see n < n by theorem b, thus establishing a contradiction.

Resolution

The paradoxical conclusion can be stated verbally as:

> The ordinal of the set of all ordinals is greater than the ordinal of the set of all ordinals.

In this statement we can identify the subject primitive term as "The ordinal of the set of all ordinals" and the predicate primitive term as "is greater than the ordinal of the set of all ordinals." Here again we see a reflexivity in the subject. If we restate the conclusion in terms of eigenvalue operators with the subject projection operator P and the predicate projection operator (1 – P) then the inner product $<\Omega_u|P(1 - P)|\Omega_u> = 0$ indicating that the statement has no truth value. Thus the statement is excluded from the set of statements with truth values in this universe of discourse calculus. The domain of the predicate does not include reflexive subjects such as the one given above.

Richard Paradox

The Richard paradox is concerned with the proposition: the set of all numerical functions is not enumerable. A commonly used argument to prove this proposition is the following. Suppose an enumeration existed symbolized by $f_n(m)$, which represents the nth function with argument m. Consider the function g defined by

$$g(n) = f_n(n) + 1$$

for any value of n. Let n_0 be the index number of g(n) in the enumeration:

$$g(n) = f_{n_0}(n) + 1$$

then

$$f_{n_0}(n) = g(n) = f_n(n) + 1 \qquad (6.27)$$

and

$$f_{n_0}(n_0) = g(n_0) = f_{n_0}(n_0) + 1 \qquad (6.28)$$

which is a contradiction. Thus the set of all numerical functions is not enumerable.

Contrarian argument: Consider the set of all definable functions. Definable is taken to mean definable in some specific language with a fixed dictionary and grammar. Since the number of words in the language is finite, then the number of expressions is enumerable. Thus the set of expressions that form the definitions of definable functions is enumerable. Thus the set of definable functions is enumerable. Since the set of numerical functions is a subset of the set of definable functions it also must be enumerable. Thus the set of all numerical functions is enumerable.

The result of the preceding two arguments is a contradiction (paradox).

Resolution

The projection operator inner product corresponding to eq. 6.27 has the form (see subsection 3.2.3 for a discussion of variables)

$$<\Omega_u| \ldots P_{f_{n_0}}(n) \ldots (1 - P_{f_n}(n)) \ldots |\Omega_u> \qquad (6.29)$$

which is non-zero (and thus the statement has a truth value) for $n \neq n_0$. However for $n = n_0$ eq. 6.29 is equal to zero[73] and thus eq. 6.28 has no truth value and therefore the paradox is avoided. The contrarian proof is not contradicted.

Gödel's Undecidability Theorem

Gödel's Undecidability Theorem, as he pointed out in his celebrated paper, is closely related to Richard's paradox (antinomy) and the Liar's paradox. Gödel provides an explicit undecidable statement, eq. 1, in his paper:

$$n \,\varepsilon\, K \equiv \overline{Bew}[R(n); n] \qquad (6.30)$$

where

1. Bew Y means Y is a provable formula (statement)
2. The bar over Bew means "not"
3. K is a class of natural numbers

---

[73] Note all projection operators commute as do all eigenvalue operators in the formulation of this chapter.

4. A class-sign is a formula[74] with one free natural number variable
5. The set of class-signs is arranged in a series using some rule with the nth class-sign denoted R(n).
6. Define [α, m] to be the formula derived upon replacing the free variable in the class-sign α by the sign of the natural number m. Thus the relation Y = [α, z] is definable in PM.
7. Since the above are all definable in PM, so is the class K that is formed from them.

Therefore there is a class-sign S such that [S; n] implies that n ε K. Thus S = R(q) for some specfic natural number q. If we assume [R(q); q] is provable, then q ε K. But, by eq. 6.30

$$\overline{\text{Bew}}[R(q); q] \tag{6.31}$$

would be true contrary to our assumption.

If we assume the contrary: the negation of

$$[R(q); q] \tag{6.32}$$

is provable, then

$$\overline{q \; \varepsilon \; K} \tag{6.33}$$

and

$$\text{Bew}[R(q); q] \tag{6.34}$$

would be true. Thus [R(q); q] and its negation would be provable.

Gödel's Theorem (more precisely any statement conforming to eq. 6.30) is excluded from a mathematical deductive system if the subject is reflexive. Eq. 6.30 can be stated in a subject predicate form as

"n is contained in K which is equivalent to …"

where "n" is the subject and "is contained …" is the predicate. In this form we can associate the projection P with "n" and (1 – P) with the predicate part "[R(n); n]" yielding

---

[74] Of the Principia Mathematica (denoted PM).

$$<\Omega_u|P \ldots (1 - P)|\Omega_u> = 0 \quad (6.35)$$

which shows the statement has no truth value.

Gödel's more succinct undecidable statement[75] 17 Gen r is similar in character but disguised in a more condensed notation. The observant reader will note the close resemblance between Gödel's undecidablity result, Richard's paradox and the Liar paradox—as Gödel himself noted.[76]

Conclusion

The preceding examples show that Operator Logic provides a mechanism to systematically exclude statements with no truth value ("undecidable" or paradoxical statements) from a universe of discourse. The following chapters show further advantages of Operator Logic including a generalization to quantum and probabilistic logic, hierarchies of universes of discourse, and the formulation of a mathematical philosophy (metaphysics) relating Platonic Ideas to the physical universe, Reality, that is conceptually similar to the theories of various Platonic schools.

---

[75] Gödel (1992) p. 60.
[76] Gödel (1992) p. 40.

7. Sentential Calculus in Operator Logic

In this section we will extend our discussion of projection operators to the determination of the truth value of simple sentences, or sentences composed of clauses consisting of a subject and a predicate.

In their classic book, Hilbert and Ackermann[77] begin with a discussion of the sentential calculus[78] (calculus of sentences[79]). They point out that sentences can be combined in a number of ways to create compound sentences using connectives such as "and", "or", "if … then", and "not". They specify five fundamental combinations of sentences using the following symbols:

Connective	Symbol
and	&
or	v
if … then (modus ponens)	→
if and only if	~
not[80]	___

The symbols can be used to form basic combinations of sentences.

Denoting sentences with upper case letters: A, B, … the five basic combinations are:

1. A & B is a compound sentence that is true if and only if both A and B are true.
2. A v B is true if and only if at least one of the sentences A or B is true.
3. A → B is false if and only if A is true and B is false.
4. A ~ B is true if and only if both A and B are true, or both A and B are false.

---

[77] Hilbert (1950).

[78] We assume a semantic universe of discourse in this, and the following, subsections.

[79] We use sentence and statement interchangeably.

[80] For lexicographic reasons we use underscore ___ for "not" rather than the more commonly used overscore ‾ symbol.

5. The sentence A̲ is false if A is true, and is true if A is false.

7.1 Truth Values of Simple Sentences and Clauses

The projection operators, that we have used up to now, have determined the decidability of sentences or clauses. A *clause* consists of a subject(s) and predicate. A sentence or clause was undecidable if the product of the subject projection operator and predicate projection operator was zero—usually in the simple form of $P(1 – P) = 0$.

Now we extend the formalism to determining the truth value of a sentence or clause using projection operators. The procedure begins by assigning a set of projections to each predicate—one truth projection for each subject in the domain of subjects for the predicate. For predicate p we will denote the truth projection as P_{ptot} (eq. 6.19). For each subject s we associate a projection P_{stot} (eq. 6.23). Then the sentence

$$<\Omega_u|A_1 + A_2|\Omega_u> \tag{7.1}$$

where A_1 is the subject and A_2 is the predicate, has the truth value

$$\text{Truth Value} = <\Omega_u|P_{1tot}P_{2tot}|\Omega_u> = <\Omega_u|P'_1P'_2|\Omega_u> \tag{7.2}$$

where the form of P_{itot} for i = 1,2 is given by eq. 6.8. For example, if the sentence is "The sky is blue" then, when we assign truth projections for all possible subjects of the predicate "is blue", we could assign

$$\begin{aligned} P_1 &= P_{\text{"The sky"}} \\ P_{1tot} &= P_1P' \\ P_2 &= P_{\text{"is blue"}} \\ P_{2tot} &= P_2P' \end{aligned} \tag{7.3}$$

and let $P' = P_1$ using eq. 6.20. Consequently, the statement has a truth value.

We now extend Classical Operator Logic by assigning numeric values to truth values:

$$\begin{aligned} \text{True} &= +1 \\ \text{Undecidable} &= 0 \\ \text{False} &= -1 \end{aligned} \tag{7.4}$$

In order to obtain numeric truth values from inner products of projection operator expressions we must introduce a dependence on the subject in the truth value projection operator expression. To accomplish this we generalize eq. 6.20 to

$$P'_{pk} = \prod_{i \varepsilon D}[(-1)^{n_{ik}} P'_i] \prod_{j \notin D} (1 - P'_j) \tag{7.5}$$

where the integers n_{ik} are 0 if subject i and predicate k form a true subject-predicate statement (or clause), and 1 if subject i and predicate k form a false subject-predicate statement (or clause). *The set of all predicates in the universe of discourse is labeled by the integer k.*

The set denoted $\{P'_s\}$ is the set of subject truth value projection operators for all subjects in the universe of discourse. The operator expression P'_{pk} satisfies

$$<\Omega_u|P'_s \, P'_{pk}|\Omega_u> = (-1)^{n_{sk}} \tag{7.6}$$

if subject s is contained in D and

$$<\Omega_u|P'_s \, P'_{pk}|\Omega_u> = 0 \tag{7.7}$$

if subject s is *not* contained in D.

For example, consider now the sentence "The sky is green." When we assign truth projections for all possible subjects of the predicate "is green" we choose to assign the subject $n_{\text{"The sky""is green"}} = 1$ so that

$$<\Omega_u|P'_{\text{"The sky"}} \, P'_{\text{"is green""The sky"}}|\Omega_u> = -1 \tag{7.8}$$

The truth value –1 indicates a false statement.

We note that the operator P'_{pk} is not a projection operator. It satisfies

$$P'_{pk} \, P'_{pk} = P'_p \tag{7.9}$$

where P'_p is defined by eq. 6.20.

For a given predicate, the choice of subject truth values (i.e. the values of the integers n_{ik}) in eq. 7.5 is a semantic issue and thus is determined by semantic considerations.

If a given sentence has multiple subjects the determination of the truth value is slightly more complex. We will consider two special cases:

$$\text{Subject1 and subject2 predicate} \tag{7.10}$$

and

$$\text{Subject1 or subject2 predicate} \tag{7.11}$$

In the case of eq. 7.10 the truth value[81] is obtained by evaluating the expression:[82]

$$\text{Truth Value} = \langle\Omega_u|P_{s1}P_{pa}|\Omega_u\rangle\langle\Omega_u|P_{s2}P_{pa}|\Omega_u\rangle[1 - \\ - 2\theta(-\langle\Omega_u|P_{s1}P_{pa}|\Omega_u\rangle - \langle\Omega_u|P_{s2}P_{pa}|\Omega_u\rangle)] \tag{7.12}$$

where the step function is

$$\theta(x) = 1 \text{ if } x > 0 \text{ and } \theta(x) = 0 \text{ if } x \leq 0 \tag{7.13}$$

and where P_{si} for i = 1, 2 are the subject projection operators for Subject1 and Subject2 respectively. P_{pa} is the predicate projection operator.

Eq. 7.10 is equivalent to

$$\text{Subject1 predicate and subject2 predicate} \tag{7.14}$$

Eq. 7.12 for the truth value follows from its equivalent eq. 7.14 taking account of the possible truth values and undecidability issue. Eq. 7.12 specifies that the subject-predicate expression must be true for both subjects for the statement to be true. If either factor in eq. 7.12 is zero

---

[81] In this example and the following examples we factor expectation values into products of expectation values of clauses since this procedure is both well defined and produces results consistent with our intuitive expectations. Interestingly, a somewhat similar procedure is followed in perturbation theory expansions in quantum field theory. In perturbation theory, terms with more than two operators in them are expanded in terms of vacuum expectation values of products of two quantum field operators.

[82] The statement is separated into an equivalent consisting of two parts each of which is evaluated using the universe state $|\Omega_u\rangle$. It is clearly the only way of obtaining truth values correctly in this formalism. Therefore we so define it.

then the truth value of the statement is zero meaning it is undecidable. If either factor, or both factors, in eq. 7.12 are false (and both decidable) then the truth value of the statement is –1 meaning false.

In the case of eq. 7.11 the truth value is obtained by evaluating the truth projection expression:

$$\text{Truth Value} = \text{Max}(<\Omega_u|P_{s1}P_{pa}|\Omega_u>, <\Omega_u|P_{s2}P_{pa}|\Omega_u>)\cdot \cdot |<\Omega_u|P_{s1}P_{pa}|\Omega_u><\Omega_u|P_{s2}P_{pa}|\Omega_u>| \quad (7.15)$$

where | … | indicates the numerical absolute value. The statement is true if the subject-predicate expression is true for either subject. The absolute value $|<\Omega_u|P_{s1}P_{pa}|\Omega_u><\Omega_u|P_{s2}P_{pa}|\Omega_u>|$ guarantees the truth value of the statement is zero if either or both subject-predicate expressions is undecidable. The function Max(x, y) is the maximum of the quantities x and y.

The generalization to the case of multiple subjects is direct.[83]

7.2 Truth Value of Compound Sentences and Clauses

In the case of compound sentences and clauses, classical Operator Logic is based on factoring them into simple subject-predicate clauses as discussed in the previous section.

In this section we will consider some general cases of compound sentences to illustrate the procedure for obtaining a truth value.

Suppose we have a sentence that can be expressed as a series of "and" clauses such as

$$\text{Statement} = \text{clause1 and clause2 and} \ldots \quad (7.16)$$

Then the truth value of the statement is given by the expression:[84]

---

[83] In the case of quantifiers such as "all" or "some" we make the quantifier part of the subject or predicate as the case may be. For example, in the sentence "All men are good" the subject is "all men" and the predicate is "are good". The truth value status of the sentence is determined by whether "all men" is a member of the domain of "are good". A similar example could be created for "some". And the appearance of "all" or "some" in a predicate also can be similarly treated. For example, "Peacemakers are all good men" has the predicate "are all good men" and the truth value status of the sentence is determined by whether "peacemakers" is in the domain of that predicate. Other quantifiers such as "there exists" and "for any" are embodiable within Operator Logic.

$$<\Omega_u|\text{statement}|\Omega_u> = |<\Omega_u|\text{clause1}|\Omega_u><\Omega_u|\text{clause2}|\Omega_u>\ldots|\{1 - $$
$$- 2\theta(|<\Omega_u|\text{clause1}|\Omega_u>| - <\Omega_u|\text{clause1}|\Omega_u> + |<\Omega_u|\text{clause2}|\Omega_u>| - <\Omega_u|\text{clause2}|\Omega_u> +$$
$$+ \ldots)\} \qquad (7.17)$$

Note the following truth value cases implied by eq. 7.17:

Case 1: all clauses true (no undecidable clauses), Truth Value = +1
Case 2: One or more false clauses (with no undecidable clauses), Truth Value = –1
Case 3: One or more undecidable clauses, Truth Value = 0

Suppose we consider a sentence that can be expressed as a series of "or" clauses such as

$$\text{Statement} = \text{clause1 or clause2 or} \ldots \qquad (7.18)$$

Then the truth value of the statement is given by the expression:[85]

$$<\Omega_u|\text{statement}|\Omega_u> = [2\theta(\theta(<\Omega_u|\text{clause1}|\Omega_u>) + \theta(<\Omega_u|\text{clause2}|\Omega_u>) + \ldots) - 1]\cdot$$
$$\cdot|<\Omega_u|\text{clause}|\Omega_u><\Omega_u|\text{clause}|\Omega_u><\Omega_u|\text{clause}|\Omega_u>\ldots| \qquad (7.19)$$

The following truth value cases are implied by eq. 7.19:

Case 1: Any clauses are true (no undecidable clauses), Truth Value = +1
Case 2: All clauses are false (with no undecidable clauses), Truth Value = –1
Case 3: One or more undecidable clauses, Truth Value = 0

Lastly, let us consider the case of an "if … then" statement:

$$\text{If A, then B} \qquad (7.20)$$

This statement is equivalent to

$$\underline{A\underline{B}} \qquad (7.21)$$

which in words is NOT(A and NOT B). In terms of Operator Logic the truth value of eqs. 7.20 and 7.21 is numerically

---

[84] Note that because the projection expressions do not change the universe state the expression $<\Omega_u|\text{clause1}|\Omega_u><\Omega_u|\text{clause2}|\Omega_u> \equiv <\Omega_u|\text{clause1clause2}|\Omega_u>$ and similarly elsewhere. The separation into "clause" factors is clearer.

[85] Note nested step functions appear in eq. 7.19.

$$- [<\Omega_u|A|\Omega_u>(- <\Omega_u|B|\Omega_u>)] = <\Omega_u|A|\Omega_u><\Omega_u|B|\Omega_u> \quad (7.22)$$

where $<\Omega_u|A|\Omega_u>$ and $<\Omega_u|B|\Omega_u>$ are evaluated in a manner similar to the previous cases (eqs. 7.17 and 7.19), or for other sentential forms of A and B.

7.3 Truth Value of the Normal Form

It is well known that all sentential calculus statements can be reduced to operations involving two logical connectives such as & and __ (not).[86] So our Operator Logic clearly suffices to handle all sentential calculus statements including undecidable statements.

Furthermore any set of sentences can be into a form called a *normal form*[87] by means of equivalence transformations.[88]

Since our Operator Logic also encompasses the predicate calculus and second level predicate calculus (See the following chapter.) we conclude we have a complete Operator Logic formalism[89] for classical Logic that decisively handles the issue of undecidability.

---

[86] Hilbert(1950) pp. 10-11.

[87] "consisting of a conjunction of disjunctions in which each component of the disjunction is either an elementary sentence or the negation of one" in the words of Hilbert (1950) p. 12.

[88] Hilbert(1950) p. 12.

[89] The rules of inference and proof procedures are analogous in Operator Logic and in conventional Logic, and so will not be considered here.

8. Why Operator Logic?

8.1 Operator Logic is the Correct Formalism for Logic

Operator Logic enables logical systems and statements to be constructed that exclude paradoxes and undecidable statements in a mathematically well defined way. In addition it has other important advantages described later. It provides a superior formalism for Logic without the known defects of conventional language/symbolic language formulations, which we now see as incomplete.

After we establish Operator Logic we can proceed to develop the intermediate mathematical-metaphysical transition to a theory of physical Reality

8.2 Advantages of Operator Logic

Operator Logic has some major advantages, which we will discuss in this chapter and in succeeding chapters:

1. Operator Logic provides a unified formulation of classical and quantum probabilistic logic.

2. Operator Logic is the only approach with a fundamental, non-intuitive basis in Nature (being similar conceptually to Quantum Theory) and thus is guaranteed to be consistent.[90] Otherwise, logic is futile in its foundations.

---

[90] G. Feinberg, Phys. Rev. **159**, 1089 (1967) quotes Dr. M. Tausner to the effect, "No [experimental] observations can be logically inconsistent." Thus no formalism truly embodying Nature can be inconsistent. This comment presumes nature to be consistent and to embody a well-defined set of laws. The regularity of nature has become more and more evident in recent centuries with the growth of scientific knowledge.

3. Raising and lowering operators can be defined that enable us to create subworlds (subuniverses) of Calculus and of Semantic Universes of Discourse.

4. Projection operators can also be used to generate subuniverses.

5. A semantic universe of discourse can be defined, which can be mapped to a physical universe realizing the Platonic concepts of Ideas (pure forms), of an intermediate mathematical connection, and of the physical universe of Reality.

These possibilities will be discussed in the following sections and chapters.

8.3 Operator Logic Raising and Lowering Operators

8.3.1 Eigenvalue Operators that are not Functions of Variables

If we consider an eigenvalue operator A_E (without a dependence on variables) for a term E, and its corresponding eigenstates |0> and |E>,

$$A_E |E> = gn(E)|E> \tag{8.1}$$
$$A_E |0> = 0 \tag{8.2}$$

then we can[91] define raising and lowering operators that transform one eigenstate into the other. We define the raising operator $a_E^\dagger$ that satisfies

$$|E> = [gn(E)]^{-½} a_E^\dagger|0> \tag{8.3}$$

with <E|E> = 1, and, the lowering operator a_E that satisfies

$$|0> = [gn(E)]^{-½} a_E|E> \tag{8.4}$$

[91] Although this is commonly done in the case of a harmonic oscillator in Quantum Mechanics, and also in Quantum Field Theory.

where † signifies hermitean conjugatation. In the present situation we do not wish repeated application of the raising operator to generate states with higher eigenvalues.[92] Therefore we define

$$a_E^\dagger|E> = 0 \qquad (8.5)$$
$$a_E|0> = 0 \qquad (8.6)$$

Consequently, raising and lowering operators must satisfy anticommutation relations with the form:

$$\begin{aligned} \{a_E^\dagger, a_E\} &= gn(E) \\ \{a_E, a_E\} &= 0 \\ \{a_E^\dagger, a_E^\dagger\} &= 0 \end{aligned} \qquad (8.7)$$

for eqs. 8.3 – 8.6 to hold.[93] where anti-commutation is defined by

$$(a, b\} = ab + ba$$

More generally if there are a number of semantic eigenvalue operators A_{E_k} where k = 1,2, …, q then the anticommutation relations are

$$\begin{aligned} \{a_{E_k}^\dagger, a_{E_{k'}}\} &= gn(E_k)\delta_{kk'} \\ \{a_{E_k}, a_{E_{k'}}\} &= 0 \\ \{a_{E_k}^\dagger, a_{E_{k'}}^\dagger\} &= 0 \end{aligned} \qquad (8.8)$$

where $\delta_{kk'}$ is the Kronecker delta. The semantic eigenvalue operators can be represented in terms of these operators by

$$A_{E_k} = a_{E_k}^\dagger a_{E_k} \qquad (8.9)$$

and the corresponding projection operators by

---

[92] Since such eigenstates would not correspond to a primitive or defined term in the universe of discourse.

[93] This definition of raising and lowering operators leads to a simpler form for eq. 8.9.

$$P_{E_k tot} \equiv P_{E_k} = P'_{E_k} \tfrac{1}{2}[1 - [gn(E_k)]^{-1} a_{E_k} a_{E_k}^{\dagger}] \qquad (8.10)$$

where P'_{E_k} is a truth value projection operator. (See section 7.1 for its definition and use.) The product of P'_{E_k} operators determines the truth value of expressions and precludes logical paradoxes.

The projection operator factor $\tfrac{1}{2}[1 - [gn(E_k)]^{-1} a_{E_k} a_{E_k}^{\dagger}]$ in eq. 8.10 projects substates from the universe state whose k^{th} eigenvalue is zero. Thus the projection operator P_{E_k} plays two roles: 1) it projects any state $|\Omega>$, whose k^{th} eigenvalue is zero, to zero; and 2) it enables statements to avoid paradoxes by zeroing statements and clauses whose subjects or predicates are not terms in the universe of discourse.[94]

Subjects that are not in the subject domain of a predicate have a projection operator $P'_{E_{k'}}$ that when multiplied by the predicate's P'_{E_k} gives zero identically. As a result such a statement's corresponding projection operator product (containing both projection operators) yields a truth value of zero indicating undecidable.

Calculus eigenvalue operators, and raising and lowering operators, have exactly the same definitions, forms and equations as the semantic analogues given above in this subsection. For definiteness we give their basic definitions:

$$\begin{aligned} \{b_{E_k}^{\dagger}, b_{E_{k'}}\} &= gn(E_k)\delta_{kk'} \\ \{b_{E_k}, b_{E_{k'}}\} &= 0 \\ \{b_{E_k}^{\dagger}, b_{E_{k'}}^{\dagger}\} &= 0 \end{aligned} \qquad (8.11)$$

with

$$B_{E_k} = b_{E_k}^{\dagger} b_{E_k} \qquad (8.12)$$

and

$$Q_{E_k} = Q'_{E_k} \tfrac{1}{2}[1 - [gn(E_k)]^{-1} b_{E_k} b_{E_k}^{\dagger}] \qquad (8.13)$$

where Q'_{E_k} is the calculus equivalent of a semantic truth projection operator. The Q'_{E_k} projection operator prevents logical paradoxes in a manner similar to the semantic case.

---

94 In the cases considered hitherto, this is not an issue since the universe state has a non-empty string factor for each term in the universe. Later when we develop subuniverses state projection operators become non-trivial. They are also relevant in Quantum Operator Logic as will be seen.

8.3.2 Eigenvalue Operators that are Functions of Variables

If we consider any eigenvalue operator that is dependent on free variables such as $A_E(x_1, x_2, \ldots)$ with corresponding eigenstates $|0, x_1, x_2, \ldots\rangle$ and $|gn(E), x_1, x_2, \ldots\rangle$ for each set of values of $x_1, x_2, \ldots$ then we can again define raising and lowering operators that transform between eigenstates. The raising operator becomes variable dependent:

$$|E, x_1, x_2, \ldots\rangle = [gn(E_k)]^{-1/2} a_E^{\dagger}(x_1, x_2, \ldots)|0\rangle \tag{8.14}$$

as does the lowering operator:

$$a_E(x_1, x_2, \ldots)|E, x'_1, x'_2, \ldots\rangle = [gn(E_k)]^{1/2}\delta(x_1, x'_1)\, \delta(x_2, x'_2)\ldots|0\rangle \tag{8.15}$$

where † signifies the hermitean conjugate, and $\delta(x_k, x'_k)$ is a Kronecker-type δ-function for discrete variables and a Dirac-type δ-function for continuous variables.

Again we do not wish repeated application of a raising operator to generate states with higher eigenvalues. Therefore

$$a_E^{\dagger}(x_1, x_2, \ldots)|E\rangle = 0 \qquad (8.17)$$
$$a_E^{\dagger}(x_1, x_2, \ldots) \ldots a_E^{\dagger}(x_1, x_2, \ldots)|0\rangle = 0 \qquad (8.16)$$
$$a_E(x_1, x_2, \ldots) \ldots a_E(x_1, x_2, \ldots)|E\rangle = 0 \qquad (8.17)$$

Consequently the reader can verify that the raising and lowering operators must satisfy the anicommutation relations:

$$\begin{aligned} &\{a_E^{\dagger}(x_1, x_2, \ldots), a_E(x'_1, x'_2, \ldots)\} = gn(E)\delta(x_1, x'_1)\delta(x_2, x'_2)\ldots \\ &\{a_E(x_1, x_2, \ldots), a_E(x'_1, x'_2, \ldots)\} = 0 \\ &\{a_E^{\dagger}(x_1, x_2, \ldots), a_E^{\dagger}(x'_1, x'_2, \ldots)\} = 0 \end{aligned} \tag{8.18}$$

More generally if there are a number of semantic eigenvalue operators A_{E_k} where k = 1,2, … , q then the anticommutation relations are

$$\begin{aligned} &\{a_{E_k}^{\dagger}(x_1, x_2, \ldots), a_{E_{k'}}(x'_1, x'_2, \ldots)\} = gn(E)\delta_{kk'}\delta(x_1, x'_1)\delta(x_2, x'_2)\ldots \\ &\{a_{E_k}(x_1, x_2, \ldots), a_{E_{k'}}(x'_1, x'_2, \ldots)\} = 0 \end{aligned} \tag{8.19}$$

$$\{a_{E_k}^{\dagger}(x_1, x_2, \ldots), a_{E_{k'}}^{\dagger}(x'_1, x'_2, \ldots)\} = 0$$

where $\delta_{kk'}$ is a Kronecker delta. The semantic eigenvalue operators can be represented in terms of these operators by

$$A_{E_k}(x_1, x_2, \ldots) = a_{E_k}^{\dagger}(x_1, x_2, \ldots)a_{E_k}(x_1, x_2, \ldots) \qquad (8.20)$$

and the projection operators are

$$P_{E_k}(x_1, x_2, \ldots) = P'_{E_k}(x_1, x_2, \ldots)\tfrac{1}{2}[1 - [gn(E_k)]^{-1}a_{E_k}(x_1, x_2, \ldots)a_{E_k}^{\dagger}(x_1, x_2, \ldots)] \qquad (8.21)$$

The calculus eigenvalue operators, and raising and lowering operators have a similar form

$$\{b_{E_k}^{\dagger}(x_1, x_2, \ldots), b_{E_{k'}}(x'_1, x'_2, \ldots)\} = gn(E_k)\delta_{kk'}\delta(x_1, x'_1)\delta(x_2, x'_2)\ldots$$
$$\{b_{E_k}(x_1, x_2, \ldots), b_{E_{k'}}(x'_1, x'_2, \ldots)\} = 0 \qquad (8.22)$$
$$\{b_{E_k}^{\dagger}(x_1, x_2, \ldots), b_{E_{k'}}^{\dagger}(x'_1, x'_2, \ldots)\} = 0$$

with

$$B_{E_k}(x_1, x_2, \ldots) = b_{E_k}^{\dagger}(x_1, x_2, \ldots)b_{E_k}(x_1, x_2, \ldots) \qquad (8.23)$$

and

$$Q_{E_k}(x_1, x_2, \ldots) = \tfrac{1}{2}[1 - [gn(E_k)]^{-1}b_{E_k}(x_1, x_2, \ldots)b_{E_k}^{\dagger}(x_1, x_2, \ldots)]Q'_{E_k}(x_1, x_2, \ldots) \qquad (8.24)$$

8.3.3 Effect of Raising and Lowering Operators on Universes of Discourse

Consider the effect of a lowering operator on a universe state such as

$$a_{E_k}|E_1, E_2, E_3, \ldots, E_k, \ldots, E_n\rangle = [gn(E_k)]^{1/2}|E_1, E_2, E_3, \ldots, \overset{k^{th}}{0}, \ldots, E_n\rangle \qquad (8.25)$$

The kth eigenvalue becomes zero, which is the Gödel number of the empty string of characters (symbols).

Now consider a semantic universe of discourse (Parallel observations and comments can be made for a calculus universe of discourse.) with terms corresponding to Gödel numbers: $E_1, E_2, \ldots, E_n$.[95] The state $|\Omega_u\rangle$ corresponding to this universe of discourse is given by eq. 6.26.

A universe of discourse state $|\Omega_u\rangle$ can be "restricted" using products of lowering operators. For example

$$a_{E_k}|\Omega_u\rangle = [gn(E_k)]^{1/2}|\Omega'_u\rangle \tag{8.26}$$

where $\langle\Omega'_u|\Omega'_u\rangle = 1$, and where the kth eigenvalue of $|\Omega'_u\rangle$ is 0:

$$|\Omega'_u\rangle = |E_1, E_2, E_3, \ldots, \overset{k^{th}}{0}, \ldots, E_n\rangle \tag{8.27}$$

The axioms, theorems and proofs of the subuniverse of discourse state $|\Omega'_u\rangle$ are those of the universe of discourse $|\Omega_u\rangle$ which do not contain the kth term.

Thus a statement such as

$$\langle\Omega'_u| \sum_{n \in \{m'\}} A_{E_n}|\Omega'_u\rangle \tag{8.28}$$

satisfies the corresponding projection operator product equation

$$F(P'_{E_1}, P'_{E_2}, \ldots) \prod_{n \in \{m'\}} P_{E_n}|\Omega'_u\rangle = 0 \tag{8.29}$$

where $F(P'_{E_1}, P'_{E_2}, \ldots)$ is a function of P'_{E_1}, P'_{E_2} and so on, if k is among the set of integers {m'}. The expectation value is[96]

$$\langle\Omega'_u|F(P'_{E_1}, P'_{E_2}, \ldots) \prod_{n \in \{m'\}} P_{E_n}|\Omega'_u\rangle = 0 \tag{8.30}$$

---

[95] Zero is the Gödel number of the empty string.

[96] Inner product.

for each statement containing the kth term. Thus the set of statements with a possible non-zero truth value in the $|\Omega'_u\rangle$ universe of discourse consists of those statements not containing the kth term. $|\Omega'_u\rangle$ is properly regarded as a subuniverse of discourse.

Whether the subuniverse of discourse is of interest is another question. In the next chapter we consider a simple case where the creation of subuniverses is of interest.

To summarize: A statement in a subuniverse defined above has the form

$$\langle\Omega'_u| \sum_{n \in \{m''\}} A_{E_n}|\Omega'_u\rangle \tag{8.31}$$

for {m"}, the subset of the integers of {m} not containing k.

The projection operator product corresponding to this statement has the form[97]

$$F(P'_{E_1}, P'_{E_2}, \ldots)\prod_{n \in \{m''\}} P_{E_n} \tag{8.32}$$

Whether a statement in the subuniverse has a truth value is determined by the expectation value

$$\langle\Omega'_u| F(P'_{E_1}, P'_{E_2}, \ldots)\prod_{n \in \{m''\}} P_{E_n}|\Omega'_u\rangle \tag{8.33}$$

If it is zero then it has no truth value; if it is non-zero then it has a truth value. The statements of this subuniverse exclude statements of the parent universe containing the kth term.

We now have a method for generating subuniverses through the application of one or more lowering operators. We shall see examples of subuniverses in the next chapter.

8.3.4 Multiple Lowering Operator Case

In the case where we apply q lowering operators (whose indices form the set {p}) to a universe state[98], then eq. 8.25 generalizes directly to

---

[97] The expression $F(P'_{E_1}, P'_{E_2}, \ldots)$ is assumed not to contain P'_{E_k}.

[98] $|\Omega_u\rangle$ is defined by eq. 6.26 and {m} is defined in the statement following it.

$$\prod_{n \in \{p\}} a_{E_n} |\Omega_u\rangle = \prod_{n \in \{p\}} [gn(E_n)]^{1/2} |\Omega''_u\rangle \qquad (8.34)$$

where $\langle\Omega''_u|\Omega''_u\rangle = 1$, and members of the set of statements with truth values in this subuniverse have the form

$$\langle\Omega''_u| \sum_{n \in \{m-p\}} A_{E_n} |\Omega''_u\rangle \qquad (8.35)$$

where {m – p} is the subset of integers in {m} that does not include any integers in {p}.

The corresponding projection operator product expectation value has the form[99]

$$\langle\Omega''_u| F(P'_{E_1}, P'_{E_2}, \ldots) \prod_{n \in \{m-p\}} P_{E_n} |\Omega''_u\rangle \qquad (8.36)$$

The truth value of the statement is determined by the value of eq. 8.36: if the value is zero then the corresponding statement has no truth value - undecidable; if the value is non-zero then the corresponding statement has a truth value.

[99] The expression $F(P'_{E_1}, P'_{E_2}, \ldots)$ is assumed not to contain projections corresponding to terms in {p}.

9. Hierarchies of Universes of Discourse

In this chapter we will define hierarchies of universes and consider some simple examples. In addition we will show how to combine universes into "superuniverses." We will use these concepts when we establish a mathematical connection to the theory of Reality – The Standard Model of Elementary Particles.

9.1 Hierarchies of Subuniverses

In the previous chapter we examined the case of a two level hierarchy. One can construct multi-level hierarchies of universes using various sets of lowering operator products at each level. Fig. 9.1 shows a three level hierarchy diagramatically. The procedure is so simple that writing the explicit equations generating the hierarchy is unnecessary in the author's view. They are similar to those given for a two level hierarchy in section 8.3.4.

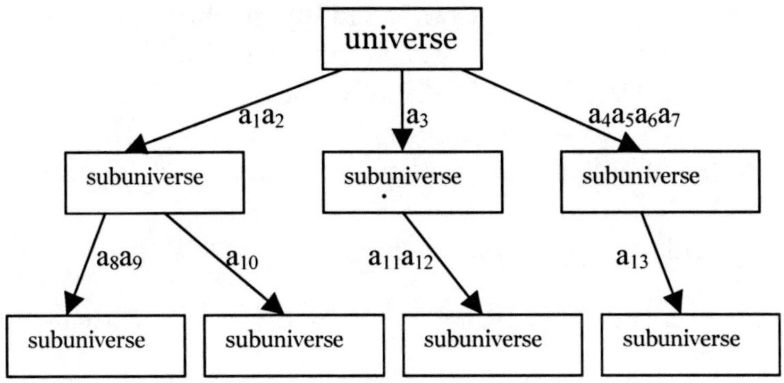

Figure 9.1. A hierarchy of universes with the lowering operators indicated for the transition to each subuniverse.

One primary purpose in defining universe hierarchies is to enable separate universes of discourse that have a significant overlap in their primitive terms, axioms and theorems to be united in a superuniverse through the reversal of the mechanism described for creating subuniverse hierarchies. We form a unified set of terms with one subset containing the primitive terms of one subuniverse and another subset containing the primitive terms of the other subuniverse. Then a combined set of axioms is constructed such that setting one set of terms to zero (empty strings) yields one of the initial subuniverses and setting a different set of terms to zero yields the other subuniverse.[100] Thus one achieves a unification of what had been two separate but similar universes of discourse.[101] The arithmetic universe considered next is an example of this process.

9.2 Subuniverses of a Simple Arithmetical Universe of Discourse

We will consider a simple universe of discourse, that is similar to Gödel's "formal system P"[102], to illustrate the utility of creating subuniverses of discourse and to illustrate combining universes to produce a super universe. The axioms in this universe are:

1. $a + b = b + a$	Commutativity of Addition
2. $a + (b + c) = (a + b) + c$	Associativity of Addition
3. $a \cdot b = b \cdot a$	Commutativity of Multiplication
4. $a \cdot (b \cdot c) = (a \cdot b) \cdot c$	Associativity of Multiplication
5. $a \cdot (b + c) = a \cdot b + a \cdot c$	Distributive Law

They are the fundamental arithmetic axioms of natural numbers (integers).

The arithmetic universe of discourse state $|\Omega_u>$ is:

$$|\Omega_u> = |\text{"a", "b", "c", "+", "}\cdot\text{", "=", "(", ")"} > \quad (9.1)$$

---

[100] Additional axioms might also be needed to unite the subuniverses. These axioms contain terms from both subuniverses and thus are only relevant for the unified subuniverses.

[101] J. C. Maxwell achieved this feat by successfully uniting electricity and magnetism in his theory of Electromagnetism in 1865.

[102] Gödel (1962) p. 41.

9.2.1 Additive Universe of Discourse

If we now apply the lowering operator for multiplication "·", namely $a_{"\cdot"}$, to $|\Omega_u\rangle$ then the universe of discourse is reduced to addition only:

1. a + b = b + a Commutativity of Addition
2. a + (b + c) = (a + b) + c Associativity of Addition

with the subuniverse state:

$$|\Omega_+\rangle = |\text{"a", "b", "c", "+", 0, "=", "(", ")"}\rangle \qquad (9.2)$$

All statements containing a multiplication become statements without a truth value when their projection operator products are evaluated between $\langle\Omega_+|$ and $|\Omega_+\rangle$. These statements are undecidable statements since they are neither true nor false and excluded from the subuniverse of decidable statements.

9.2.2 Multiplicative Universe of Discourse

The multiplicative universe of discourse state $|\Omega_\bullet\rangle$ is obtained by applying the lowering operator for addition "+", namely $a_{"+"}$, to $|\Omega_u\rangle$. The subuniverse state $|\Omega_\bullet\rangle$ is:

$$|\Omega_\bullet\rangle = |\text{"a", "b", "c", 0, "·", "=", "(", ")"}\rangle \qquad (9.3)$$

The axioms in this subuniverse are the purely multiplicative axioms:

1. a·b = b·a Commutativity of Multiplication
2. a·(b·c) = (a·b)·c Associativity of Multiplication

Thus we obtain the multiplicative subuniverse. All statements containing an addition become undecidable (truth value = 0) when their projection operator products are evaluated between $\langle\Omega_\bullet|$ and $|\Omega_\bullet\rangle$.

9.3 Subuniverses of Euclidean Geometry

Euclidean geometry was plagued for millenia by concerns over the fifth postulate. In the 19th Century Bolyai and Lobachevsky showed that omitting the fifth postulate opened the door to non-Euclidean geometries that were valid on certain curved surfaces. Euclidean geometry has five axioms plus five common notions including the notion of congruence. The primitive terms and axioms are:

Primitive terms:
point, line, straight line, circle, angle, congruent (in common notions)

Axioms:
1. Given two points there is a straight line that joins them.
2. Any straight line can be prolonged indefinitely.
3. A circle can be constructed when its center, and a point on it, are specified.
4. All right angles are equal.
5. If a straight line falling on two straight lines makes the interior angles on the same side sum to less than two right angles, the two straight lines if prolonged indefinitely, meet on that side on which the angles are less than two right angles.

They also include some common English words.

If we consider the semantic universe of discourse of Euclidean geometry, then the interpreted primitive terms, the axioms, and common notions comprise the knowledge basis of the Euclidean universe of discourse.

If we define a semantic subuniverse by setting "congruent" to the empty string using a lowering operator, and reinterpreting the primitive term "straight line" so that axiom 5 is not necessarily true, then we obtain a family of geometries that includes Euclidean geometry and non-Euclidean geometries.

We conclude a subuniverse can be more general then its parent universe.

9.4 A Subuniverse of Electromagnetism

A more general form of the Maxwell equations (We will call it the pseudo-Maxwell equations.) in the absence of sources and currents can be defined by

$$\nabla \times \mathbf{E} = -\beta \partial \mathbf{B}/\partial t$$
$$\nabla \cdot \mathbf{E} = 0 \tag{9.4}$$

$$\nabla \cdot \mathbf{B} = 0$$
$$\nabla \times \mathbf{B} = -\beta\partial \mathbf{E}/\partial t$$

where β is a constant and × represents the vector cross product while · represents a vector inner product. If β = 1 then eqs. 9.4 are exactly the Maxwell equations of electrodynamics in the absence of sources and currents.

We now define the nine primitive terms of eqs. 9.4:

"∇ ×" "∇·" "=" "–" "β" "∂/∂t" "**E**" "**B**" "0"

which we associate with the eigenvalue operators $A_1, A_2, \ldots, A_9$. The universe state is

$$|\Omega_u\rangle = |\,\text{“}\nabla\times\text{”}, \text{“}\nabla\cdot\text{”}, \text{“=”}, \text{“–”}, \text{“}\beta\text{”}, \text{“}\partial/\partial t\text{”}, \text{“}\mathbf{E}\text{”}, \text{“}\mathbf{B}\text{”}, \text{“0”}\rangle \qquad (9.5)$$

We apply the lowering operator to make the eigenvalue of A_5 (the term "β") zero producing the resulting universe state $|\Omega_{u1}\rangle$:

$$[gn(\text{“}\beta\text{”})]^{1/2}|\Omega_{u1}\rangle = a_{\text{“}\beta\text{”}}|\Omega_u\rangle$$

$$= |\,\text{“}\nabla\times\text{”}, \text{“}\nabla\cdot\text{”}, \text{“=”}, \text{“–”}, 0, \text{“}\partial/\partial t\text{”}, \text{“}\mathbf{E}\text{”}, \text{“}\mathbf{B}\text{”}, \text{“0”}\rangle \qquad (9.6)$$

Thus the first and fourth Maxwell equations become:

$$\langle\Omega_{u1}|(A_1 + A_7 + A_3 + A_9)|\Omega_{u1}\rangle \equiv \nabla\times\mathbf{E} = 0$$
$$\qquad (9.7)$$
$$\langle\Omega_{u1}|(A_1 + A_8 + A_3 + A_9)|\Omega_{u1}\rangle \equiv \nabla\times\mathbf{B} = 0$$

The complete set of the subuniverse "Maxwell" equations is

$$\begin{aligned} \nabla\times\mathbf{E} &= 0 \\ \nabla\cdot\mathbf{E} &= 0 \\ \nabla\cdot\mathbf{B} &= 0 \\ \nabla\times\mathbf{B} &= 0 \end{aligned} \qquad (9.8)$$

Eqs. 9.8 represent a "new" theory in which the electric and magnetic fields are decoupled and become separate "subtheories." An important result of this theory is the absence of electromagnetic waves.

The creation of subuniverses, and the reverse process of combining subuniverses to produce a superuniverse, provide a mechanism for relating theories, and a mechanism for unifying similar theories into a joint theory in some area of physics or mathematics.

9.5 Products of Universes

All universes, as we have defined them, are linear vector spaces. Thus it is natural to consider direct products of universes. We will start by considering the case of the direct product of the additive and multiplicative subuniverses defined earlier in this chapter, namely $|\Omega_+\rangle|\Omega_\bullet\rangle$. Then the axioms

1. a + b = b + a	Commutativity of Addition
2. a + (b + c) = (a + b) + c	Associativity of Addition
3. a·b = b·a	Commutativity of Multiplication
4. a·(b·c) = (a·b)·c	Associativity of Multiplication

emerge directly for the product of subuniverses. For example, axiom 1 is generated from

$$\langle\Omega_\bullet|\langle\Omega_+|(A_{+\text{"a"}}I_\bullet + A_{+\text{"+"}}I_\bullet + A_{+\text{"b"}}I_\bullet + A_{+\text{"="}}I_\bullet + A_{+\text{"b"}}I_\bullet + A_{+\text{"+"}}I_\bullet + A_{+\text{"a"}}I_\bullet)|\Omega_+\rangle|\Omega_\bullet\rangle$$

where $I_\bullet$ is the identity operator in the multiplicative subuniverse. The other axioms are similarly generated. The plus subscript on the operators A_+ indicates they are operators in the additive subuniverse. Since we are taking the direct product of linear vector spaces the eigenvalue operators are paired from each space as shown.

The fifth axiom

5. a·(b + c) = a·b + a·c	Distributive Law

which combines addition and multiplication can also be added to the set of axioms. The statement of the fifth axiom is slightly more complex since both subuniverses "interact" through this axiom:

$$\langle\Omega_\bullet|\langle\Omega_+|(A_{+\text{"a"}}I_\bullet + A_{\bullet\text{"."}}I_+ + A_{+\text{"("}}I_\bullet + A_{+\text{"b"}}I_\bullet + A_{+\text{"+"}}I_\bullet + A_{+\text{"c"}}I_\bullet + A_{+\text{")"}}I_\bullet + A_{+\text{"="}}I_\bullet +$$
$$+ A_{+\text{"a"}}I_+ + A_{\bullet\text{"."}}I_+ + A_{+\text{"b"}}I_\bullet + A_{+\text{"+"}}I_\bullet + A_{+\text{"a"}}I_\bullet + A_{\bullet\text{"."}}I_+ + A_{+\text{"c"}}I_\bullet)|\Omega_+\rangle|\Omega_\bullet\rangle$$

The "•" subscript on the operators $A_\bullet$ indicates they are operators in the multiplicative subuniverse. The axiom can be restated in alternate ways as well since we identify the terms "a", "b", and "c" in $|\Omega_+\rangle$ with the corresponding terms in $|\Omega_\bullet\rangle$.

This example illustrates direct products of universes. It also shows that an issue, that may arise, is duplicate terms in the universes. As shown, this issue is usually easily resolved.

The procedure of taking the direct product of two universes can result in the creation of a "superuniverse" that combines the features of both universes. Additional axioms can be added to the combined universe to provide inter-relationships between the universes. Axiom 5 above is an example of an axiom that interrelates (causes an interaction between) the component universes. Thus a combination of universes can lead to more than their initial combined content.

10. Matrix Form of Operator Logic

The eigenvalue operators A_{E_k} and B_{E_k} in the semantic and calculus universes of discourse each have two eigenvalues: a Gödel number $gn(E_k)$ and zero (which is the Gödel number of the empty string). For each k, the operator A_{E_k} or B_{E_k}, and each operator's two eigenstates $|0>$ and $|E_k>$, can be expressed in terms of SU(2) operators and eigenstates in a 2×2 matrix formulation. This representation is also used to describe spin ½ particles such as electrons and quarks in physics theories. Later when we develop the Platonic realm of Ideas and their relation to Reality this correspondence will turn out to be of importance.

10.1 Spinor Representation of Operator Logic

The SU(2) spinor representation is based on the Pauli matrices:

$$\sigma_1 = \begin{bmatrix} 0 & 1 \\ 1 & 0 \end{bmatrix} \qquad \sigma_2 = \begin{bmatrix} 0 & -i \\ i & 0 \end{bmatrix} \qquad \sigma_3 = \begin{bmatrix} 1 & 0 \\ 0 & -1 \end{bmatrix} \tag{10.1}$$

Operator Logic can be expressed in a spinor representation. Consider an eigenvalue operator A_{E_k} for some value k.[103] We can represent this operator and its eigenvectors as a matrix and as column vectors (spinors):

$$A_{E_k} \rightarrow \begin{bmatrix} gn(E_k) & 0 \\ 0 & 0 \end{bmatrix} \qquad |E_k> \rightarrow \begin{bmatrix} 1 \\ 0 \end{bmatrix} = s_{k\uparrow} \qquad |0> \rightarrow \begin{bmatrix} 0 \\ 1 \end{bmatrix} = s_{k\downarrow} \tag{10.2}$$

The raising and lowering operators of the spinor representation are

---

[103] Calculus Logic Operators have a completely analogous formalism.

$$c_{E_k}^{\dagger} = \begin{bmatrix} 0 & 1 \\ 0 & 0 \end{bmatrix} \qquad c_{E_k} = \begin{bmatrix} 0 & 0 \\ 1 & 0 \end{bmatrix} \tag{10.3}$$

where † signifies hermitean conjugation, and where

$$c_{E_k}^{\dagger} \equiv [gn(E_k)]^{-1/2} a_{E_k}^{\dagger} \tag{10.4}$$

$$c_{E_k} \equiv [gn(E_k)]^{-1/2} a_{E_k}$$

with the anti-commutation relations

$$c_{E_k}^{\dagger} c_{E_{k'}} + c_{E_{k'}} c_{E_k}^{\dagger} \equiv \{c_{E_k}^{\dagger}, c_{E_{k'}}\} = \delta_{kk'}$$
$$\{c_{E_k}, c_{E_{k'}}\} = 0 \tag{10.5}$$
$$\{c_{E_k}^{\dagger}, c_{E_{k'}}^{\dagger}\} = 0$$

where $\delta_{kk'}$ is the Kronecker delta. Note the semantic eigenvalue operator A_{E_k} then becomes

$$A_{E_k} = gn(E_k) c_{E_k}^{\dagger} c_{E_k} \tag{10.6}$$

and the corresponding projection operator is

$$P_{E_k} = P'_{E_k} \tfrac{1}{2}[1 - c_{E_k} c_{E_k}^{\dagger}] \tag{10.7}$$

where P'_{E_k} is a truth projection operator. (See section 7.1 for details on truth projection operators.)

The universe of discourse $|\Omega_u\rangle$ with q factors is generated by the direct product of the q individual eigenstates $s_{i\dagger}$ of each primitive term in the universe:

$$|\Omega_u\rangle = \prod_i s_{i\dagger} \tag{10.8}$$

Note the universe state is normalized to one, $\langle\Omega_u|\Omega_u\rangle = 1$. A statement has the form:

$$\text{statement} = \langle\Omega_u|A_1 + A_2 + A_3 + \ldots |\Omega_u\rangle \quad (10.9)$$

where the matrices A_i have the form specified in eq. 10.2. The corresponding truth value is given by

$$\text{Truth Value} = \langle\Omega_u|\, F(P'_1, P'_2, \ldots)P_1\, P_2\, P_3 \ldots |\Omega_u\rangle \quad (10.10)$$

A simple example of this formalism is a two term universe. In this universe of discourse

$$|\Omega_u\rangle = s_{1\uparrow}\, s_{2\uparrow}$$

The statement (with a symbolic + sign)

$$\begin{aligned}\text{statement} &= \langle\Omega_u|A_1 + A_2|\Omega_u\rangle \\ &= \text{gn}(E_1) + \text{gn}(E_2)\end{aligned}$$

can be transformed into a human language by mapping the Gödel numbers to their human language equivalent, dropping the + sign, and putting the statement into a grammatically correct form for that language. This might include add “a” and “the” (in English), and possibly reordering the words. (For example in German words often appear in a different order than English.)

10.2 A Symmetry of the Spinor Matrix Representation

Eq. 10.2 implies a discrete symmetry is present in the matrix form of a universe of discourse[104]. If we define the eigenvalue and projection operators

$$A''_{E_k} = \sigma_1^{-1} A_{E_k} \sigma_1 \quad (10.11)$$
$$P''_{E_k} = \sigma_1^{-1} P_{E_k} \sigma_1 \quad (10.12)$$
$$P'''_{E_k} = \sigma_1^{-1} P'_{E_k} \sigma_1 \quad (10.13)$$

for all k using $\sigma_1^{-1} = \sigma_1$[105], and if we define a new universe state[106]

---

[104] Similar comments apply to the matrix form of a calculus universe of discourse.

[105] A σ_1 transformation is applied for each k.

[106] There is a σ_1 transformation applied to each factor in $|\Omega_u\rangle$.

$$|\Omega_u\rangle = \sigma_1^{-1}|\Omega''_u\rangle \tag{10.14}$$

then the transformed statements and their truth values in the new matrix representation are the same as that of the original matrix representation.

A statement and its truth value is transformed to

$$\text{statement} = \langle\Omega_u|A_1 + A_2 + A_3 \ldots |\Omega_u\rangle = \langle\Omega''_u|A''_1 + A''_2 + \ldots |\Omega''_u\rangle \tag{10.15}$$

and

$$\text{TV} = \text{Truth Value} = \langle\Omega_u|F(P'_{E_1}, P'_{E_2}, \ldots)\prod_{n \in \{m'\}} P_{E_n} |\Omega_u\rangle$$

$$= \langle\Omega''_u|F(P''_{E_1}, P''_{E_2}, \ldots)\prod_{n \in \{m'\}} P''_{E_n} |\Omega''_u\rangle \tag{10.16}$$

Thus we have invariance under the discrete transformation σ_1.

10.3 Direct Sum of Spinor Representations of Universes – Dirac Matrices

In this section we will consider the direct sum of the spinor representations of two universes that are generally different universes. In part, our goal is to construct a space whose eigenvectors are similar to 4-component Dirac spinors. We will use these eigenvectors later to develop the relation of the Platonic realm of Ideas to Reality.

The universes that we will define may have different terms, axioms or proof schema. We will first define the direct sum and then introduce terms, which join the two universes of discourse to create a combined universe. We denote the constructs of the first universe as

$$A_{E1k} \quad P_{E1k} \quad c_{E1k} \quad c_{E1k}^{\dagger} \quad s_{1k\uparrow} \quad s_{1k\downarrow} \quad |\Omega_{1u}\rangle \tag{10.17}$$

and the constructs of the second universe as

$$A_{E2k} \quad P_{E2k} \quad c_{E2k} \quad c_{E2k}^{\dagger} \quad s_{2k\uparrow} \quad s_{2k\downarrow} \quad |\Omega_{2u}\rangle \tag{10.18}$$

From these constructs we form four column vectors with four columns:

$$s_{\uparrow\uparrow k} = \begin{bmatrix} s_{1k\uparrow} \\ s_{2k\uparrow} \end{bmatrix} \quad s_{\uparrow\downarrow k} = \begin{bmatrix} s_{1k\uparrow} \\ s_{2k\downarrow} \end{bmatrix} \quad s_{\downarrow\uparrow k} = \begin{bmatrix} s_{1k\downarrow} \\ s_{2k\uparrow} \end{bmatrix} \quad s_{\downarrow\downarrow k} = \begin{bmatrix} s_{1k\downarrow} \\ s_{2k\downarrow} \end{bmatrix} \tag{10.19}$$

and 4×4 matrices composed of 2×2 pairs of matrix entries[107]

$$A_{E_k} = \begin{bmatrix} A_{E_{1k}} & 0 \\ 0 & A_{E_{2k}} \end{bmatrix} \qquad P_{E_k} = \begin{bmatrix} P_{E_{1k}} & 0 \\ 0 & P_{E_{2k}} \end{bmatrix} \tag{10.20}$$

$$c_{E_k} = \begin{bmatrix} c_{E_{1k}} & 0 \\ 0 & c_{E_{2k}} \end{bmatrix} \qquad d_{E_k} = \begin{bmatrix} c_{E_{1k}} & 0 \\ 0 & c_{E_{2k}} \end{bmatrix} \tag{10.21}$$

and also similarly for $c_{E_k}^{\dagger}$ and $d_{E_k}^{\dagger}$. The direct sum universe state is

$$|\,\Omega_u\rangle = \begin{bmatrix} |\Omega_{1u}\rangle \\ |\Omega_{2u}\rangle \end{bmatrix} \tag{10.22}$$

10.3.1 Independent Universes

We can express all the statements in the first universe by projecting out the parts associated with the second universe. The universe projection operator that plays this role is

$$U_1 = ½(I_4 + \gamma^0) \tag{10.23}$$

---

[107] If the number of terms in universe one and universe two differ then one pads eq. 10.20 and 10.21 with 2×2 zero submatrices for the values of k with unpaired submatrices.

where I_4 is the 4×4 identity matrix and γ^0 is one of the four Dirac γ matrices.[108] The four Dirac γ matrices and their product $\gamma^5 = \gamma_5 = i\gamma^0\gamma^1\gamma^2\gamma^3$ are:

$$\gamma^0 = \begin{bmatrix} I & 0 \\ 0 & -I \end{bmatrix} \quad \gamma^i = \begin{bmatrix} 0 & \sigma_i \\ -\sigma_i & 0 \end{bmatrix} \quad \gamma^5 = \begin{bmatrix} 0 & I \\ I & 0 \end{bmatrix} \tag{10.24}$$

for i = 1, 2, 3 where the submatrices are the 2×2 Pauli matrices, and I is the 2×2 identity matrix. We will use the γ^5 matrix in the next subsection.

A statement strictly of the first universe has the form

$$\text{Universe 1 statement} = \langle\Omega_u|U_1(A_1 + A_2 + A_3 \ldots\ )|\Omega_u\rangle \tag{10.25}$$

where the eigenvalue operators A_i have the form of eq. 10.20 and its corresponding truth value is given by

$$\text{Universe 1 Truth Value} = \langle\Omega_u|U_1F(P'_1, P'_2, \ldots)P_1\ P_2\ P_3 \ldots|\Omega_u\rangle \tag{10.26}$$

where the projection operators P_i and P'_i have the form of eq. 10.20.

Similarly if we define the universe projection operator

$$U_2 = ½(I_4 - \gamma^0) \tag{10.27}$$

then we see that a statement strictly of the second universe has the form

$$\text{Universe 2 statement} = \langle\Omega_u|U_2(A_1 + A_2 + A_3 \ldots\ )|\Omega_u\rangle \tag{10.28}$$

where the eigenvalue operators A_i have the form of eq. 10.20 and its corresponding truth value is given by

$$\text{Universe 2 Truth Value} = \langle\Omega_u|U_2F(P'_1, P'_2, \ldots)P_1P_2P_3 \ldots|\Omega_u\rangle \tag{10.29}$$

---

[108] We introduce Dirac γ matrices with a Platonic view of connecting Operator Logic to Elementary Particle Physics later in this volume.

where the projection operators P_i and P'_i have the form of eq. 10.20. Thus both universes are embodied in the direct sum.

10.3.2 Mixing Between the Universes

In the preceeding subsection both universes in the direct sum maintained their distinct identity with separate sets of statements. In this section we will consider statements that have terms from both universes. This possibility is of interest because it is analogous to the mixing of two component particle spinors in the free Dirac equation in Physics that is caused by a particle's mass term. Again we are looking to the development of the Platonic Ideas – Reality relationship later. This possibility of mixing universes is of interest in its own right as well.

In order to create mixed statements with terms that come from both universes we have to define eigenvalue and projection matrices that perform that task:[109]

$$\hat{A}_{E_k} = \gamma^5 A_{E_k} \qquad \text{and} \qquad \hat{S} = \gamma^5 P_{E_k} \tag{10.30}$$

using the matrices in eq. 10.20. The new matrices have the form

$$\hat{A}_{E_k} = \begin{bmatrix} 0 & A_{E_{2k}} \\ A_{E_{1k}} & 0 \end{bmatrix} \qquad \hat{S}_{E_k} = \begin{bmatrix} 0 & P_{E_{2k}} \\ P_{E_{1k}} & 0 \end{bmatrix} \tag{10.31}$$

Some examples of statements that mix both universes are

$$\text{Mixed Universe statement} = \langle\Omega_u|U_1(A_1 + \ldots + \hat{A}_i + \ldots + A_j + \ldots + \hat{A}_k + \ldots)|\Omega_u\rangle \tag{10.32}$$

and

$$\text{Mixed Universe statement} = \langle\Omega_u|U_2(A_1 + \ldots + \hat{A}_i + \ldots + A_j + \ldots + \hat{A}_k + \ldots)|\Omega_u\rangle \tag{10.33}$$

with their corresponding truth values having the form

---

[109] There is also an alternate approach based on the use of a mass term, which we will discuss later.

Mixed Universe Truth Value 1 =

$$= \langle\Omega_u|U_1F(P'_1, P'_2, \ldots, S'_1, S'_2, \ldots)P_1 \ldots \hat{S}_i \ldots P_j \ldots \hat{S}_k \ldots |\Omega_u\rangle \tag{10.34}$$

Mixed Universe Truth Value 2 =

$$= \langle\Omega_u|U_2F(P'_1, P'_2, \ldots, S'_1, S'_2, \ldots)P_1 \ldots \hat{S}_i \ldots P_j \ldots \hat{S}_k \ldots |\Omega_u\rangle \tag{10.35}$$

respectively where the A_k in eq. 10.32 and 10.33 correspond to the projections P_k in eqs. 10.34 and 10.35; and the $\hat{A}_k$ in eq. 10.32 and 10.33 correspond to the projections $\hat{S}_k$ in eqs. 10.32 and 10.33.

A reader who is a logician may wonder why we have developed the concept of mixed universe statements. Again we note this concept is analogous to the Dirac equation for spin ½ elementary particles and will be used later as we attempt to construct a realization of Plato's theory of Ideas and Reality. The analogy is also evident in the two component form of the free spin ½ particle Dirac equation. The Dirac wave function ψ has four components that can be viewed as a pair of spinors:

$$\psi = \begin{bmatrix} \psi_+ \\ \psi_- \end{bmatrix} \tag{10.36}$$

that satisfy the two-component form of the Dirac equation:

$$i\partial\psi_+/\partial t = -i\boldsymbol{\sigma}\cdot\nabla\psi_+ - m\psi_- \tag{10.37}$$

$$i\partial\psi_-/\partial t = i\boldsymbol{\sigma}\cdot\nabla\psi_- - m\psi_+ \tag{10.38}$$

where m is the particle's mass, $\boldsymbol{\sigma}$ is a 3-vector whose components are the Pauli matrices, and ∇ is the grad differential operator. If we view these equations as statements, which is what they are, then the mass terms mix the "universe" of ψ_+ spinors with the "universe" of ψ_- spinors. Thus the analogy. We will discuss this in more detail at a later point.[110]

---

[110] Eqs. 10.38 and 10.39 also show we could create a spinor representation for spin ½ particles directly. Therefore forming the direct sum (eqs. 10.19 – 10.23) was not necessary but it was helpful because it led directly to a Dirac matrix formulation that is conventional in spin ½ particle physics.

The fact that γ_5 appears in the Operator Logic mixing terms (eq. 10.30) is also significant since γ_5 also appears in the 4 component Dirac equation formalism for left-handed and right-handed neutrinos. (Neutrinos are also spin ½ particles.)

10.4 Vector Operator Logic

In this section we will define a vector matrix representation of Operator Logic. It is possible that statements in a universe of discourse have terms that appear in logically related pairs. It is also not uncommon that a universe of discourse (or a deductive system) consists of a number of distinct sectors that are intertwined by the addition of one or more axioms that reference terms and operations in other sectors. Section 9.2 provides an example of this situation with two axioms for addition only, two axioms for multiplication only, and a third axiom that combines addition and multiplication operations.

The most important reason for considering Vector Operator Logic is our desire to build the Platonic connection between Ideas and Reality. The elementary particles, with which we are familiar, are spin ½ fermions and spin 1 vector bosons (plus spin 2 gravitons). Thus Vector Operator Logic representations may be relevant to connect Ideas with Reality. (We will forego presenting spin 2 Operator Logic in this book.)

We have seen the spinor matrix formulation (section 10.1) of universes of discourse earlier. It is also possible to develop a vector formalism for Operator Logic that might be appropriate for some types of universes. The matrix Vector Operator Logic formalism that we will consider in this section is similar to the formalism for spin 1 physical systems.[111]

We will use the 3-dimensional (vector) representation of SU(2). The three generator matrices, T_i, of SU(2) can be placed in a form where T_3 is diagonal:

[111] It is also possible to develop spin 3/2, spin 2, and higher spin formalisms for universes of discourse. Their development parallels the spin ½ and spin 1 (vector) cases discussed here.

$$T_3 = \begin{bmatrix} 1 & 0 & 0 \\ 0 & 0 & 0 \\ 0 & 0 & -1 \end{bmatrix} \quad T_+ = \begin{bmatrix} 0 & \sqrt{2} & 0 \\ 0 & 0 & \sqrt{2} \\ 0 & 0 & 0 \end{bmatrix} \quad T_- = \begin{bmatrix} 0 & 0 & 0 \\ \sqrt{2} & 0 & 0 \\ 0 & \sqrt{2} & 0 \end{bmatrix} \tag{10.39}$$

The kth eigenvalue and projection operators can be assumed to have the form:

$$A_{E_k} = \begin{bmatrix} gn(E_{k1}) & 0 & 0 \\ 0 & gn(E_{k2}) & 0 \\ 0 & 0 & 0 \end{bmatrix} \qquad P_{E_k} = \begin{bmatrix} P_{Ek1} & 0 & 0 \\ 0 & P_{Ek2} & 0 \\ 0 & 0 & 0 \end{bmatrix} \tag{10.40}$$

where A_{Ek1} and A_{Ek2} are eigenvalue operators for the specific Gödel eigenvalues $gn(E_{k1})$ and $gn(E_{k2})$ respectively of specific terms of the universe, and P_{Ek1} and P_{Ek2} are their corresponding projection operators. *Thus each eigenvalue operator represents two different terms in general.*

For each integer value k, the three eigenvectors are

$$s_{k1} = \begin{bmatrix} 1 \\ 0 \\ 0 \end{bmatrix} \qquad s_{k2} = \begin{bmatrix} 0 \\ 1 \\ 0 \end{bmatrix} \qquad s_{k3} = \begin{bmatrix} 0 \\ 0 \\ 1 \end{bmatrix} \tag{10.41}$$

Assuming there are q eigenvalue operators (each containing two terms in general) in the universe then an individual eigenstate will be a direct product with the form:

$$|\{m\}> = \prod_{\substack{k=1 \\ i \in \{m\}}}^{q} s_{ki} \tag{10.42}$$

where {m} is an ordered set of q values from the set i = 1, 2, 3. For example, if q = 5 and {m} = {1, 3, 1, 2, 2} then

$$|\{m\}> = s_{11}s_{23}s_{31}s_{42}s_{52}$$

is an outer product of 5 eigenvectors.

If there are q eigenvalue operators then the universe state $|\Omega_u\rangle$ is the product of q eigenvectors:

$$|\Omega_u\rangle = 2^{-q/2} \prod_{k=1}^{q} (s_{k1} + s_{k2}) \qquad (10.43)$$

The universal state is normalized to unity:[112]

$$\langle\Omega_u|\Omega_u\rangle = 1 \qquad (10.44)$$

10.4.1 Vector Statements

Vector universes can generate two statements if one does not mix the first and second rows of the eigenvalue operators. The "top" statement statement and its truth value have the forms

$$\text{statement} = \langle\Omega_u|U_1(A_i + A_j + A_k \ldots)|\Omega_u\rangle \qquad (10.45)$$

and

$$\text{Truth Value} = \langle\Omega_u|U_1F(P'_1, P'_2, \ldots)(P_iP_jP_k \ldots)|\Omega_u\rangle \qquad (10.46)$$

where U_1 is a projection operator defined below. The "second" statement statement and its truth value have the forms

$$\text{statement} = \langle\Omega_u|U_2(A_i + A_j + A_k \ldots)|\Omega_u\rangle \qquad (10.47)$$

and

$$\text{Truth Value} = \langle\Omega_u|U_2F(P'_i, P'_j, \ldots)P_iP_jP_k \ldots|\Omega_u\rangle \qquad (10.48)$$

The projection operators are:

$$U_1 = \begin{bmatrix} 1 & 0 & 0 \\ 0 & 0 & 0 \\ 0 & 0 & 0 \end{bmatrix} \qquad U_2 = \begin{bmatrix} 0 & 0 & 0 \\ 0 & 1 & 0 \\ 0 & 0 & 0 \end{bmatrix} \qquad (10.49)$$

---

[112] Using the factor of $2^{-q/2}$.

10.4.2 S_3 Symmetry of Vector Universe of Discourse

The matrix formulation of Vector Operator Logic has a discrete S_3 symmetry[113] under permutations of row and column numbers. For example the interchange of row 1 and column 1 with row 2 and column 2 throughout does not change any statement. More generally if we define a symmetric group transformation S of the eigenvalue and projection operator matrices

$$A''_{E_k} = S^{-1}A_{E_k}S \tag{10.50}$$
$$P''_{E_k} = S^{-1}P_{E_k}S \tag{10.51}$$

for all k, and if we define a new universe state where S^{-1} is understood to be applied to each multiplicand in the direct products of eigenvectors:

$$|\Omega_u\rangle = S^{-1}|\Omega''_u\rangle \tag{10.52}$$

then the transformed statements and their truth values in the new matrix representation are the same as that of the original matrix representation.

Under the symmetric group operation of interchanging rows and columns, 1 and 2, a statement and its truth value is transformed to

$$\begin{aligned}\text{statement} &= \langle\Omega_u|U_1(A_1 + A_2 + A_3 \ldots)|\Omega_u\rangle \\ &= \langle\Omega''_u|U_2(A''_1 + A''_2 + \ldots)|\Omega''_u\rangle\end{aligned} \tag{10.53}$$

and

$$\begin{aligned}\text{Truth Value} &= \langle\Omega_u|U_1(P_1\, P_2\, P_3 \ldots)|\Omega_u\rangle \\ &= \langle\Omega''_u|U_2P''_1\, P''_2\, P''_3 \ldots |\Omega''_u\rangle\end{aligned} \tag{10.54}$$

Thus Vector Operator Logic has an S_3 invariance.[114] The eigenvalue operators can be redefined to produce new operators that break S_3 symmetry.

10.4.3 Raising and Lowering Operators to Define Vector Subuniverses

The raising and lowering operators that are used in the vector representation, are T_+ and T_- respectively. (See eq. 10.39). T_{k-} (the

---

[113] The symmetric group of degree 3 having six elements.

[114] The projections U_1 and U_2 must also be transformed under an S_3 transformation to maintain the invariance: $U_i'' = S^{-1}U_iS$ for i = 1, 2.

lowering operator for the kth term can be used to create a subuniverse with the kth term lowered. Products of T_-'s can create subuniverses with several terms lowered. For example $T_{k-}T_{l-}T_{m-}T_{n-}$ lowers four terms: the kth, l^th^, m^th^, and n^th^ terms.

The action of the kth raising and lowering operator for the kth term's eigenvectors is

$$\begin{aligned} T_{k+}\, s_{k1} &= 0 \\ T_{k+}\, s_{k2} &= \sqrt{2}\, s_{k1} \\ T_{k+}\, s_{k3} &= \sqrt{2}\, s_{k2} \\ \\ T_{k-}\, s_{k1} &= \sqrt{2}\, s_{k2} \\ T_{k-}\, s_{k2} &= \sqrt{2}\, s_{k3} \\ T_{k-}\, s_{k3} &= 0 \end{aligned} \tag{10.55}$$

where 0 represents the 3-vector composed of zeroes. We note

$$\begin{aligned} T_{k+}T_{k+}T_{k+} &\equiv T_{k+}^{\;3} = 0 \\ T_{k-}T_{k-}T_{k-} &\equiv T_{k-}^{\;3} = 0 \end{aligned} \tag{10.56}$$

We will consider an example of creating a subuniverse in the Vector Operator Logic. Applying T_{1-} to $|\Omega_u\rangle$ yields

$$|\Omega'_u\rangle = T_{1-}|\Omega_u\rangle \tag{10.57}$$

with

$$\langle\Omega'_u|\Omega'_u\rangle = 1 \tag{10.58}$$

Using eigenvalue operators with the form of eq. 10.40 the statement

$$\text{statement} = \langle\Omega'_u|U_1(A_1 + A_2)|\Omega'_u\rangle \tag{10.59}$$

yields

$$\text{statement} = \text{gn}(E_{21}) \tag{10.60}$$

after evaluating the matrices A_1 and A_2 between the universe state eq. 10.57.

If we define another subuniverse by

$$|\Omega"_u\rangle = T_{2-}T_{1-}|\Omega_u\rangle \tag{10.61}$$

we note

$$\langle\Omega"_u|\Omega"_u\rangle = 1 \tag{10.62}$$

The statement

$$\text{statement} = \langle\Omega"_u|U_2(A_1 + A_2)|\Omega"_u\rangle \tag{10.63}$$

has the value[115]

$$\text{statement} = gn(E_{11}) + gn(E_{21}) \tag{10.64}$$

These examples illustrate the creation of subuniverses in Vector Operator Logic.

Vector Operator Logics may provide an appropriate formalism for some deductive systems. Spin 3/2, spin 2 and higher spin matrix formulations are also possible as noted earlier.

[115] Note we use U_2 in this example (for illustrative purposes) as opposed to the previous example where we used U_1.

11. Probabilistic and Quantum Operator Logics

The Operator Logic that we have developed in the preceding chapters is not a probabilistic or quantum logic. The eigenvalue operators and the projection operators of all the terms in a universe of discourse commute with each other. The expectation values/inner products defined within this formalism have a well-defined value: true, false or undecidable (+1, –1 or 0).

In this chapter we develop the formalisms for Probabilistic Operator Logic and Quantum Operator Logic. These logics have expectation values that are probabilities and thus can be called probabilistic or quantum respectively depending on the semantic universe in which they are defined.

The material in this chapter is not needed to develop Relativistic Quantum Metaphysics in the succeeding chapters. However it seemed important to the author to discuss probabilistic Operator Logics as a natural extension of the development of deterministic Operator Logic in the preceeding chapters. *Deterministic Operator Logic will be crucial in the development of Relativistic Quantum Metaphysics.*[116]

However Quantum Operator Logic may be of importance in analyzing the cause(s) (rationale) of the formation of our universe with the physical laws and features that it has. We will consider this possibility in a later chapter.

---

[116] The reader may wonder why Quantum Operator Logic is not required for Relativistic Quantum Metaphysics. The reason is that the "Quantum" in Quantum Operator Logic refers to a new form of quantum behavior in the probability of the truth of statements while the word "Quantum" in Relativistic Quantum Metaphysics refers to the quantum nature of Reality such as the Heisenberg Uncertainty Principle which applies to the spatial position and momentum of a particle as well as other natural quantum phenomena. **Quantum Operator Logic concerns the quantum probability of statements – not the quantization of physical phenomena.**

11.1 Quantum Operator Logic is Not Fuzzy Logic

Fuzzy Logic in its simplest form is a type of classical probabilistic logic that is based on assigning probabilities to the members of the domain of a predicate that leads to a probability for each statement consisting of the predicate and one or more of the subjects of its domain.

We discuss Fuzzy Logic within the framework of our operator formalism. We call it Probabilistic Operator Logic.

Quantum Operator Logic which is the main subject of this chapter is probabilistic in nature but is not Fuzzy Logic. It describes a quantum probabilistic universe of discourse where terms may not commute and statements have a truth value that is a probability.

11.2 Propositions are Analogous to Experiments

There is a correspondence between the stages of an experiment and the sequence of parts of a statement. In part this analogy is based on Hilbert's[117] view of a predicate, "the use of the term 'predicate' is the one usual in philosophy, namely, that by which one subject can be more particularly characterized." Correspondingly the analysis of an experiment shows it generally to be a series of filtrations which particularize the input to an experiment through a sequence of stages to reach a final stage of output.[118] A filtration in an experiment takes an input state and lets a specified part of it proceed to the next stage of the experiment – stage by stage. A simple example is the case of an input stream of particles of varying velocities. A filtration may select only those particles of a certain velocity to go through to the next stage of the experiment. Analogously the statement "The car is green" takes the general set of all cars and particularizes (filters) it to a green car. An example of a multiple particularization (multi-stage) is "The car is green and has a silver star on it." This statement first particularizes the subject with the predicate "is green" and then particularizes it with the predicate "has a silver star on it."

We thus come to the conclusion that there is a formal analogy between physical experiments and logic propositions. In a sense this

[117] Hilbert (1950) p. 44.

[118] See Gottfried (1989), Messiah (1965), and Mackey (1963) as well as certain papers of Schwinger referenced in Gottfried (1989).

observation is the beginning of the relation between the Platonic sphere of Ideas and physical Reality. *Ideas and Reality "map" into each other.*

11.3 Overview of the Relation Between Operator Logic, Probabilistic Operator Logic & Quantum Operator Logic

All three of these formalisms assume that any proposition can be has one of three forms:

1. A simple subject- predicate form.
2. A combination of clauses of the simple subject-predicate form that are joined by connectives such as "and" and "or".
3. A mathematical equation.

11.3.1 Operator Logic Formulation

In the previous chapters we developed the formalism of Operator Logic. In Operator Logic the eigenvalue operators commute with each other – subject state projection operators, predicate state projection operators, predicate truth value projection operators, and subject truth value projection operators.[119] As a result we obtain deterministic propositions and deterministic well defined truth values for propositions.

11.3.2 Probabilistic Operator Logic Formulation

Probabilistic Operator Logic is a generalization of Operator Logic in which the eigenvalue operators commute with each other. Predicate state projection operators, and subject state projection operators also all commute with each other. As a result simultaneous eigenstates of all terms in a universe of discourse are allowed.

However, predicate *truth value* projection operators and subject *truth value* projection operators do not form a commuting set. For a given predicate each subject in the predicate's domain has a probability value associated with it that determines the truth value of a subject-predicate statement or clause. These probabilities are classical probabilities – not quantum probabilities.[120] The sum of the probability values of all the subjects in each predicate's domain is one.

---

[119] See the discussion of eqs. 6.19 – 6.23.

[120] The probability of each subject in a predicate's domain is determined by semantic considerations in the case of a semantic universe of discourse and arbitrarily in the case of a calculus universe of discourse.

11.3.3 Quantum Operator Logic Formulation

Quantum Operator Logic addresses a theoretical physics issue as well as more general issues. Logic, Operator Logic and Probabilistic Operator Logic all concern c-number terms – terms that are simply words in a language or mathematics – that commute with each other even in our Operator Logic formulations. The areas of Quantum Mechanics and Quantum Field Theory have statements in the form of equations that contain q-number terms – terms that are non-commuting operators. Because of the inherent quantum nature of the terms in these theories the truth or falsity of statements (equations) containing them cannot be truly addressed by Logic, Operator Logic or Probabilistic Operator Logic.

The normal approach of physicists is to postulate a lagrangian and derive equations through variational techniques ignoring the q-number nature of the terms.[121] While this approach works in physics[122] it is not satisfactory from the viewpoint of logic.

In this chapter we will define a Quantum Operator Logic for equations that enables the truth or falsity of q-number equations (statements) to be determined. Quantum Operator Logic lies at the basis of Relativistic Quantum Metaphysics.

11.4 Probabilistic Operator Logic Formalism

The Probabilistic Operator Logic formalism is an extension of the formalism developed for classical Operator Logic in earlier chapters to the case where the eigenvalue operators of the terms in a universe commute with each other[123] but *the set of truth value projection operators of the universe of discourse do not commute with each other in general.*

We will define the state of a probabilistic universe of discourse in the same manner as we did in the case of the classical universe of discourse using Dirac's bra-ket notation for states and operators. We will

---

[121] To some extent this issue is mitigated by path integral formulations of quantum theories that use c-number terms but lead to results identical to q-number quantum theories (if the q-number quantum theories are correctly calculated.)
[122] And avoids the issue of false equations generally.
[123] This enables the eigenvalue operators all terms in a universe of discourse to have a simultaneous eigenvector, namely, the universe state as defined previously.

assume we have q primitive terms with which any statement in the universe can be expressed.

Thus the universe state for a q term probabilistic universe of discourse (either semantic or calculus) will be

$$|\Omega_u> = \prod_{i=1}^{q} |E_i> \tag{11.1}$$

$$\equiv |E_1, E_2, E_3, \ldots, E_q>$$

with the set of q Gödel numbers E_i each corresponding to a term gi(E_i) as in eq. 6.26. Note $|\Omega_u>$ is normalized to one: $<\Omega_u|\Omega_u> = 1$.

In defining eq. 11.1 we are assuming that the set of q eigenvalue operators for the primitive terms, which we denote $\{A_i\}$, is compatible—the eigenvalue operators all commute with each other.

The elements of the set of state projection operators have the form[124]

$$P_{E_k} = ½[1 - c_{E_k} c_{E_k}^{\dagger}] \tag{11.2}$$

and also all commute with each other.

The operators $c_{E_k}^{\dagger}$ and c_{E_k} are raising and lowering operators defined by eq. 10.3 – 10.5. (We will not use the spinor formalism here to avoid mathematical complexity.)

The elements of the set of truth projection operators are denoted P'_{E_k}. The set of truth projection operators P'_{E_k} form an incompatible set—some, or all, of them do not commute with each other in general. For a predicate term A_{E_k} the truth value projection operator factor P'_{E_k} has the form[125]

$$P'_{E_k} = \sum_j a_{E_j E_k} |E_j><E_k| \tag{11.3}$$

where subject states are denoted E_j, and predicate states are denoted E_k; and where the sum is over all subjects in the universe of discourse. The

[124] These operators, together with the eigenvalue operators, conform to the general character of linear vector space operators. Truth projection operators are well-defined linear vector space operators.

[125] Eq. 11.3 is a sum of outer products of eigenvectors.

number $a_{E_jE_k}$ is the probability that the statement "gi(E_j) gi(E_k)" is true. Each combination of subject, E_j, and predicate, E_k, in a statement or clause has the probability $a_{E_jE_k}$ of being true. If "gi(E_j) gi(E_k)" is undecidable or is nonsense or is outside the domain of subjects of the predicate, *then we define* $a_{E_jE_k}$ *such that* $0 > a_{E_jE_k} \geq -1$.[126] *The introduction of negative probabilities is necessary to distinguish totally false statements from undecidable statements as we will see below.*[127]

For a subject term A_{E_j} the truth projection operator is

$$P'_{E_j} = |E_j\rangle\langle E_j| \tag{11.4}$$

The probability of the truth of a statement:

$$\text{statement} = \langle\Omega_u|A_1 + A_2 + A_3 + \ldots\ |\Omega_u\rangle \tag{11.5}$$

is determined by the evaluation of its probability

$$\text{Probability} = \langle\Omega_u|P_{1tot}\ P_{2tot}\ P_{3tot}\ \ldots\ |\Omega_u\rangle \tag{11.6}$$

where the $P_{itot} = P'_iP_i$ is the projection operator corresponding to the eigenvalue operator A_i.

A clause or subject-predicate statement has the form

$$\begin{aligned}\text{statement} &= \langle\Omega_u|A_{subject} + A_{predicate}|\Omega_u\rangle \\ &= \text{gn(subject)gn(predicate)}\end{aligned} \tag{11.7}$$

namely, a sequence of two Gödel numbers equivalent to two terms in the universe of discourse. The corresponding probability of the subject-predicate statement or clause in Probabilistic Operator Logic is

$$\begin{aligned}\text{Probability} &= \langle\Omega_u|P_{subject}P_{predicate}|\Omega_u\rangle \\ &= \langle\Omega_u|P'_{subject}P'_{predicate}|\Omega_u\rangle \\ &= \langle\Omega_u|\ |E_{subject}\rangle\langle E_{subject}|a_{subject,predicate}|E_{subject}\rangle\langle E_{predicate}|\ |\Omega_u\rangle \\ &= a_{subject,predicate}\end{aligned} \tag{11.8}$$

---

[126] If the probability of undecidability is not known then one can define $a_{EjEk} = -1$.
[127] Note this differs from deterministic Operator Logic. See eq. 7.4.

where $P'_{subject}$ and $P'_{predicate}$ are the eigenvalue operators' truth value projections.

Probabilities are assigned to all subjects in the domain of a predicate according to some semantic (or calculus) rule.

At this point we see we have a sentential calculus and a rule to reduce all compound statements to equivalent series of clauses as seen previously.

We note again the sentential connectives and their interpretation

Connective	Symbol
and	&
or	v
if … then (modus ponens)	→
if and only if	~
not	$\underline{\quad}$

Denoting sentences with upper case letters: A, B, … the five basic sentential combinations have the probabilities:

1. A & B is a compound sentence whose probability is the product of the probabilities of A and B.
2. A v B is a compound sentence whose probability is the sum of the probabilities of A and B.
3. A → B is a compound sentence whose probability = 1 – (probability of A)·(1 – (probability of B)). This is based on the non-probabilistic case: A → B is false if and only if A is true and B is false, which is equivalent to $A\underline{B}$ or in words is equivalent to NOT(A and NOT B).
4. A ~ B is a compound sentence whose probability = 1 – |(probability of A) – (probability of B)|. This is based on the non-probabilistic case: A ~ B is true if and only if both A and B are true, or both A and B are false.
5. The sentence $\underline{A}$ has the probability = 1 – (probability of A). This is based on the non-probabilistic case: the sentence $\underline{A}$ is false if A is true, and is true if A is false.

(11.9)

In specifying these probability rules we assume that if a statement has the probability $p \geq 0$ of being true then $p_f = 1 - p$ is the probability that the statement is false.

Rule 11.1: *Given a simple or compound statement with one or more clauses the probability of the statement follows the rules stated in eq. 11.9 which is an expression in the probabilities p_i of the clauses that we denote $F(p_1, p_2, \ldots)$ multiplied by the minimum value in the set of p_i divided by its absolute value:*

$$\text{Probability} = F(p_1, p_2, \ldots)\text{Min}(p_1, p_2, \ldots)/|\text{Min}(p_1, p_2, \ldots)| \qquad (11.10)$$

If the probability is positive or zero it is the probability of the statement being true.[128]

11.4.1 Probabilistic Example "if … then …" Modus Ponens

We will now examine the case of a probabilistic modus ponens (if … then …) statement. Suppose we have a statement

$$A \rightarrow B \qquad \text{which is equivalent to} \qquad \underline{A\underline{B}} \qquad (11.11)$$

and A can be expressed as a sequence of "anded" clauses:

$$A = \text{clauseA1 and clauseA2 and} \ldots \qquad (11.12)$$

and B can also be expressed as a sequence of "anded" clauses:

$$B = \text{clauseB1 and clauseB2 and} \ldots \qquad (11.13)$$

Then the probability of the statement eq. 11.10 according to eq. 11.9 is given by the expression:

$$\text{probability} = 1 - (\text{probability of A})\cdot(1 - (\text{probability of B})) \qquad (11.14)$$

---

[128] If any clause is undecidable then the factor Min(p1, p2, …)/|Min(p1, p2, …)| is negative and the probability has a negative value indicating the statement is undecidable or has a probability of undecidability specified by eq. 11.10. See section 11.4.3 for the interpretation of negative probabilities.

where the probability for A is

$$<\Omega_u|A|\Omega_u> = \{ p_{A1}\, p_{A2} \ldots\}\text{Min}(p_{A1}, p_{A2}, \ldots)/|\text{Min}(p_{A1}, p_{A2}, \ldots)| \quad (11.15)$$

with $p_{Ai} = <\Omega_u|\text{clauseAi}|\Omega_u>$; and where the probability for B is

$$<\Omega_u|B|\Omega_u> =\{ p_{B1}\, p_{B2} \ldots\}\text{Min}(p_{B1}, p_{B2}, \ldots)/|\text{Min}(p_{B1}, p_{B2}, \ldots)| \quad (11.16)$$

with $p_{Bi} = <\Omega_u|\text{clauseBi}|\Omega_u>$. The probability of A is

$$<\Omega_u|A|\Omega_u> \quad (11.17)$$

and the probability of B is

$$<\Omega_u|B|\Omega_u> \quad (11.18)$$

If, for the sake of example,

$$<\Omega_u|A|\Omega_u> = 3/4 \quad (11.19)$$

and

$$<\Omega_u|B|\Omega_u> = 1/4 \quad (11.20)$$

then the probability of A → B

$$p(A \rightarrow B) = 1 - 3/4(1 - 1/4) = 7/16 \quad (11.21)$$

For the special case where the probability of A equals one and the probability of B is one then p(A → B) = 1. If the probability of A equals one and if the probability of B is zero then p(A → B) = 0. Both these special cases conform to our expectations based on classical Logic.[129] If the probability is negative then the statement is undecidable.

11.4.2 Probabilistic Operator Logic Example

Consider a mechanical process that causes a stream of particles to fill a circle of a certain radius on a screen. The probabilities of the

[129] Note that this formalism differs from Operator Logic. Compare the above discussion to the discussion of section 7.1.

various possible radii of the circle are calculated and found to yield the $a_{subject,predicate}$ values for the cases listed below.

Predicate = "is the radius"

Domain:

Subject1 = "Five cm" $a_{subject,predicate}$ = ¼
Subject2 = "Ten cm" $a_{subject,predicate}$ = ¼
Subject3 = "Fifteen cm" $a_{subject,predicate}$ = ¼
Subject4 = "Twenty cm" $a_{subject,predicate}$ = ¼

Based on this probabilistic universe of discourse we find the probabilities of each statement to be:

p("Five cm ", "is the radius") = ¼
p("Ten cm ", "is the radius") = ¼
p("Fifteen cm ", "is the radius") = ¼
p("Twenty cm ", "is the radius") = ¼

The probabilities sum to one as they should.

11.4.3 The Probability of Undecidability

Rule 11.1 defined probability and its interpretation in the case of non-negative probabilities. Negative probabilities indicate a statement is undecidable: completely undecidable, or to some degree undecidable. Since decidable statements have a probability of truth or falsity, consistency suggests, in principle, decidability should also be probabilistic in nature. In the case of deterministic Logic decidable statements are either true or false; statements are either decidable or undecidable. In a probabilistic logic decidability and truth should both be probabilistic.

Therefore it is reasonable to define the absolute value of a negative probability as the probability that a statement is undecidable. One can envision situations where a proposition has a certain probability of undecidability. The common view is to regard a proposition as either decidable or undecidable. However, there could be situations, where terms are not fully defined or there are other sources of ambiguity, in which one can only estimate the probability that a proposition is undecidable.[130]

The thought process for the probability of undecidability of a proposition is analagous to the thought process for the probability of the truth of a proposition.

The issue of the probability of undecidability is also relevant for Quantum Operator Logic although we will not address it in this book except to say the previous statements in this subsection also apply in Quantum Operator Logic.

11.5 Quantum Operator Logic

Quantum Operator Logic deals with propositions containing terms that are q-numbers.[131] The label q-number comes from physics and designates a term that is inherently an operator.[132] Quantum Mechanics is based on q-numbers. For example, the position of a particle is represented in quantum mechanics as an operator – a q-number. The momentum of a particle is similarly an operator. These operators satisfy the Heisenberg Uncertainty Principle

$$[x, p] \equiv xp - px = i\hbar \qquad (11.22)$$

where $\hbar = h/(2\pi)$ with h being Planck's constant. Quantum Mechanics is best understood as an operator formalism. Operator expressions are evaluated between quantum states (inner products) to yield expectation values (probability amplitudes whose absolute values squared are probabilities.)

Quantum Electrodynamics, the most accurately measured theory in physics, is formulated in terms of operators for the electromagnetic fields. Unified Quantum Field theories such as ElectroWeak Theory[133]

---

[130] In science it is not uncommon to do experiments to measure something where the success of the measurement may be undecidable or only decidable with a certain probability.

[131] We use the label "Quantum" in this context because the propositions, which are usually equations, contain q-numbers – operators. The label does not limit the formalism to quantum physics.

[132] The label c-number, used earlier, also comes from physics and designates a term that is not an operator and has a strictly numerical value.

[133] The union of electromagnetism and the Weak Interactions.

and The Standard Model of Elementary Particles[134] are also formulated in terms of operators for the various particle fields.

The operator equations of Quantum Mechanics and the aformentioned Quantum Field Theories are not propositions whose truth or falsity can be determined in Logic, Operator Logic or Probabilistic Operator Logic. This issue has been overlooked because Quantum Mechanical Theories of specific phenomena and the various Quantum Field Theories have been normally defined using a lagrangian and a variational principle.[135] If the lagrangian is correct the quantum equations derived from it are true and the issue of truth or falsity does not arise. So the logic analysis of operator equations has hitherto not been investigated.[136] In the author's opinion the logic analysis of operator equations, viewed as propositions, is of interest for physics, logic and metaphysics.

In the case of metaphysics we are confronted with a very successful theory of Reality, The Standard Model. Its basis in metaphysics is not susceptible to significant metaphysical analysis because the "language" of the theory, its quantum operator equations, cannot be analyzed with the logic that has hitherto been developed. Metaphysics, as Leibniz observed, is based on logic. So we must develop a logic suitable for operator propositions. We call this logic Quantum Operator Logic. Having this logic in hand we can address the deeper issues of the foundations of The Standard Model in a meaningful way.

11.5.1 Quantum Operator Logic Formalism

The fundamental consideration of Quantum Operator Logic is the non-commutativity of some or all of the terms in a universe of discourse. The consequence of non-commuting terms is non-commuting eigenvalue operators as well as non-commuting projection operators in general.

In the previously discussed formalisms of Operator Logic and Probabilistic Operator Logic we could define a single state for a universe of discourse (See eq. 6.1). *In the present case we must define a set of*

---

[134] The union of electromagnetism, the Weak Interactions and the Strong Interactions.

[135] The variational principles of physics come from the Calculus of Variations which was originally formulated by Leibniz and others.

[136] Of course experimental verification has been used to determine whether a theory is true or not. But that is not logic analysis.

universe of discourse states: one state for each set of commuting eigenvalue operators.

For example if term1 and term2 are operators that do not commute with each other while term3, term4, ... commute with each other and with term1 and term2[137] then there are two universe states:

$$|\Omega_{u1}\rangle = |E_{term1}, E_{term3}, E_{term4}, \ldots\rangle \tag{11.23}$$

and

$$|\Omega_{u2}\rangle = |E_{term2}, E_{term3}, E_{term4}, \ldots\rangle \tag{11.24}$$

In general, the set of universe states consists of all states that are simultaneous eigenstates of the maximally commuting subsets of the set of all terms in the universe of discourse. Since the number of terms in a universe is finite by assumption the number of maximally commuting subsets of terms is finite. The form of the ith term is

$$|\Omega_{ui}\rangle = \prod_{j\in\{m_i\}} |E_j\rangle \tag{11.25}$$

where $\{m_i\}$ is the ith maximally commuting set of terms and thus eigenvalue operators.

We will now consider the example of a universe of discourse composed of a coordinate operator x and its corresponding momentum operator p where x and p satisfy eq. 11.22 and thus do not commute. We define the corresponding eigenvalue operators as $A_{"x"}$ and $A_{"p"}$ with Gödel number eigenvalues $E_{"x"}$ and $E_{"p"}$ respectively. We also define the eigenvalue operator $A_{"i\hbar"}$ with eigenvalue $E_{"i\hbar"}$. $A_{"i\hbar"}$ commutes with both $A_{"x"}$ and $A_{"p"}$. It will be convenient to define the set of universes of discourse

$$|\Omega_{u"x"}\rangle = |E_{"x"}, E_{"i\hbar"}\rangle \quad \text{and} \quad |\Omega_{u"p"}\rangle = |E_{"p"}, E_{"i\hbar"}\rangle \tag{11.26}$$

and form a two-vector (spinor) from them which we will call *term space*

---

[137] To keep the example simple we assume [term1, term2] equals a c-number and thus this commutator commutes with term3, term4, ... This is required by the commutator identity [A,[B,C]] + [C,[A,B]] + [B,[C,A]] = 0.

$$| \Omega_u> = \begin{bmatrix} |\Omega_{u\text{"x"}}> \\ \\ |\Omega_{u\text{"p"}}> \end{bmatrix} \tag{11.27}$$

This form will also be seen to be suggestive of the appearance of group symmetries in The Standard Model.

We will now introduce a matrix notation that enables us to evaluate equations (statements) involving non-commuting terms. We define the two diagonal matrices I_t in term space as having the matrix elements:

$$I_t = [\delta_{ti}] \tag{11.28}$$

where δ_{ik} is the Kronecker delta function with value one if i = k and zero otherwise. Thus I_t has only one non-zero entry, namely, a one at row t and column t – the diagonal matrix element for the tth term.

We now consider two examples that illustrate the basic nature of the Quantum Operator Logic formalism. First consider the additive commutation condition which in terms of the x and p operators is

$$x + p = p + x \tag{11.29}$$

This operator equation can be expressed in terms of Quantum Operator Logic eigenvalue operators

$$A_{tot\text{"x"}} = A_{\text{"x"}}\, I_{\text{"x"}} \tag{11.30}$$

and

$$A_{tot\text{"p"}} = A_{\text{"p"}}\, I_{\text{"p"}} \tag{11.31}$$

where

$$A_{\text{"x"}}\, |\Omega_{u\text{"x"}}> = gn(\text{"x"})|\Omega_{u\text{"x"}}> \tag{11.32}$$

and

$$A_{\text{"p"}}\, |\Omega_{u\text{"p"}}> = gn(\text{"p"})|\Omega_{u\text{"p"}}> \tag{11.33}$$

The eigenvalue operator equation leading to eq. 11.29 is

$$A_{tot\text{"x"}} + A_{tot\text{"p"}} = A_{tot\text{"p"}} + A_{tot\text{"x"}} \tag{11.34}$$

Taking the expectation value (inner product) of eq. 11.34 using the universe of discourse two-vector (eq. 11.27)

$$<\Omega_u|(A_{tot"x"} + A_{tot"p"} = A_{tot"p"} + A_{tot"x"})|\ \Omega_u>$$

yields (after mapping Gödel numbers to their corresponding terms)

$$x + p = p + x \qquad (11.35)$$

upon realizing that the matrices contain operators and the order of operators matters. We thus find eq. 11.29 follows.[138] Note that we do not express the plus sign or the equal sign as operators although we might have done so.

We now consider the case of the commutation relation eq. 11.22. We wish to evaluate the expectation value (inner product) of the commutator $[A_{tot"x"}, A_{tot"p"}]$ between universe two-vectors (eq. 11.27). The only sensible way to proceed is to take the expectation value of each eigenvalue operator independently:

$$\begin{aligned}<\Omega_u|[A_{tot"x"}, A_{tot"p"}]|\Omega_u> &= \text{Tr}\ \{<\Omega_u|A_{tot"x"}|\Omega_u><\Omega_u|A_{tot"p"}|\Omega_u> - \\ &\quad - <\Omega_u|\ A_{tot"p"}|\Omega_u><\Omega_u|A_{tot"x"}|\Omega_u>\} \\ &= E_{"x"}\ E_{"p"} - E_{"p"}\ E_{"x"} \end{aligned} \qquad (11.36)$$

Upon mapping the Gödel numbers to the corresponding operators we obtain

$$<\Omega_u|[A_{tot"x"}, A_{tot"p"}]|\Omega_u> = xp - px \qquad (11.37)$$

which gives

$$\begin{aligned}<\Omega_u|[A_{tot"x"}, A_{tot"p"}]|\Omega_u> &= xp - px = <\Omega_u|[A_{tot"x"}, A_{tot"p"}]|\Omega_u> \\ &= <\Omega_u|A_{tot"i\hbar"}|\Omega_u> = \text{Tr}\ <\Omega_u|IA_{"i\hbar"}/2|\Omega_u> \\ &= i\hbar \end{aligned} \qquad (11.38)$$

---

[138] Addition is normally considered a commutative operation but it is not required to be commutative. Thus commutativity is often an axiom and sometimes a theorem or result.

where I is the 2×2 identity matrix and $A_{\text{"}i\hbar\text{"}}$ is the eigenvalue operator for the c-number quantity $i\hbar$. Thus we obtain eq. 11.22 through a suitably evaluated Quantum Operator Logic statement:

$$<\Omega_u|[A_{tot\text{"}x\text{"}}, A_{tot\text{"}p\text{"}}]|\Omega_u> = <\Omega_u|A_{tot\text{"}i\hbar\text{"}}|\Omega_u> \qquad (11.39)$$

The evaluation of the sequence of eigenvalue operators between universe states in the preceding examples requires:

1. Taking the expectation values of each eigenvalue operator in an operator product separately to obtain its Gödel number.
2. Using a matrix notation with 2-vectors and 2×2 matrices (or an equivalent).
3. Evaluating the inner product of a statement.
4. Mapping Gödel numbers to corresponding operator terms in a fashion similar to Operator Logic.
5. Using the mathematical operators for multiplication, addition, subtraction, and so on after substituting corresponding operators for Gödel numbers.

Thus we find the Quantum Operator Logic equation

$$[A_{tot\text{"}x\text{"}}, A_{tot\text{"}p\text{"}}] = A_{\text{"}i\hbar\text{"}} \qquad (11.40)$$

when evaluated in the above formalism implies eq. 11.22.

11.5.2 Operator Term Space

The non-commutativity of operators in Quantum Operator Logic statements leads to a set of universe states (eq. 11.25). Each universe state has a set of maximally commuting subsets, C, of the set of operator terms. Since the number of operator terms is finite by assumption, the number of sets in C is finite. We can construct a universe state for each member of C. The number of such universe states is finite. They can be used to form a vector space. Eq. 11.27 in the preceding subsection is an example of a two dimensional vector space of universe states.

It is easy to find larger, more complex examples, particularly in physics. For example the group SU(2) plays a role in many areas of quantum physics. Let us consider the general case of its three generators S^i. They satisfy the commutation relations:

$$[S^i, S^j] = i\varepsilon_{ijk} S^k \qquad (11.41)$$

where ε_{ijk} is the unit totally anti-symmetric tensor with $\varepsilon_{123} = 1$. Since there are three non-commuting operators, there are three subsets in C, and we can form a three-dimensional universe vector space:

$$|\,\Omega_{u3}\rangle = \begin{bmatrix} |\Omega_{u\text{"}S^1\text{"}}\rangle \\ |\Omega_{u\text{"}S^2\text{"}}\rangle \\ |\Omega_{u\text{"}S^3\text{"}}\rangle \end{bmatrix} \qquad (11.42)$$

We introduce three eigenvalue operators $A_{\text{"}S^i\text{"}}$ that correspond to the three SU(2) operators where $A_{\text{"}S^i\text{"}}$ has the Gödel number eigenvalues E_i = gn("S^i") or zero. We define projection operators (similarly to eq. 11.28) as the diagonal matrices I_t in term space with the matrix elements:

$$I_{3t} = [\delta_{ti}] \qquad (11.43)$$

for t = 1, 2, 3 where δ_{ik} is the Kronecker delta function.

Eigenvalue operators in term space have the more general form

$$A_{tot\text{"}S^i\text{"}} = A_{\text{"}S^i\text{"}}\, I_{3i} \qquad (11.44)$$

for i = 1, 2, 3.

Then eq. 11.41 is equivalent to the Quantum Operator Logic statement

$$[A_{tot\text{"}S^i\text{"}}, A_{tot\text{"}S^j\text{"}}] = i\varepsilon_{ijk} A_{tot\text{"}S^k\text{"}} \qquad (11.45)$$

If we evaluate

$$\langle\Omega_{u3}|[A_{tot\text{"}S^i\text{"}}, A_{tot\text{"}S^j\text{"}}] = i\varepsilon_{ijk} A_{tot\text{"}S^k\text{"}}\, |\Omega_{u3}\rangle \qquad (11.46)$$

in the same manner as we did in the case of eqs. 11.36 – 11.40 then we find eq. 11.46 leads to eq. 11.41.

Similarly in the case of the SU(3) Lie algebra, which is also of great importance in elementary particle theory, we find that the eight generators of SU(3) lead us to define a seven dimensional space since two of the generators commute with each other. The commuting generators are usually designated F^3 and F^8. The *universe vector* is thus

$$|\,\Omega_{u7}\rangle = \begin{bmatrix} |\Omega_{u\text{"F1"}}\rangle \\ |\Omega_{u\text{"F2"}}\rangle \\ |\Omega_{u\text{"F3", "F8"}}\rangle \\ |\Omega_{u\text{"F4"}}\rangle \\ |\Omega_{u\text{"F5"}}\rangle \\ |\Omega_{u\text{"F6"}}\rangle \\ |\Omega_{u\text{"F7"}}\rangle \end{bmatrix} \tag{11.47}$$

where

$$|\Omega_{u\text{"F3", "F8"}}\rangle = |E_{\text{"F3"}}, E_{\text{"F8"}}\rangle \tag{11.48}$$

We define projection operators (similarly to eq. 11.28) as the diagonal matrices I_t in term space with the matrix elements:

$$I_{7t} = [\delta_{ti}] \tag{11.49}$$

for t = 1,2, …, 7 where δ_{ik} is the Kronecker delta function.

Eigenvalue operators in term space have the general form

$$A_{tot\text{"Fi"}} = A_{\text{"Fi"}}\, I_{7i} \tag{11.50}$$

for i = 1, 2, …, 7 and

$$A_{tot\text{"F8"}} = A_{\text{"F8"}} I_{73} \tag{11.51}$$

Thus

$$A_{tot\text{"Fi"}}|\Omega_{u7}\rangle = E_{\text{"Fi"}}\, I_{7i}|\Omega_{u7}\rangle \tag{11.52}$$

for i = 1, …,7 excepting i = 3, and

$$A_{tot\text{"F3"}}|\Omega_{u7}\rangle = E_{\text{"F3"}} I_{73}|\Omega_{u7}\rangle = gn(\text{"}F^3\text{"}) I_{73}|\Omega_{u7}\rangle \tag{11.53}$$

$$A_{tot\text{"F8"}}|\Omega_{u7}\rangle = E_{\text{"F8"}}\, I_{73}|\Omega_{u7}\rangle = gn(\text{"}F^8\text{"}) I_{73}|\Omega_{u7}\rangle \tag{11.54}$$

The commutation relations for SU(3) are

$$[F^i, F^j] = ic_{ijk} F^k \tag{11.55}$$

Again following the procedure outlined above we see that the corresponding Quantum Operator Logic statement is

$$[A_{toti}, A_{totj}] = ic_{ijk} A_{totk} \tag{11.56}$$

11.5.3 The Term Space of Quantum Operator Logic

In the preceding subsections we have seen how to develop and evaluate Quantum Operator Logic statements. These statements are typically equations and constructs in quantum mechanics and quantum field theories in physics. Statements (propositions) in ordinary language usually do not contain operator (q-number) terms.[139] Such statements are handled within the framework of Operator Logic.

We will now consider a theory that consists of statements (axioms, theorems, and equations) that are constructed from a set of generally non-commuting operator terms. The examples in the preceding subsections illustrate the main features of the Quantum Operator Logic formulation of this type of theory.

One might ask why we should develop a reformulation of perfectly acceptable quantum theories? I believe the answer lies in the hope of finding a more fundamental theory of Reality – particularly in the case of The Standard Model of Elementary Particles. This theory is the closest experimentally verified theory of fundamental Reality.

We found the concept of *term space* was necessary to develop a Quantum Operator Logic formulation of a set of operator statements since one cannot formulate a simultaneous eigenstate of non-commuting operators. Thus we were led to define a universe of discourse state that is a vector[140] of eigenstates of maximally commuting subsets of terms $\{m_i\}$ of the form of eq. 11.25:

$$|\Omega_{ui}\rangle = \prod_{j \in \{m_i\}} |E_{ij}\rangle \tag{11.25'}$$

[139] They could also contain a set of operator terms – all of which commute with each other.

[140] Eqs. 11.42 and 11.47 are examples of Quantum Operator Logic universe of discourse states.

the universe state of the ith maximally commuting subset $\{m_i\}$ of set of operators where each E_{ij} is the non-zero Gödel number of a term operator. Since the number of terms in a universe is finite by assumption the number of maximally commuting subsets of terms is finite. If there are q maximally commuting subsets then the universe of discourse state vector has q components:

$$|\Omega_u\rangle = \begin{bmatrix} |\Omega_{u1}\rangle \\ \cdots \\ |\Omega_{uq}\rangle \end{bmatrix} \qquad (11.57)$$

as does its hermitean conjugate row vector $\langle\Omega_u|$. We call the q dimensional vector composed of these q universe states a vector in *term space*. To access each of the individual universe states it is necessary to define projection operators that are diagonal matrices I_i in term space with the matrix elements:

$$I_i = [\delta_{ij}] \qquad (11.58)$$

for i, j = 1,2, …, q where δ_{ij} is the Kronecker delta function.

The eigenvalue operators $A_{\text{"}T^{ij}\text{"}}$ satisfy

$$A_{\text{"}T^{ij}\text{"}}\,|E_{ij}\rangle = E_{ij}|E_{ij}\rangle = gn(\text{"}T^{ij}\text{"})|E_{ij}\rangle \qquad (11.59)$$

The form of a full eigenvalue operator in term space is

$$A_{tot\text{"}T^{ij}\text{"}} = A_{\text{"}T^{ij}\text{"}}\, I_i \qquad (11.60)$$

where j designates an operator within the ith maximally commuting subset $\{m_i\}$. T^{ij} is the name of the jth operator term in the ith subset of commuting term operators with its eigenvalue specified by eq. 11.59. The eigenvalue equation in term space is

$$A_{tot\text{"}T^{ij}\text{"}}|\Omega_u\rangle = E_{ij}\, I_i\, |\Omega_u\rangle = gn(\text{"}T^{ij}\text{"})I_i\, |\Omega_u\rangle \qquad (11.61)$$

for i = 1, …, q.

Non-operator terms (c-number terms such as numeric constants and mathematical operation signs) that appear in Quantum Operator Logic statements are basically treated like terms in Operator Logic. We can choose to add these terms to the largest subset of commuting operators (although this is not required). The steps of defining corresponding eigenvalue operators whose eigenvalues are the Gödel numbers of the c-number terms and then using them to express a statement in terms of eigenvalue operators is similar to the Operator Logic case – see chapter 6 – particularly section 6.1 and eq. 6.24.

To each c-number term C_a we associate an eigenvalue operator A_{C_a} and an eigenstate $|E_{C_a}>$ satisfying

$$A_{C_a}|E_{C_a}> = E_{C_a}|E_{C_a}> = gn(C_a)|E_{C_a}> \qquad (11.62)$$

We can (arbitrarily) adjoin the c-number term eigenstates to the eigenvector of the largest maximally commuting subset of operators denoted $\{m_{i_1}\}$

$$|\Omega_{ui_1}> = \prod_{j\in\{m_{i_1}\}} |E_{i_1j}> \prod_{k\in\{m_C\}} |E_{Ck}> \qquad (11.25'')$$

where $\{m_C\}$ is the set of all c-number terms in the universe of discourse. We define

$$A_{totC_a} = A_{C_a} I_{i_1} \qquad (11.63)$$

Thus we have a Quantum Operator Logic that encompasses a universe of discourse with both q-number and c-number terms.

As we saw in the previous section we can map between operator statements (equations) and eigenvalue operator statements in term space:

$$S(T^1, T^2, \ldots, T^n, C_1, C_2, \ldots, C_m) \leftrightarrow S'(A_{tot"T^1"}, A_{tot"T^2"}, \ldots, A_{tot"T^n"}, A_{totC_1}, A_{totC_2}, \ldots, A_{totC_m}) \qquad (11.64)$$

which are evaluated as expectation values of the universe state eq. 11.57:

$$\langle\Omega_u|\, S'(A_{tot\text{"}T^1\text{"}}, A_{tot\text{"}T^2\text{"}}, \ldots, A_{tot\text{"}T^n\text{"}}, A_{totC_1}, A_{totC_2}, \ldots, A_{totC_m})|\Omega_u\rangle \quad (11.65)$$

to obtain the the operator statement $S(T^1, T^2, \ldots, T^n, C_1, C_2, \ldots, C_m)$.

We have succeeded in mapping operator statements (equations) of a theory into a term space of Quantum Operator Logic. At this point it seems appropriate to inquire as to the benefits of such a map. One primary goal seems evident: to uncover a deeper layer of Reality using the formalism and methodology of Quantum Operator Logic that we can the map back to the original operator theory. The key will be to develop a semantic theory for the origin and truth[141] of the statements in term space. We initiate a discussion of a semantic theory in term space in the next subsection.

The existence of a term space due to the operator nature of the original theory is provocative. We will now examine some features of term space. If we apply the projection operators I_i to the universe state we can generate q unit vectors in the various "directions" of term space

$$v_i = I_i\, |\Omega_u\rangle \quad (11.66)$$

where i = 1, 2, … , q and where

$$v_i^\dagger v_j = \delta_{ij} \quad (11.67)$$

by the projection I_i appearing in eq. 11.66. If we sum the vectors v_i we obtain the universe of discourse state

$$|\Omega_u\rangle = \sum_{i=1}^{q} v_i \quad (11.68)$$

We can define raising and lowering operators for the eigenstates within the product of eigenstates constituting each unit vector v_i. In a manner similar to subsections 8.3.1 and 8.3.2. For the jth eigenstate in the product of eigenstates that constitutes v_i we can define a lowering operator $a_{E_{ij}}$ by

$$|0\rangle = [gn(E_{ij})]^{-1/2} a_{E_{ij}} |E_{ij}\rangle \quad (11.69)$$

[141] Or probability amplitude of truth.

and, the corresponding raising operator $a_{E_{ij}}^{\dagger}$ with

$$|E_{ij}> = [gn(E_{ij})]^{-½} a_{E_{ij}}^{\dagger}|0> \tag{11.70}$$

where † signifies hermitean conjugatation. Thus

$$v_{i,j} = [gn(E_{ij})]^{-½} a_{E_{ij}} v_i \tag{11.71}$$

is the vector obtained from v_i by setting the jth eigenvalue to zero. As in sections 8.3.1 and 8.3.2 we do not wish repeated application of raising operators to generate states with higher eigenvalues.[142] Therefore we define

$$a_{E_{ij}}^{\dagger}|E_{ij}> = 0 \tag{11.72}$$

$$a_{E_{ij}}|0> = 0 \tag{11.73}$$

for each individual eigenstate. Consequently the raising and lowering operators must satisfy the *anticommutation* relations with the form:

$$\begin{aligned}\{a_{E_{ij}}^{\dagger}, a_{E_{km}}\} &= \delta_{ik}\,\delta_{jm}\,gn(E_{ij})\\ \{a_{E_{ij}}, a_{E_{km}}\} &= 0\\ \{a_{E_{ij}}^{\dagger}, a_{E_{km}}^{\dagger}\} &= 0\end{aligned} \tag{11.74}$$

for eqs. 11.72 and 11.73.

Eigenvalue operators can be represented in terms of these operators by

$$A_{E_{k_{ij}}} = a_{E_{k_{ij}}}^{\dagger} a_{E_{k_{ij}}} \tag{11.75}$$

and the corresponding projection operators by

$$P_{E_{k_{ij}}} = P'_{E_{k_{ij}}}\, ½[1 - [gn(E_{k_{ij}})]^{-1} a_{E_{k_{ij}}} a_{E_{k_{ij}}}^{\dagger}] \tag{11.76}$$

---

[142] Since such eigenstates would not correspond to a primitive or defined term in the universe of discourse.

where P'_{E_k} is a truth projection operator. (Truth projection operators are discussed in the next subsection.)

Lastly it is possible to define rotations in this q dimensional space having the group structure of U(q) – the unitary group in q dimensions. These rotations would lead to rotations amongst the eigenvalue operators and thus to their Gödel number eigenvalues. Then the mapping from Gödel numbers to operators in statements would be similarly rotated. However the statements would remain the same. U(q) rotations may eventually play some role in understanding the underpinnings of The Standard Model of Elementary Particles.

11.5.4 Quantum Operator Logic Projection Operators

In the previous subsections of 11.5 we have focussed on the mapping from operator terms in the operator equations of a theory to eigenvalue operators in term space. We have only briefly examined projection operators in term space (eq. 11.76) and have not addressed the issue of the truth value projection operators that figured prominently in our study of Operator Logic.

Determining truth value projection operators is essentially a semantic question. Bacon's famous opening line of his essay "On Truth" begins "What is truth …" and we must confront the same issue. In Operator Logic and Probabilistic Operator Logic we resolved the issue for subject-predicate statements and statements reducible to combinations of subject-predicate clauses by defining a domain for each predicate in a universe of discourse.[143] The specification of a domain is a semantic issue for semantic universes of discourse and a matter of definition for a calculus universe of discourse.

Thus one approach to the determination of truth value projection operators in Quantum Operator Logic would be to separate each operator equation (statement) into a "subject" part and a "predicate" part in such a way as to minimize the number of predicates,[144] and then to determine the domain of subjects for each predicate. An example of this approach

---

[143] With a probability for each subject-predicate pair in the case of Probabilistic Operator Logic. In the case of Quantum Operator Logic following one approach we could define a probability amplitude for each subject-predicate pair and calculate probabilities from the probability amplitudes. Statements that had negative probabilities would again be undecidable as discussed in subsection 11.4.3.

[144] For the sake of simplicity.

is to regard the SU(3) commutation relations (eq. 11.55) as a set of statements where the subject is "$[F^i, F^j]$" and the predicate is "$= ic_{ijk} F^k$". The domain of this predicate is the subject "$[F^i, F^j]$".

This approach is viable but it assumes that one specifies the set of domains and thus requires a foreknowledge of all the domains. One obtains a well-defined universe of discourse within the framework of Quantum Operator Logic but one only gets that which one puts in. Perhaps in some cases that is all one really wants—a reformulation of a set of operator equations (statements) in the formalism of Quantum Operator Logic.

Another more interesting approach, in the view of this author, to the case of the metaphysics of ultimate Reality is to grow a Quantum Operator Logic formulation from a "secure" foundation using a set of principles that lead to a physical theory of ultimate Reality. Among the principles that seem to be most relevant are Ockham's Razor[145] and Leibniz's Principle of Perfection.[146] We will consider this approach in the next chapter. Its constructive approach to the creation of the theory of Reality is more intellectually satisfying then simply positing a theory either as a set of postulates or a lagrangian for the universe.

11.6 Quantum Operator Logic Subuniverses Generated by Lowering Operators

In the preceding sections we have considered Quantum Operator Logic universes of discourse. We now consider the possibility of subuniverses as we did similarly in the case of Operator Logic earlier.

The creation of subuniverses of discourse is described in the case of Operator Logic in chapter 9. The creation of a quantum subuniverse follows almost the same procedure of deleting primitive terms from a universe of discourse by using lowering operators as in those sections.

Suppose we wish to delete the operator term with eigenvalue operator A_{E_m}, eigenstate $|E_m\rangle$ and lowering operator a_{E_m} from the

---

[145] William of Ockham – Law of Parsimony – "Pluralitas non est ponenda sine necessitate" or "Plurality should not be posited without necessity." First stated by Durand De Saint-Pourçain (1270-1334 AD). In simple terms the principle states the simplest solution to a problem is most likely to be the correct solution.

[146] This principle can be phrased for our purposes as, "The universe is based on the smallest set of properties or features that lead to the greatest variety of phenomena." This principle reflects the minimum/maximum criteria of the Calculus of Variations that play a central role in many physics theories. Leibniz was one of the founders of the Calculus of Variations.

universe of discourse. The universe of discourse eigenstates are defined by eqs. 11.25, 11.25', 11.25", and 11.57. We must first identify the subset of the states $\{|\Omega_{ui}>\}$ containing the eigenstate $|E_m>$ as a factor. We will denote these states as $|\Omega_{umi}>$ and define a new vector

$$|\Omega_{um}> = \begin{bmatrix} |\Omega_{um1}> \\ \cdots \\ |\Omega_{ump}> \end{bmatrix} \quad (11.77)$$

where $p \le q$. Thus we have a p dimensional vector and can apply the lowering operator

$$a_{E_m}|\Omega_{um}> = \begin{bmatrix} |\Omega_{um1}'> \\ \cdots \\ |\Omega_{ump}'> \end{bmatrix} \equiv |\Omega_{um}'> \quad (11.78)$$

The application of the lowering operator produces a state such that

$$A_{E_m}|\Omega_{um}'> = 0 \quad (11.79)$$

If we now construct a q dimensional vector from eq. 11.57 by substituting a member of the set $\{|\Omega_{umi}'>\}$ for the corresponding member in the vector specified by eq. 11.57 then we obtain a vector $|\Omega_u'>$ that when used to evaluate the expectation value of an equation (statement)

$$<\Omega_u'|S|\Omega_u'> \quad (11.80)$$

does not contain the term corresponding to A_{E_m} but does contain all the other terms that were originally in the equation. Thus we have generated a subuniverse of discourse with the term corresponding to A_{E_m} omitted from all statements. This procedure could be applied repeatedly to obtain "sub-sub- … -sub universes generating specialized subtheories from an initial general theory.

12. From Ideas to Reality: First Principles

In this chapter we will develop a framework for relating the realm of Ideas as conceived by the Platonists, but updated, through a path of construction to ultimate physical Reality as we currently understand it. The approach will be based on certain principles which are closer to physical thought than the principles that have been the subject of investigation and discussion by metaphysicians from early times to the present.

12.1 Six General Principles

In section 2.3 we discussed seven principles upon which Leibniz based his metaphysics. Considering the date of his work – the seventeenth century – and the primitive state of our knowledge of physical Reality, his principles furnished a framework for thought and discussion but did not have a connection to physical Reality as we now understand it. The principles he enunciated were simply too general to provide a basis for an understanding of physical Reality. Metaphysical thought since Leibniz still suffers from a lack of specificity about physical Reality although some efforts have come to grips with issues related to Reality.

In this section we will specify six principles which furnish a starting point and a guide to the development of a fundamental basis for physical Reality and the construction of a specific physical theory of ultimate Reality – The Standard Model. The principles are subject to dispute but seem reasonable, and sufficiently detailed, to provide useful guidance in the construction of a theory of physical Reality. They also are general in nature. Thus they cannot be said to be merely an alternate definition of the physical theory that emerges.

The six principles are:

1. Logic, particularly Operator Logic and Quantum Operator Logic as the case may be, applies to metaphysical and physical derivations and discussions. (This principle is clearly assumed by metaphysicians.)
2. In the case of several alternative choices Ockham's Razor[147] should be used to determine the correct choice.[148]
3. We assume the universe is based on the smallest set of properties or features that lead to the greatest variety of phenomena.[149] Alternately put, we choose the most minimal conditions necessary to lead to the known phenomena of nature.
4. Whatever physical entities exist in the universe they can only make their presence known by interacting (exerting forces) with other entities. Thus each physical entity interacts with at least one other type of entity.[150]
5. The behavior and properties of physical entities must be reproducible whenever the same circumstances occur. Behavior can be deterministic or can be specified in terms of probabilities. Consequently the behavior and properties must be governed directly or indirectly by mathematics.
6. We assume that space and time exist and that the properties of entities are local in the sense that the properties of an entity depend on the point in space and time where it is located.[151]

12.2 Logic and Metaphysics

The principles stated in the previous section in themselves do not constitute a physical theory. In order to progress further we need to use logic and appropriate constructs.

---

[147] See subsection 11.5.4.

[148] One very clear reason for this principle is that physics is very difficult and the simplest choice is usually the choice that enables further progress to occur. An example is Copernicus' theory that the planets circled the sun. This theory was nicely adapted to support Newton's work on the theory of gravitation. On the other hand, the Ptolemaic theory is more complex (and also not consistent with Newton's theory of gravitation.) Copernicus' theory was far simpler than Ptolemaic theory.

[149] A version of Leibniz's Principle of Perfection.

[150] Consequently there is no knowable physical entity with no interactions with other types of entities.

[151] This principle is the basis of local rotational invariance in Yang-Mills theories or sectors of physical theories.

Metaphysics deals with many areas including physical Reality, the nature of God, attempts at proofs of God's existence, Theology, and so on. We will focus entirely on physical Reality as stated earlier. We will begin with a foundation built on Logic – the only fundamental starting point of which we can be certain. Without Logic the construction and study of the implications of theories is impossible. From Logic we will develop constructs that lead us to a physical theory of matter and energy. In this author's view, this approach is sounder and more intellectually satisfying than simply positing a physical theory with a set of axioms or simply stating a lagrangian for "everything."

Physicists have spent many years developing a comprehensive "Theory of Everything" called SuperString Theory. This theory purports to be ultimate Reality. However at present there is no experimental evidence for it and no conclusive derivation of The Standard Model from it. It assumes that everything is ultimately composed of ultra-small mathematical strings. However it does not answer the question: Why is it *ultimate* Reality? One simply posits certain types of mathematical strings and then derives their consequences. But where and why do strings arise? We conclude that a SuperString Theory even if found to be experimentally true merely reduces physics to a lower level of Reality but, in turn, raises questions as to its origin in a yet deeper Reality. Some metaphysicians have proposed that there is no ultimate Reality but only ever-increasing lower levels of Reality. We do not believe this to be true.

We suggest that a better alternative, if it is viable,[152] *is to found the ultimate theory of Reality on the bedrock of Logic. For if Logic is not a true science, then no science or mathematics is possible. And choosing Logic as the bedrock of ultimate Reality gives us an unimpeachable foundation below which no deeper foundation is possible. Our experience of Reality in the past few hundred years very convincingly demonstrates that science and mathematics "works."*[153]

---

[152] And we believe it is.

[153] It is the author's belief that the enormous progress in the sciences – particularly in the past two hundred years – has mistakenly led us to believe that we are approaching a fairly complete knowledge of the physical and biological sciences. In the author's view nothing could be further from the truth. Mankind has barely begun to scratch the surface of the sciences and can look forward to a glorious further awakening over the next millenium as the scope and depth of its knowledge increases.

12.3 Platonic Conception of the Relation of Ideas to Reality

Plato and subsequent Platonic philosophers postulated that there existed a realm of Ideas independent of the human mind but consisting of the concepts and thought processes (Logic) existent in the mind.[154] Each thing, both material and conceptual, in Reality had an abstract counterpart in the realm of Ideas that embodied its features. Thus there was an Idea of a plow, an Idea of Justice, and so on. The realm of Ideas was connected through mathematics (Number) to the realm of Reality with which we are familiar. The nature of the connection was not known to the Platonic philosophers since the relevant mathematics and physics were not known.

Plato and subsequent philosophers were not familiar with the modern view of Reality, which has only become apparent in the quantum revolution of the twentieth century. The concepts of quantum physics apply to non-relativistic quantum mechanics and relativistic quantum field theory.

Quantum Theory has a formal method of viewing the world. It describes the process of quantum experimental measurement as a series of filtrations (or refinements that select certain properties or states) that occur at various stages of an experiment. Since all observations are experiments, it applies in principle to all the ways that we obtain information about the universe in which we live. In the macroscopic world quantum effects are usually negligible. In the very small, quantum effects often dominate physical processes. Quantum Theory applies in principle in all cases.

So the gropings of Plato and other philosophers towards a total view of Reality had an element of truth in that there is a fundamental set of Ideas that govern Reality, but was wrong—for good reason—Reality does not map directly to Ideas in the one-to-one fashion envisioned by the Platonists.

Instead, the set of Ideas is the set of laws of physics: Quantum Theory, the Theory of Relativity and elementary particle theory. The phenomena of Reality behave according to the laws of physics. The mathematics connecting Ideas to Reality will be shown to start from Logic and end in The Standard Model of Elementary Particles. Thus

[154] Plato discusses the concept of Ideas in the dialogue Parmedides as well as other dialogues.

Platonic thought was correct in a general sense; but the specifics were incorrect due to the primitive state of physical science until recently.

12.4 Operator Logic and Quantum Operator Logic Exist Independent of Our Knowledge of Them

We have constructed Operator Logic and Quantum Operator Logic in the "language" of linear vector spaces, and, more specifically, based on the view that a proposition is analogous to a quantum experiment in that it proceeds to particularize subjects in the same manner that a quantum experiment filters quantum states to achieve a particular final state. Linear vector spaces appear in many areas of physical Reality. Since Reality cannot be internally inconsistent, incomplete or "incorrect", Operator Logics are well founded and exist independently of our knowledge of them.[155] Our understanding of quantum physics only emerged in the twentieth century yet the universe has followed the laws of quantum theory since its beginning. Thus we must attribute a conceptual reality to the concepts and laws of Quantum Theory. The extreme precision of the theory of Quantum Electrodynamics strongly indicates that Quantum Theory is correct.

Consequently the only reasonable philosophical stance is Platonist—the concepts and laws of Quantum Theory have a true existence outside of our knowledge of them.[156]

12.5 The Realm of Ideas

The preceding section shows that our formalisms for Operator Logic and Quantum Operator Logic exist independently of our awareness of them. They exist! One may ask what other Ideas exist in the Realm of Ideas. Confining ourselves to the physical, we see that the Realm of Ideas must contain Operator Logic, Quantum Operator Logic, Mathematics, and the fundamental theory of physical Reality. The

---

[155] The deepest part of our current physical understanding of the universe is Quantum Theory. All branches of physics ultimately conform to it. It is extremely unlikely that Quantum Theory will change.

[156] This statement should not be confused with the quantum mechanical requirement of an observer to make quantum mechanical measurements. Quantum physical processes can proceed without an observer although experimental results, being probabilistic, can only be determined by an observer.

consideration of other possible parts of the Realm of Ideas in Ethics and Theology are beyond the scope of this book.

12.6 Role of the Observer in the Realm of Reality

Section 12.4 addresses the issue of whether the Realm of Ideas exists independently of our knowledge of it and its contents. One might also ask the same question of the Realm of Reality: Do events happening in Reality, and the manner in which they occur, require an Observer to detect them? Or do events in Reality happen irrespective of our presence? The answer to these questions lies in Quantum Theory.

In Quantum Theory there is an implicit Observer who "sets up" experiments, performs eigenvalue measurements (filters) at various stages in an experiment, and measures the end result(s) of an experiment.

12.6.1 The Observer in Quantum Theory

The role of the Observer in Quantum Theory is extremely significant. One indication of the importance of the Observer in quantum theory is the effect of an observation (without filtration) of the state of an intermediate stage of an experiment compared to the case of the non-observation of the state of the intermediate stage of an experiment.

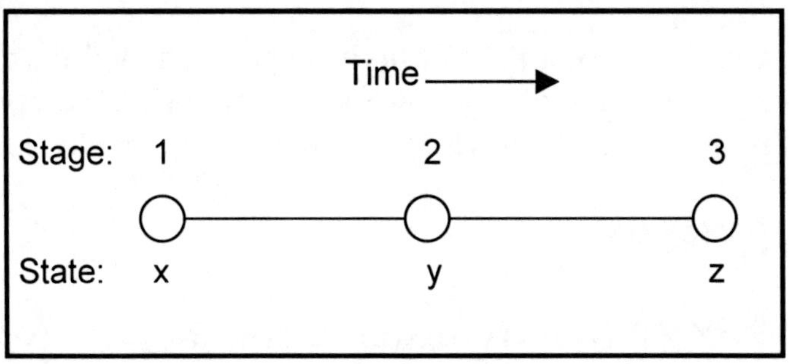

Figure 12.1. A three stage experiment beginning with a system in the state with eigenvalue x of observable X, having an intermediate state 2 with the eigenvalue y of observable Y, and ending in state 3 with eigenvalue z of observable Z. Time proceeds from left to right in the diagram.

Consider the experiment diagrammed in Fig. 12.1. The probability amplitude of the transition of the initial physical state from having eigenvalue x of X to having eigenvalue y of Y and thence to having eigenvalue z of Z is

$$<Z(z)|Y(y)><Y(y)|X(x)>$$

and thus the corresponding quantum probability is

$$|<Z(z)|Y(y)>|^2|<Y(y)|X(x)>|^2 \qquad (12.1)$$

If we do not filter at stage 2 of the experiment then we must sum over the Y eigenvalue states to obtain the probability of the transition from x to z:

$$p_Y(z, x) = \sum_y |<Z(z)|Y(y)>|^2|<Y(y)|X(x)>|^2 \qquad (12.2)$$

$$= \sum_y <Z(z)|Y(y)><Y(y)|Z(z)><Y(y)|X(x)><X(x)|Y(y)> \qquad (12.3)$$

Note the probability $p_Y(z, x)$ is dependent on the summation over intermediate (stage 2) Y eigenstates in the particular way specified by eq. 12.3.

If stage 2 were not present in the experiment and the experiment ran from stage 1 directly to stage 3 without intermediate observation then a different probability, which we will denote p(z, x), would result.

$$p(z, x) = |<Z(z)|X(x)>|^2 \qquad (12.4)$$

$$= \sum_{y_1} <Z(z)|Y(y_1)><Y(y_1)|X(x)> \sum_{y_2} <X(x)|Y(y_2)>< Y(y_2)|Z(z)>$$

$$= \sum_{y_1, y_2} <Z(z)|Y(y_1)><Y(y_1)|X(x)><X(x)|Y(y_2)>< Y(y_2)|Z(z)> \qquad (12.5)$$

which is manifestly different from $p_Y(z, x)$ as given in eq. 12.3.

Thus we conclude that the mere presence of an observation point (stage 2 in the present example) changes the probability for the experiment even if a filtration is not made.

We thus have concrete proof (substantiated by extensive experiments) that the passive observation of the state of an experiment in progress affects the probability of the outcome of the experiment.

Since an observation (passive, or active (i.e. a filtration)) necessarily requires an observer we see that observers influence the evolution of experiments, and thus have an important role in quantum theory. Since the mathematics of Operator Logic, and Quantum Operator Logic, is analogous in many respects to Quantum Theory we conclude that a parallel observer role exists, in principle, for Operator Logic and Quantum Operator Logic.

12.6.2 The Observer in Operator Logic & Quantum Operator Logic

In Operator Logic and Quantum Operator Logic an implicit "observer" (a logician or theorist) determines the truth (semantic universe of discourse) or provability (calculus universe of discourse) of a statement.

An observer has three major roles in Operator Logic and Quantum Operator Logic:

1. Create statements.
2. Determine the truth (provability) of statements.
3. Create derivations of statements (theorems).

The effect of the observer on the truth (provability) of statements depends on whether the statement is a c-number statement (Operator Logic) or a q-number statement (Quantum Operator Logic). In the case of Operator Logic the observer has no effect on the truth value (true or false, provable or not, or probability of truth) of the statement. In the case of Quantum Operator Logic, if the observer determines the state of a proposition (statement) at an intermediate point(s) in the proposition (statement) then the probability of the statement is affected in the same manner as we saw in the previous subsection in the discussion of Quantum Measurement Theory.

12.7 Space and Time

In previous section we have begun the process of constructing the transition from the Realm of Ideas to the Realm of Reality. An important idea in the Realm of Ideas is the concept of time. Time can be viewed as a necessary (although usually implicit) component of Logic, Operator Logic, and Quantum Operator Logic. Space is also an important Idea in the Realm of Ideas as we will see later.

12.7.1 The Necessity of Time, and an Arrow of Time, in Operator Logic and Quantum Operator Logic

A number of logicians have noted that a concept of time is implicit in conventional Logic. For example, proofs of theorems proceed step by step from initial postulates and theorems to a theorem's proof. Embedded in that process is a notion of discrete time steps, and a direction of the time steps. The directionality of the time steps[157] specifies an "arrow of time." The question of the Arrow of Time – why time proceeds forward and not backward – has been a subject of much discussion over the years. In the present situation Logic, Operator Logic, Quantum Operator Logic automatically embody an arrow of time.

Not only is this true for proofs but it is also true for statements. Although the order of the parts of a statement is language dependent the order is specific and consecutive within a given language and thus has a time order as well.

So we conclude that the various Logics that we have considered all embody discrete time steps and a definite concept of time ordering – an Arrow of Time.

Having ascertained that discrete time, and time ordering, is implicit in our Logics we now define physical time as the continuous limit of discrete time (with the understanding that physical time may be discrete and may possibly consist of very small time steps of time intervals of the order of the Planck time scale 5.39×10^{-44} seconds.) Discrete time intervals of that order of magnitude are not detectable experimentally at present or in the foreseeable future. Thus the assumption of continuous time, with an arrow of time, is satisfactory.

---

[157] After all one does not proceed "backward" from a theorem through the proof steps to the initial postulats.

12.7.2 Why add Space to the Logics?

Space is necessarily a part of the Realm of Ideas because propositions often depend on a spatial location. (See subsection 8.3.2 where the variables of eigenvalue operators x_1, x_2, ... could specify spatial locations.) Thus we must add spatial (and time) dimensions to our specification of the Realm of Reality as well as the Realm of Ideas. Later we will connect the spin ½ matrix formulation of Operator Logic augmented with space-time dimensions to the Dirac equation for spin ½ particles. Thus we map Operator Logic spinors and spinors in physical Reality. Fermion particles (spin ½ particles) in physical Reality emerge from a map from Operator Logic spinors.

Now we address the issue of the number of space-time dimensions. Clearly if spinors exist in Reality, as we know they do, then they must be "spinning" in spatial dimensions. The number of components of a spinor is related to the total number of time and space dimensions.[158] For the case of an even number of dimensions d a spinor has $2^{d/2}$ components. For the case of an odd number of dimensions d a spinor has $2^{(d-1)/2}$ components. Based on these formulas we find the results in the following table.

Total Number of Space-Time Dimensions d	Number of Spinor Components
1	1
2	2
3	2
4	4

Table 12.1. The number of spinor components for various numbers of space-time dimensions.

The case of d = 1 is immediately ruled out because Operator Logic supports, at minimum, 2 component spinors or 4 component spinors. The case d = 2 is also ruled out because spinor particles in a one-dimensional space reduce to scalar particles, and Reality has true spinors. The case d = 3 is ruled out because in two spatial dimensions there is no

[158] Weinberg (1995) p. 216.

difference between left-handedness and right-handedness. Thus the minimal number of spatial dimensions that yield true physical spinors and support "handedness" is three spatial dimensions. This case meets Leibniz's criteria: principle 3 of section 12.1. The simplest features associated with space are spin (represented by spinors) and handedness. They yield a rich spectrum of particle types and interaction types (maximal complexity).

Thus we have a rationale for the extension of Operator Logic to include one time and three spatial dimensions.

12.8 Being and Existence of the Material World

Having established the Ideas of Quantum Theory, Time and Space and their counterparts in Reality we now turn to the Idea of Being.

The question of being or existence has been a subject of discussion in Philosophy and Metaphysics for millenia. In the absence of "experimental" information the discussions have centered on the definitions of being and the implications of these definitions for the "properties" of being.

For many scholars the state of Philosophy and Metaphysics was considered satisfactory in the 20th century with respect to issues such as Being. For example, Hans Bethe, perhaps the dominant figure in theoretical physics from the 1930's through the 1950's, and a Nobel Prize winner, stated that at the beginning of his "graduate" studies in the 1920's he considered the state of Philosophy and Metaphysics, and concluded that they were satisfactory. He then decided to become a physicist where he felt that he could make significant contributions (which he did by discovering the solar energy carbon cycle, and making notable contributions to nuclear physics and quantum field theory as well as guiding a generation of physicists including R. P. Feynman and M. Gell-Mann). In our view, Professor Bethe's views on Metaphysics did not anticipate the impact of Quantum Theory and Relativity.

In this section we will consider a phenomenological (Realistic) view of Being and nothingness based on the experimental observations of the creation and annihilation of particles over the past eighty years. The experimental picture that we will use dates from the 1930's when the occurrence of particle creation and annihilation was first recognized. The mathematics of quantum field theory adequately describes particle creation and annihilation as seen experimentally. Since particle creation is the “creation” of Being and since particle annihilation is the “destruction” of Being one could say that the issues of Being and non-Being can be resolved by experimental observations and their theoretical analysis – thus making metaphysical speculations experimentally accountable. *Experiment can provide direct guidance on the nature of Being.*

Having said that, we recognize that the mathematics of creation and annihilation of particles somehow doesn’t fully answer the question, “What is Being?” from a human perspective. Part of the problem is that we don’t exactly “know” what a particle is. We know particles have particle-like properties, and wave-like properties as well. But statements about properties do not address the issue of what Being is in itself. Can we say particles are composed of a substance or substances? If so, what substance or substances? Or are particles merely form without substance?

12.8.1 The Substance of Particles

Philosophers have long argued (two and a half millenia) about the nature of the substances of Reality. These discussions have largely centered around the definition of substances followed by an analysis based on one definition or another. The question of their reality was also a recurring issue. In the earliest discussions the world was thought to consist of varying combinations of the four elements: earth, air, fire and water. So early philosophical discussions took the lead from “experimental” observation because these four elements are what we encounter in nature, and one can argue that various materials seen in nature are combinations of the four elements. More recently philosophers such as Bishop Berkeley have raised the question whether the world is real or perhaps some form of evolving “thought” in the mind of God, or a similar insubstantiality.

In this subsection we address the issue of Being based on experimental observations – particularly particle observations made from the 1930’s onward. *We look to experiment/reality, as the first philosophers did, to uncover the nature of substances.*

The first major qualitative experimental reality that we find is that elementary particles can transform into each other in certain ways that are theoretically well described by the Standard Model of

Elementary Particles.[159] Some typical transformations are diagrammed in Fig. 12.2.

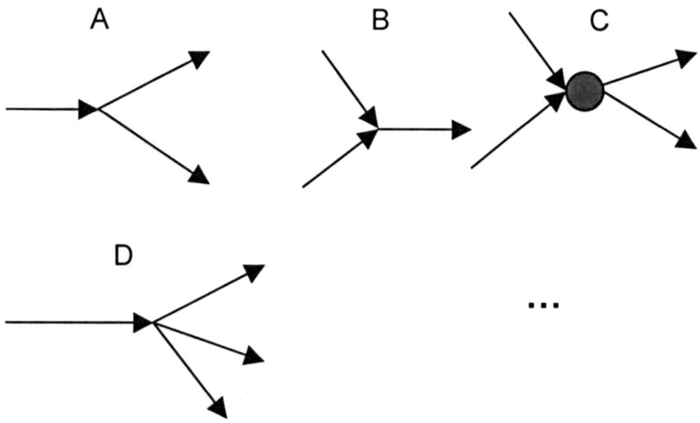

Figure 12.2. Diagrams showing some ways in which particles interact with each other and transform to other types of particles. In diagram A a particle "splits" into two particles. In B two particles combine to produce an "outgoing" particle. In C two particles interact in some complex way represented by the filled circle to produce two outgoing particles which may or may not be the same particles. In D a particle transforms (decays) into three particles.

Because of the nature of the Standard Model, particles of matter (quarks and leptons) and particles of "energy" (photons and massive vector bosons) can transform into each other in all sorts of ways if the incoming particle(s) have sufficient energy and momentum to generate the outgoing particles.

Since, under the right energy and momentum conditions, it is possible for any incoming set of particles to transform into a set of

---

[159] There are some uncertainties with regard to a small number of special transformations, and the possibility of new particles and new types of transformations at higher energies is not excluded. But the overwhelming majority of known particle physics phenomena are very well described by The Standard Model.

different outgoing particles[160] it seems reasonable to assume that the substance of *all particles in nature is the same and differ only in their form.* This assumption follows from the principle of Ockham's Razor, which states that usually the simplest explanation of a phenomenon is usually the correct one.[161]

So instead of assuming each particle is composed of a different substance the well-substantiated Standard Model, which contains transformations (interactions) between all the known particles, suggests that particles are made of the same substance (or perhaps no substance) and that their differences are due to their internal form.[162] There is no experimental data that could clarify the nature and properties of this substance. There is a further problem in that the dominant form of matter in our universe is Dark Matter and the dominant form of energy is Dark Energy. Dark Matter and Dark Energy interact with "normal" matter and energy through the gravitational interaction, which is very weak.[163] So the substance of which Dark Matter and Dark Energy are made are open questions.

It is possible that all types of matter and energy are not made of any substance but instead consist of form imposed on nothingness. Then Dark Matter and Dark Energy would be "dark" only because they did not interact with normal matter and energy except possibly through very weak forces.

It is difficult to imagine form imposed on nothingness yet this possibility seems reasonable. There is also the question of how nothingness relates to the vacuum. In quantum field theory the vacuum is a quantum thing and possesses properties. It possesses enormous energy. Vacuum fluctuations can occur in which particle – antiparticle pairs "pop out" of the vacuum for extremely short periods of time. The vacuum can "exert" Casimir forces that can be measured in the laboratory. So the vacuum is an extremely dynamic "substance" in quantum field theory.

---

[160] With the proviso that charge and internal quantum numbers such as strangeness, color, and so on are properly conserved (or not conserved according to the dictates of The Standard Model).

[161] Principle 2 of section 12.1. Richard P. Feynman expressed a similar view – that the simplest solution is often the correct one.

[162] As stated earlier this concept is also implicit in SuperString Theories. All particles consist of extremely small mathematical strings in these theories and differ only in form. Thus the substance would be mathematical strings.

[163] And possibly other ways.

If we identify the vacuum of quantum field theory with nothingness then nothingness is an extremely dynamic thing as well. Perhaps its major distinguishing feature is its non-conservation. Nothingness is unlimited in quantity as opposed to substances of everyday experience which have a fixed mass or size. One thinks of the "sacred" law of 19th century physics, "Mass is conserved and cannot be created or destroyed," which was overturned by Einstein's theory of Relativity and the transformation of matter into energy in fusion and fission. Nothingness being unlimited can easily be interpreted as the substance of particles and its unlimited nature enables transitions between the numbers and types of particles.

12.8.2 The Form of Particles

If, as the previous subsection suggested, particles are composed of one substance, and particles are differentiated from one another by their form, then the connection of the realm of Ideas with Reality becomes less difficult since form is based on Ideas – theoretical concepts. We have learned that the form of particles is in part based on space-time features such as their spin, their energy, their mass, and their momentum. In addition, their form is based on their electric charge and their internal quantum numbers. These quantities are determined by their group structure. Their group structure is SU(2)⊗U(1) (ElectroWeak symmetry) and color SU(3) (quark and gluon particle Strong interaction symmetry). There may be additional symmetries that will appear in the future as higher energy accelerators discover new phenomena.

But, from what we have learned, it is clear that space-time symmetry and internal group symmetries determine the form of particles.

12.8.3 Being as Form

The consequences of the considerations in the previous two subsections suggest that Being is form and that there is one unlimited substance in nature from which the particles of Reality and thus Reality is constructed.

If Being is form, then there is no difficulty in combining Being with Operator Logic to form Reality using the knowledge base of space-time features and internal symmetries that we have acquired experimentally.

Thus we have realized the Platonic scenario of a realm of Ideas connected mathematically to Reality.

12.8.4 Origin of Being

The components of Reality are tangible—perceivable by our senses (directly or indirectly), and the components of Reality interact with each other. But how did the components of Reality acquire existence?

Remarkably, there is strong evidence that the universe emerged from a point (or a "small" neighborhood of a point) at its beginning, The Big Bang. So the question of Being may degenerate to the question of "Being" at a single point without extension, without space or time, and perhaps without content. A number of theorists have suggested that the Big Bang is a quantum fluctuation—something emerging from nothing (the "vacuum") in such a way that the sum total of the emergent fluctuation has zero energy and thus is still nothing if considered in toto.[164]

In view of the major uncertainties, both physically and philosophically, in the understanding of existence and Being the question of the origin of Being is primarily work for the future. This section provides a preliminary discussion of Being based on our knowledge of particle physics – a subject which confronts existence and non-existence (Being and non-Being) on an everyday experimental and theoretical basis.

12.9 The Origin and Mechanism of Physical Laws

When we were developing the various Logics in earlier chapters we assumed a corpus of predicates and subjects, and also domains for each predicate. The specification of predicate domains was based on an implicit *knowledge base* that was used to determine the subjects that were in the domain of a given predicate and/or to determine the probability or probability amplitude that a subject was in the domain of a particular predicate. Thus a knowledge base is an essential ingredient of Operator Logic and Quantum Operator Logic.

---

[164] Being then becomes an illusory artifact of our consciousness. Combinations of forms (particles) constitute matter and the forces between these particles en masse give matter its solidity or liquidity as the case may be. The forces between chunks of matter when they collide are the cumulative effect of the forces between the particles constituting each mass.

In the case of physical Reality a physics knowledge base was implicit and operated between the various stages of an experiment. It affected the filtrations that took place as the experiment progressed. Physical quantities such as mass, energy, momentum and so on, and the laws that governed them, constitute a knowledge base that is implicit in the application of Quantum Theory experiment.

To make the transition from the realm of Ideas to Reality not only do we need the existence of physical things but we also need a source for physical laws and a mechanism for their execution. In human societies legislatures frame laws and the police enforce them. In Reality we must determine the origin of physical laws and the mechanism by which they are "enforced" on the components of Reality.

The typical methods, by which physical laws are determined, are 1) a fundamental lagrangian which through "canonical" methods leads to the equations embodying physical laws; or 2) a set of fundamental equations from which all physical laws are derived.

In either of these cases the question arises: "Where did the fundamental lagrangian or equations come from?" One could say that their origin is not knowable within a scientific framework. (Perhaps a deity specified them.) Or one could say they were the result of some as yet unknown process of self-organization (order emerging from chaos).

Once we have the laws then the question arises of how things in Reality are coerced into obeying them. Who are the police that enforce physical laws? The importance of this question is very apparent today in the case of quantum entanglement, and especially in the Einstein-Rosen-Podolsky thought experiment. In this experiment doing something to a particle that is quantum entangled with another particle at a great distance away—affects the distant particle. How? Clearly there is an enforcement issue for physical laws that is not as yet understood. Interestingly, the enforcement must be instantaneous and thus infinitely faster than the speed of light. One mechanism that could resolve this problem is the tachyon propagation of quantum effects.[165] Tachyons can travel at infinite speed and thus instantaneously "enforce" physical laws.

Thus we add the need for a source of physical laws and a mechanism for their enforcement to the need for the origin of Reality.

---

[165] See Blaha (2008) for a discussion of tachyon quantum field theory and the possible role tachyons play in the Standard Model of elementary particles.

13. Précis of the Transition from Ideas to Reality

Ultimate Reality is based on Logic, properly understood.

13.1 The Aspects of Reality

The four aspects of Reality, space, time, matter, and radiation, are intimately related. Without change in matter and/or radiation, time becomes unmeasurable and irrelevant. Without matter and/or radiation, distance and thus space becomes unmeasurable and irrelevant. So the presence of matter and/or radiation are necessary to make time and space meaningful and thus real.

In turn matter and radiation modify the properties of time and space as Einstein's Theory of General Relativity shows. Thus the components of Reality are highly interrelated.

13.2 Matter

Our understanding of matter has evolved tremendously in the past two millenia. For much of that time we viewed matter as concrete, substantial "stuff" that we could kick, as Dr. Johnson did to refute Bishop Berkeley's claim that matter was not "real". Dr. Johnson reputedly said, " I refute it thus!" as he kicked a rock. Starting in the nineteenth century it became clear that matter was composed of molecules, which in turn were composed of atoms, which in turn were composed of electrons circling a nucleus containing protons and neutrons (the solar system view of the atom.) In the nineteen seventies it became clear that protons, neutrons and other particles were composed of quark particles. Matter appears, at present,[166] to be composed of eighteen types of quarks and six

[166] There is also an unknown form of matter called Dark Matter that constitutes about 95% of the matter in the universe. The properties and nature of this form of matter are not known at present although there are a number of speculative theoretical proposals.

particles[167] called leptons. The major qualitative, distinguishing property between quarks and leptons is that quarks experience a type of force called the strong interaction while leptons do not. Rather than describe quarks and leptons in detail we suggest the interested reader read one of the many popular books on elementary particles.

This author has proposed a set of postulates from which the theory of these particles (the Standard Model of Elementary Particles) may be "derived."[168] We will examine the starting point of this derivation/construction in this chapter. Blaha (2008) contains the derivation, which we will provide in an expanded form later in this book.

13.3 Matter is Insubstantial

We now wish to get to the heart of the nature of matter. What is a particle made of? It appears that a particle has form without substance (as we commonly understand substance). Its form is particulate in part and wave-like in part. We might think of it as "nothingness" upon which a semi-permanent (or permanent) form is imposed with the quality of "Being." Being consists of existence for some interval of time and implies observability. An entirely unobservable entity could have a persistent form, but, lacking observability, could not be considered real since there would be no manifestations of it in our Reality. Fortunately, for us Quantum Theory requires an observer, and observability in some sense, for all real entities and events. Thus all known particles interact and that makes them part of our Reality.[169]

Particles exist and possess Being. Particles can interact with each other to truly create new particles or to truly annihilate into radiation.[170] So the forms imposed on nothingness (particles) can transform to other forms but do so in such a way that certain features of these forms are preserved. These features satisfy what we call conservation laws such as the conservation of energy.

It is rather remarkable that particles can undergo true creation and annihilation because these features represent transformations between

---

[167] Electrons, muons, τ particles – each with a corresponding neutrino.

[168] Blaha (2008).

[169] This includes Dark Matter, which interacts with normal matter through gravitation and perhaps through other forces very weakly.

[170] Radiation is composed of particles as well: photons, ElectroWeak bosons, Strong Interaction gluons, and gravitons. Gravitons have not as yet been detected due to the weakness of the force of gravity.

Being and *non-Being*. Earlier we developed Operator Logic and Quantum Operator Logic based on linear vector space operators similar to those of Quantum Theory. Noting that one thus has a framework applicable to both Logic and Reality, we raised the issue of the "addition" of the quality of Being (existence) to "create" Reality from Ideas. Seeing the process of true creation and annihilation in the laboratory it is clear that Being is an acquirable property. As such, the Big Bang, which seems to be the origin of the universe, can be accepted as a transition from nothingness to Being through an expansion from a point to space-time differentiated regions with a variety of forms of particles. Thus we perceive the nature of the process although the precise details remain to be determined.[171]

13.4 The Knowledge Base of Reality

As we have seen, a calculus universe of discourse consists of a set of axioms written in terms of undefined primitive terms and the theorems derived therefrom. A semantic universe of discourse attributes meanings/interpretations to the primitive terms thereby giving meaning to the axioms and consequent theorems. A theory expressed in terms of Operator Logic usually is based on a set of axioms for either type of universe of discourse. A physical interpretation of the primitive terms is also required for a semantic universe of discourse.

We will call the set of primitive term definitions and the set of axioms the *knowledge base* of the semantic universe of discourse of Reality.

A knowledge base for physical Reality (at least the part of it with which we are familiar) was first expounded in a series of books by Blaha culminating in Blaha (2008). Blaha (2008) lists a set of axioms that lead directly, and *exactly*, to most of the established features of the Standard Model of Elementary Particles. The major known features of this theory of elementary particles were determined in a period of forty-five years (1930 – 1975 approximately). These features include parity violation, peculiar symmetries, and the somewhat complicated nature of the particle spectrum. These features are *exact* results of the axioms in Blaha (2008) unlike other theoretic attempts to explain the Standard Model. Other attempts view the Standard Model as a low energy approximation

---

[171] Blaha (2004) describes a theory of a quantum Big Bang.

of a larger theory and have no inherent justification for parity violation but rather define their theories to incorporate it.

In Blaha's derivation every basic particle of matter, whether quark or lepton, has spin ½. We have seen in chapter 10 that the basic algebra of Operator Logic eigenvalue operators, and that of the raising and lowering operators, is the same as the algebra of free spin ½ particles. Operator Logic is part of the Realm of Ideas. Our goal is to build a theory of Reality on the scaffolding of Operator Logic. The remainder of this chapter will be devoted to outlining the general idea of the construction of the theory of Reality. Blaha (2008) provides a detailed discussion. In building this theory we will have implemented the Platonic goal of mathematically connecting the Realm of Ideas and the Realm of Reality.

We will start with the direct sum of two spinor universes of discourse as given in section 10.3. In particular the 4-vectors of four rows defined in section 10.3 were

$$s_{\uparrow\uparrow k} = \begin{bmatrix} s_{1k\uparrow} \\ \\ s_{2k\uparrow} \end{bmatrix} \quad s_{\uparrow\downarrow k} = \begin{bmatrix} s_{1k\uparrow} \\ \\ s_{2k\downarrow} \end{bmatrix} \quad s_{\downarrow\uparrow k} = \begin{bmatrix} s_{1k\downarrow} \\ \\ s_{2k\uparrow} \end{bmatrix} \quad s_{\downarrow\downarrow k} = \begin{bmatrix} s_{1k\downarrow} \\ \\ s_{2k\downarrow} \end{bmatrix} \tag{10.19}$$

We form linear combinations of these vectors to establish contact with the standard treatment of spin ½ free quantum fields in physics:

$$u(k, +½) = \begin{bmatrix} 1 \\ 0 \\ 0 \\ 0 \end{bmatrix} \quad u(k, -½) = \begin{bmatrix} 0 \\ 1 \\ 0 \\ 0 \end{bmatrix} \tag{13.1}$$

and

$$v(k, -½) = \begin{bmatrix} 0 \\ 0 \\ 1 \\ 0 \end{bmatrix} \quad v(k, +½) = \begin{bmatrix} 0 \\ 0 \\ 0 \\ 1 \end{bmatrix} \tag{13.2}$$

where the first arguments of u and v, which have the value k = 0, will be identified as the momentum of a spin ½ particle shortly. The second arguments of u and v are the values of particle spin: +½ represents an "up" spin state and –½ represents a "down" spin state.[172] Note the relations to the spinors in eq. 10.19 such as

$$u(k, +\tfrac{1}{2}) + v(k, -\tfrac{1}{2}) = s_{\uparrow\uparrow k} \qquad (13.3)$$

etc.

Having defined the spinor states for a particle in eqs. 13.1 and 13.2 we now introduce space and time, which of necessity must be in the knowledge base of Reality. Quite possibly the three real space dimensions and the one real time dimension of our experience are the residue of a larger space-time. But we experience three space dimensions so they must appear in the knowledge base. We introduce the coordinates $x = (t, \mathbf{x})$ where $\mathbf{x}$ is a 3-vector and t is the time, and the momentum $p = (p^0, \mathbf{p})$ where p^0 is the energy and $\mathbf{p}$ the 3-momentum of a particle.

Then we form a Fourier representation of a spin ½ particle wave function:

$$\psi(x) = \sum_{\pm s} \int d^3p N(p)[b(p,s)u(p, s)e^{-ip\cdot x} + d^\dagger(p,s)v(p, s)e^{+ip\cdot x}] \qquad (13.4)$$

where N(p) is a normalization factor, b(p,s) is lowering operator that annihilates a particle of momentum p and spin s, $d^\dagger(p,s)$ is a creation operator, u(p, s) and v(p, s) are column vectors like those in eqs. 13.1 and 13.2 that have been "boosted" to momentum p, and $p\cdot x = p^0x^0 - \mathbf{p}\cdot\mathbf{x}$ is the inner product of 4-vectors.[173] The raising and lowering operators b(p,s) and $d^\dagger(p,s)$ (and their hermitean conjugates $b^\dagger(p,s)$ and d(p,s)) are mathematically similar to the raising and lowering operators of eq. 10.21. They satisfy the anticommutation relations

---

[172] Again we choose not to deviate from our line of discussion to discuss the details of particle spin and refer the interested reader to popular books on elementary particles. The point we are seeking to establish is that the direct sum of the matrix representation of logical universes of discourse mathematically leads to the mathematics of spin ½ Dirac particles in a simple and direct manner. And that gives us the mathematical connection of Ideas to Reality of which the Platonists spoke.

[173] This equation is discussed in detail in the many good books on quantum field theory.

$$\{b(q,s), b^\dagger(p,s')\} = \delta_{ss'}\delta^3(\mathbf{q} - \mathbf{p}) \qquad (13.5)$$
$$\{d(q,s), d^\dagger(p,s')\} = \delta_{ss'}\delta^3(\mathbf{q} - \mathbf{p})$$

using Dirac δ-functions instead of Kronecker δ-functions due to the continuous nature of the momentum variables.

Thus we see spin ½ particle wave functions as emanating from the spinors, and raising and lowering operators, of the spinor formulation of Operator Logic. We will see how this happens in some detail in section 13.6 below and in more detail in subsequent chapters.

Another interesting point that emerges from this discussion is the nature of spin ½ particle states such as

$$|p, s\rangle = b^\dagger(p, s)|0\rangle \qquad (13.6)$$

This state is interpreted in Reality as a one particle state. It also has an analogous interpretation in Operator Logic as creating a one term universe of discourse – a construct which is in part linguistic and in part logic.

13.5 Particles as Monads

The combination of a physical interpretation and an Operator Logic interpretation is somewhat similar to Leibniz's concept of a *monad*. However it differs in that no spiritual aspect is evident[174] and also in that states, such as that defined in eq. 13.6, do not have "perception" in the usual sense of the word. But they do have something analogous to perception through their interactions in the sense that they are affected by interactions, and thus may be said to 'perceive" other monads (particles) interacting with them.

13.6 Dirac-like Equations for Spin ½ Matter

Section 13.4 indicated in a general way the manner in which we can develop a Dirac-like wave function within the framework of the spinor matrix formulation of Operator Logic. In this section we will show how Dirac-like equations can be obtained by "boosting" the Dirac

[174] A religious reader might say that since everything in Reality emanates from God an implicit spiritual aspect is present. Possible spiritual extensions of this development are beyond the scope of this book.

equation for a particle at rest to a particle of momentum $p = (p^0, \mathbf{p})$. The novel feature of this derivation is that the particle mass turns out to be a scale factor with the dimension of [mass] times a Gödel number.

Our starting point is A_{E_k} in eq. 10.20 which expanded to show its 4×4 rows and columns is

$$A_E = \begin{bmatrix} gn(E_1) & 0 & 0 & 0 \\ 0 & 0 & 0 & 0 \\ 0 & 0 & gn(E_2) & 0 \\ 0 & 0 & 0 & 0 \end{bmatrix} \tag{13.7}$$

where we surpress the subscript k. A_E can be put into a more physically relevant form using the unitary transformation[175]

$$U = \begin{bmatrix} 1 & 0 & 0 & 0 \\ 0 & 0 & 1 & 0 \\ 0 & 1 & 0 & 0 \\ 0 & 0 & 0 & 1 \end{bmatrix} = U^\dagger = U^{-1} \tag{13.8}$$

$$A_E' = UA_EU^{-1} = \begin{bmatrix} gn(E_1) & 0 & 0 & 0 \\ 0 & gn(E_2) & 0 & 0 \\ 0 & 0 & 0 & 0 \\ 0 & 0 & 0 & 0 \end{bmatrix} \tag{13.9}$$

If we now let $E_2 = E_1$ and $A = A_E'$ then

$$A = \begin{bmatrix} gn(E_1) & 0 & 0 & 0 \\ 0 & gn(E_1) & 0 & 0 \\ 0 & 0 & 0 & 0 \\ 0 & 0 & 0 & 0 \end{bmatrix} \tag{13.10}$$

$$= gn(E_1)(\gamma^0 + I_4) \tag{13.11}$$

---

[175] This transformation poses no physics or logic issues since 4×4 Dirac γ matrices are equivalent up to a unitary transformation. See p. 18 of Bjorken (1965) and also R. H. Good Jr., Re. Mod. Phys. **27**, 187 (1955).

where I_4 is the 4×4 identity matrix and γ^0 is one of the four Dirac γ matrices defined in eq. 10.24. If we multiply eq. 13.11 by a mass scale m_0 and define the particle mass[176]

$$m = gn(E_1)m_0 \qquad (13.12)$$

then we have

$$m_0A = m\gamma^0 + mI_4 \qquad (13.13)$$

If we apply eq. 13.13 to a Dirac spinor U such as those in eqs. 13.1 or 13.2, introduce an exponentiated time factor, and set eq. 13.13 to zero we obtain the Dirac equation for a particle at rest:

$$(m\gamma^0 + mI_4)Ue^{-imt} = 0 \qquad (13.14)$$

where t is the time variable.[177] If we now perform a "Lorentz boost" – a Lorentz transformation to a reference frame moving with a velocity v with respect to the reference frame in which the particle is at rest and described by eq. 13.14 – we obtain[178]

$$(\not{p} - m)e^{-ip\cdot x}U(p) = 0 \qquad (13.15)$$

where the exponential factor, mt, is also "boosted" to $p\cdot x = p^0 x^0 - \mathbf{p\cdot x}$, and where $\not{p} = \gamma^0 p^0 - \boldsymbol{\gamma}\cdot\mathbf{p}$. The 4-momentum $p = (p^0, \mathbf{p})$ satisfies the mass condition

$$(p^0)^2 - |\mathbf{p}|^2 = m^2 \qquad (13.16)$$

---

[176] The reader may wonder whether the known quark and lepton masses – particularly their ratios have the Gödel number form of $2^a3^b5^c7^d$… where a, b, c, d, … are integers. Unfortunately the masses of most particles are not well known, and we expect the Higgs Mechanism and other mechanisms that may modify particle masses to ruin the simple Gödel ratios. The known mass ratios of the charged leptons are not exact Gödel ratios. There were observations of factors of 3 in the mass spectrum of "constituent" quarks some years ago with the mass ratios of s::c::b quarks being 3. m_s = 0.5 Gev, m_c = 1.5 Gev, and m_b = 4.5 Gev. (See "On A Possible Similarity Between The Heavy Lepton And Heavy Constituent Quark Mass Spectra", S. Blaha, Phys.Lett. **B84**, 116 (1979)) Since then the masses of these quarks have been adjusted to experimentally more acceptable values and the almost exact ratio of 3 has disappeared.

[177] From the viewpoint of Operator Logic eq. 13.14 can be viewed as a proposition or statement in the Operator Logic formalism.

[178] Blaha (2008) p. 29.

where m is given by eq. 13.12. A Fourier transformation of eq. 13.15 yields the free, coordinate space Dirac equation:[179]

$$(i\gamma^{\mu}\partial/\partial x^{\mu} - m)\psi(x) = 0 \qquad (13.17)$$

with the general solution given by eq. 13.4. The Dirac equation; generalized to include internal symmetries associated with the ElectroWeak Interaction and the Strong Interaction, incorporating the three known generations of quarks and leptons, and with interaction terms introduced; has the form of the matter sector of the Standard Model of Elementary Particles.

Thus we have found a mathematical path from Operator Logic implemented by a spin ½ matrix formalism to the basic equations of matter. The Platonists, of course, had no knowledge of Operator Logic, or The Standard Model, or the mathematical path that we have outlined between them. But they were able to develop a conceptual outline of the realm of Ideas and a conceptual mathematical bridge to Reality.

In subsequent chapters we will develop an expanded version of Blaha (2008) and (2009) that leads to the major parts of the form of The Standard Model.

The steps from Operator Logic to the matter sector of the Standard Model may be briefly summarized:

1) The γ matrices being 4×4 matrices imply a 4-dimensional space-time (Weinberg (1995) p. 216).
2) Assume the complex Lorentz subgroup L(C) of GL(4) is the space-time group of particles we find four types of spin ½ Dirac-like equations follow through boosts from the rest state to a state with one real time variable. These correspond to the four known types of spin ½ particles: up-quark, down-quark, electron-like and neutrino-like.
3) The four free Dirac-like equations are naturally grouped in doublets to obtain L(C)[180] covariant dynamic equations, and parity violation naturally appears as does color SU(3).

---

[179] Three other Dirac-like equations can be obtained using complex Lorentz boosts. The set of four equations corresponds naturally to the four known types of spin ½ particles and leads naturally to the form of the Standard Model.

[180] L_c is the complex Lorentz group.

4) Upon the introduction of weak interaction coupling terms and particle mass splitting, L(C) is broken to Lorentz covariance.[181] With the addition of another GL(4) group the three known generations of spin ½ particles and mass mixing follow yielding the known particle matter sector of the Standard Model.

Thus we have found the mathematical bridge between the Realm of Ideas and the Realm of Reality.

[181] Lorentz covariance of the real Lorentz group.

14. The Nature of the Physical Derivation/Construction of a Fundamental Theory of Physics

The further we pursue these inquiries, the fewer become the primitive truths to which we reduce everything; and this simplification is inself a goal worth pursuing. … The aim of proof is, in fact, not merely to place the truth of a proposition beyond all doubt, but also to afford us an insight into the dependence of truths upon one another.
Gottlob Frege – *The Foundations of Arithmetic*

14.1 What is the Form of the Fundamental Theory of Physics?

Many individuals have speculated on the ultimate form of the fundamental theory of physics, often called the "Theory of Everything". Some say it is a form of string theory. Others have suggested a theory based on a discrete lattice-like space-time with, perhaps, extra dimensions. Yet others have suggested a self-organizing theory from a primordial chaos. And, lastly, a more conservative group has suggested theories based on quantum field theory.

In all cases the fundamental theory of physics must take the form of a set of postulates that use primitive terms that are either loosely defined or "intuitive" or both. The implications of the postulates are explored and experimentally testable consequences ultimately derived.

It is clear from the experimental success of the Standard Model of Elementary Particles (modulo a few discrepancies and some major open questions[182]) that the Standard Model must be "derivable" or "constructable", at least approximately, for currently accessible energies from THE fundamental theory.

[182] Such as how Dark Matter and Dark Energy relate to The Standard Model.

14.2 A Theory for its Time

In the time of Charlemagne Western European scholars obtained access to the literature and philosophy of classical Greek times. A mini-Renaissance, the Carolingian Renaissance, followed that expired after a relatively short time because the intellectual climate of the times in Western Europe "was not ready for it." In contrast the Italian Renaissance flowered, and was culturally diffused throughout a receptive Western Europe, beginning the modern age of culture and science. For a cultural or intellectual advance to succeed, and not wither, the intellectual/cultural climate must be in a receptive state.

In the case of physics a similar situation holds. Consider the case of Newton's theory of motion, which survived unchanged for hundreds of years. If someone in that period had pointed out that Newton's transformation between reference frames in relative motion with respect to each other was a special case of the Lorentz transformation, the general reaction would have been vast disinterest since there was no experimental means to test the dynamics of relativistic velocities. Neither was there any theoretical motivation. And one could also argue that there were many possible transformation laws that approximated Newton's transformation at low velocities.[183] Thus the times were not ripe for the appearance of the Theory of Special Relativity.

More generally, if we trace the development of physics from Aristotelian times to the present then it is clear that, at all times, the major changes in physical theories were unanticipated and generally driven by experimental results. This was particularly true in the twentieth century. It is also true in each historical period of physics, that when a successful theory was constructed, it described experimental results as they were understood in that period. The transitions between physical theories evidenced a continuity that was reflected in the "old" theory being an approximation to the new theory in the region of the "old' theory's approximate validity.

The drift of the above considerations is clear. Having been driven to the Standard Model by over seventy years of experimental and theoretical work, it is incumbent on theoretical physics to make sense of its form and origin.

---

[183] There was also a belief in that period that the speed of light was infinite.

Countless negative comments about the peculiarity of its form have appeared since the general form of the Standard Model was established around 1975. Attempts to develop a more fundamental theory of which the Standard Model is an approximation, have been largely in the form of SuperString theories. Lacking experimental guidance, and limited by the complexity and variety of the mathematics of SuperString theories, progress along these lines has been somewhat limited and not experimentally substantiated.

Thus we still face the issue of explaining the Standard Model. It certainly accounts for the vast bulk of experimental data. And so we must ask for a rationale/derivation of the Standard Model within the context of the current time rather than attempt to leapfrog to a future physics without any significant experimental guidance.

We have two choices to explain the Standard Model. The first choice—although it seems not to be an appealing choice—is to take the Standard Model lagrangian (with perhaps a few refinements and a unification with quantum gravity) as The Fundamental Theory and simply accept its peculiarities. The other choice is to develop a reasonable set of postulates from which the known particles and interactions of nature follow as well as the peculiar form of the Standard Model lagrangian.

Recently a series of books[184] by this author have appeared that have developed a set of postulates that lead to the known particles, interactions and the form of the Standard Model lagrangian. This book extends Blaha (2009) with some changes and improvements. The great virtue of this derivation, which is in fact a construction,[185] is that it leads directly to the Standard Model (possibly with a fourth generation) but without a plethora of new, unseen particles.

14.3 Derivation vs. Construction of a Physical Theory

Let us assume that we have a physics theory based on a lagrangian such as the Standard Model that describes physical phenomena such as the accumulated experimental data on elementary particles and their interactions.

---

[184] Blaha (2006), (2007b), (2008), (2009), and (2010a).

[185] A construction starts with a set of axioms and proceeds to develop a theory based on the addition of features that are consistent with the set of axioms.

Perhaps the lagrangian in itself describes the phenomena and a deeper layer of physical theory is not possible. (A simple example of this case is a classical particle in a conservative potential.)

But in other cases it is possible that the lagrangian is the result of a more fundamental physics. (An example of this situation is the Landau-Ginzberg theory of superconductivity, which is based on the more fundamental theory of Cooper pairing.)

It is almost universally believed that the Standard Model is a phenomenological theory that is a consequence of a more fundamental theory of elementary particles. The open question, of course, is the nature of the more fundamental theory. We will define a set of postulates that lead to the Standard Model. But we realize from simple logic that the ability to derive a theory from a more fundamental theory does *not* prove the more fundamental theory is the one and only correct theory.

So in seeking to define a more fundamental theory that is *most likely* to be the true physical basis of an experimentally verified (phenomenological) theory, it seems reasonable to assume the six principles stated previously in section 12.1.

Further, it will be seen that the development of the specific phenomenological theory of interest (The Standard Model of Elementary Particles) can be best viewed as a construction from first principles rather than as a strict derivation from axioms. A *construction* begins with a set of axioms, and then develops a theory based on the axioms, and on the addition of features that are consistent with, and share the spirit of, the set of axioms.

Euclid's geometry is an example of a construction although most students have viewed it as strictly derived from its five axioms. As logicians have pointed out the geometrical figures used to prove many geometrical theorems embody extensions of the five axioms, and are not implied by the axioms. Thus ***Euclid's gemoetry is a construction*** – not a strictly derived theory.

In the sprit of Euclid's approach we will construct The Standard Model's form from a set of initial postulates. These postulates will be extended as we proceed by reasonable constructs that are consistent with, and reflect the spirit of the initial postulates. As a result of the construction the superficial peculiarities of form of The Standard Model will be seen to follow from the initial postulates and to thus be based on a deeper level of Reality than appears in The Standard Model.

14.4 What Purpose does a Construction or Derivation Serve?

As the eminent logician Frege indicated in the introductory quotes there are several benefits in the reduction of a theory to a more fundamental theory (set of postulates). These benefits include:

i) The more fundamental theory is usually simpler and more comprehensive.
ii) The derivation of the phenomenological theory shows the origin of the various parts of the derived theory, and the interdependence of the parts of the derived theory and their derivation from a particular postulate or set of postulates.
iii) The more fundamental theory enables us to consider a deeper level of physics and perhaps find a path to a yet deeper level.
iv) The more fundamental theory enables us to consider alternative postulate sets and the universes to which they lead. (A good example of this possibility is the controversy over Euclid's Fifth Postulate and the non-Euclidean geometries that emerged from the controversy.)

14.5 The Rigor of a Derivation

In defining a set of postulates that lead to the Standard Model, either exactly or as an approximation, the question of the rigor of the derivation unavoidably appears. First there is the question of the rigor of differential and integral calculus which is still an issue despite the apparent rigorous development of calculus in the nineteenth century by Dedekind and others. To show the issue is still alive we simply mention the question, "If a point has no width or breadth, what does it mean to say the "next point" on a line?" This question inevitably leads to issues when one considers the definition of a derivative as a limit. Since "bare" particles are "point-like" the issue also surfaces in physics.

Secondly there is the issue of the path integral formulation of the Standard Model which (together with the Faddeev-Popov Mechanism) generates the correct perturbation theory in the ElectroWeak sector and is generally believed to generate correct results for Standard Model particle physics. The path integral formalism has not been put on a rigorous basis and it appears unlikely to be made rigorous for the foreseeable future.

These obstacles to a completely rigorous development mean that we must follow a procedure similar to the "traditional" physics approach of doing things "rigorously": reasonable moderation in rigor. In this approach we have the history of physics since Newton and Leibniz to support us. For approximately three hundred years physicists used the differential and integral calculus successfully. These areas of mathematics were not put on a quasi-rigorous footing in the view of many mathematicians until the mid-nineteenth century and there remains more to be done to obtain a truly rigorous calculus.

Thus a quasi-rigorous physicist's approach to proofs and derivations appears to be fully justified.

14.6 Consistency and Completeness of a Set of Postulates

Since our goal is to define a set of postulates at a deeper level from which we can derive the Standard Model it is prudent to inquire about the consistency and completeness of the set of postulates so defined. It would be the height of hubris to believe that the set of postulates that we define completes the study of the fundamental nature of the universe(s). So we will take these postulates to be a step in the direction of deeper knowledge but realize that these postulates now become the subject of deeper theoretical and experimental investigation. There may be more phenomena at higher energies that will extend the domain of particle physics beyond the Standard Model. Indeed the vast majority of particle physicists expects new phenomena as higher energy accelerators appear. And the mysteries of Dark Matter, Dark Energy, and other unusual cosmological features, also suggest there is much more to learn since we have found that the very small is intimately connected to the very large.

However taking the postulates seriously as an interim deeper theory of particle physics we have two "simple" issues to address: the consistency of the set of postulates and the completeness of the set of postulates.

The consistency issue can be resolved by two remarks:

i) The consistency of a set of postulates cannot be mathematically proved, in principle, within the framework of the theory it defines according to the celebrated consistency theorem of Gödel.

ii) From a physicist's point of view the theory is consistent if it predicts a unique result for all possible experiments within the domain of applicability of the theory. This criteria is, of course, impossible to meet since the number of possible experiments is infinite. (Presumably, the predictions are verified by experiment as well.)

The path integral formulation of the Standard Model appears to fulfill the second consistency criteria since it provides, in principle, a unique, well-defined approach to calculating any experimental result.

The completeness of the set of postulates is determined by the theoretic results it implies. If we assume the Standard Model is a complete description of particle physics (and, of course, we do not) then if the postulates lead to The Standard Model then the postulates can be considered complete from a physical point of view. If new phenomena are found that require the Standard Model be extended, then the set of postulates may well have to be extended as well: the extended set of postulates should imply the extended Standard Model. (The possibility exists that some of the postulates might have to be modified as well.)

14.7 The Difference Between a Mathematical-Deductive System and a Fundamental Scientific Theory

The eminent philosopher and logician R. B. Braithwaite[186] has said, "The irreducible difference between the propositions of logic and mathematics and those of a natural science are that the former are logically necessary and the latter contingent."[187] This observation is true, in the author's opinion, in the case of interim scientific theories that are dependent upon experiment for clarification and further growth. However, a theory, which purports to be a complete "Theory of Everything", falls into the other category—a mathematical theory of a purely mathematical deductive type—that needs, in principle no further

[186] Braithwaite (1960) p. 353.

[187] Braithwaite is not using "necessary" and "contingent" in the sense of modality where one distinguishes between the necessary and the contingent. Rather it appears he is using these words to distinguish between mathematics which has a logical structure that follows from fundamental logic and mathematical principles, and natural science which is provisional and deduced from experiment.

experimentation and thus is strictly mathematical-deductive in nature.[188] In this book we claim to have a theory that is deeper than the Standard Model and that implies the Standard Model. This theory, framed in terms of postulates, is a step in the direction of a mathematical-deductive theory and thus is treated as such. However it is clearly not a complete "Theory of Everything" and does not pretend to be.

It is only a step—a significant step—in the author's view that obviates certain other theoretical attempts and brings us closer to a truly fundamental theory—of great simplicity—and depth—that is based on the geometry of the universe.

14.8 Modality and Physical Theories

Superficially the issues raised in the study of modality: necessary vs. contingent would appear to be relevant to the physical sciences. However a consideration of the test for contingency shows it to be untenable for physical science: a property is *contingent* if it is not required in all possible worlds (universes); a property is *necessary* if it is required in all possible worlds (universes). For physics Modality requires physical knowledge of all possible worlds (universes). This knowledge is unattainable experimentally or theoretically.

We can only create a fundamental physical theory for our universe at best. Other universes are, by definition, beyond our ken. Thus we must conclude that all physical properties—all features of the fundamental physical theory—are contingent due to our inability to investigate all possible universes. We cannot create a set of postulates that form the basis of Reality and declare any of the postulates to be necessary.[189]

Thus in the derivation/construction of The Standard Model in the chapters that follow all postulates are contingent. Modality is a moot issue in the study of Reality for the foreseeable future.[190]

---

[188] It should be noted that plane geometry was a deductive theory until Euclid's five postulates were recognized as implying all the theorems of plane geometry. At this point geometry transitioned to a mathematical-deductive theory. Thus one can say Euclid's geometry is a "Theory of Everything" for plane geometry. The ultimate goal of elementary particle theory is to accomplish the same feat for elementary particle physics.

[189] Unless they are trivially true.

[190] Note that a world (universe) is totally separate and distinct from any other world (universe). If one says that one can tunnel from one universe to another, then one must view these universes as

If in some distant epoch Mankind can develop the mathematical and physical knowledge to declare a property is necessary then it can only result from a deep analysis that shows any world without the property is "inconsistent." Gödel's Inconsistency Theorem[191] indicates that a proof that a system is internally consistent is not possible in general. Consequently a proof that a system is internally *in*consistent is undecidable. (It can be proven to be consistent and can be proven to be inconsistent within the system.) In addition, a totally empty universe without quantum phenomena stands as an example of a property-less universe (world). *Therefore no physical property is necessary using the many worlds analysis of Modal Logic.*

---

branches of one all-encompassing universe. In this sense we cannot learn of other worlds (universes) through experiments. They are by definition totally separate from our universe.

[191] The theorem, in brief, states, "If a formal system A is consistent, then its consistency is not provable within A."

15. Time, Space, and the Complex Lorentz Group

15.1 Are Time and Space Contingent or Necessary?

In section 12.7 we gave arguments for the existence of time and a direction for time based on the consecutive nature and directionality of proofs in logic. Since logical statements may be true in one locale and not true in another the existence of space is also required. Further upon examination of space and the types of spinors that it supports we found that one time dimension and three space dimensions were required[192] at minimum to have spinors of the form with which we are familiar.

So we associated space and time with a generic term in Operator Logic in section 13.6 and found that the simplest 4×4 Operator Logic statement (eq. 13.14) could be transformed into the Dirac equation (eq. 13.17) by performing a Lorentz (boost) transformation on the statement.

Thus we find that we can begin the construction of a theory of Reality by taking a simple, generic Operator Logic statement, introducing space and time, and proceding to apply a Lorentz transformation.

Earlier we discussed modal considerations (subsection 3.2.4): is a property necessary or contingent? In the case of time it appears that time is a necessary property of any universe (world) or logical proofs would not be possible. To paraphrase Descartes: I prove, therefore time exists. In the case of space one could argue that space is a contigent property if the common conception of the Big Bang is correct. The universe at the time (and before) of the Big Bang consisted of one point and had no extension. Thus we have an example of a spaceless universe demonstrating the contingency of space.

Although space may be a contingent property of a universe—our universe at present does have extension (space). So in the construction of

[192] Table 12.1 and the discussion following it.

our theory of Reality we introduce a space of three dimensions. The issue that now arises is the structure of space and its relation to time.

15.2 Transformations Between Coordinate Systems

The measurement of time and space is simple in practice but raises weighty questions when their underlying basis is concerned. We shall begin by simply measuring spatial distances by marking them off with a ruler and measuring time with a clock. Earlier we determined that four dimensions: one space dimension and three space dimensions were required for a non-trivial universe (section 12.7). We now create rectangular coordinate systems with x, y, and z axes as pictured in Fig. 15.1 below. We then assert the postulate:

Postulate 15.1. Any observer can define a set of time and space coordinates called a coordinate system in which the observer is at rest. One can define a transformation that relates the coordinate systems of two observers traveling at a constant velocity with respect to each other.

The idea of relating the coordinates of two coordinate systems is based on the fact that an observer in each coordinate can specify the coordinates of an object located at some point, and that the coordinates of the object in one coordinate system have to be related the coordinates of the object in the other coordinate system.

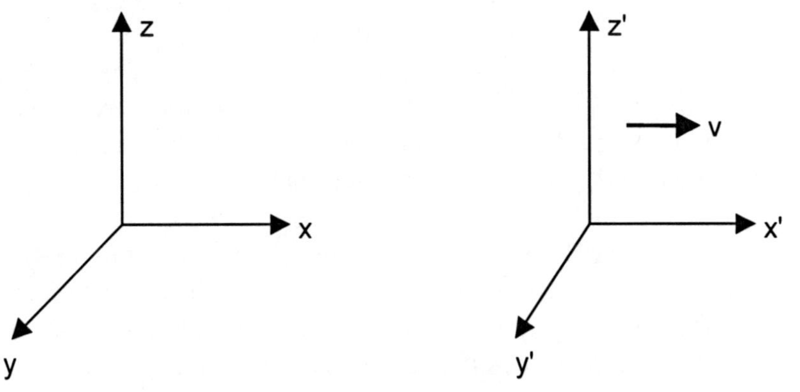

Figure 15.1. Depiction of two coordinate systems. The “primed” coordinate system is moving with velocity **v** in the positive x direction with respect to the unprimed coordinate system.

If space is flat the relation between the respective coordinates is linear. (One could reverse the logic of that statement by defining a flat space to be one in which the coordinates of a point in any coordinate system are linearly related to the coordinates of any other coordinate system moving at a constant velocity with respect to it.) Thus we can express the relation between the coordinates in the "unprimed" system to the coordinates in the "primed" system as the transformation between coordinate systems

$$\mathbf{a}' = A\mathbf{a} + \mathbf{B}t + \mathbf{C} \qquad (15.1)$$
$$t' = Dt + \mathbf{E}\cdot\mathbf{a}$$

where A is a 3×3 matrix, **B**, **C** and **E** are 3-vectors, and D is a number (scalar value).

Having restricted the set of transformations between coordinate systems to the form of eq. 15.1 we now assert postulates that restrict the form of the transformation to Lorentz transformations and transformations similar to Lorentz transformations.

Postulate 15.2. The speed of light, c, is the same in all coordinate systems.

Postulate 15.3. The invariant interval or distance $d\tau$ is defined by

$$d\tau^2 = g_{\mu\nu}dx^\mu dx^\nu \qquad (15.2)$$

It is invariant under a change of coordinate systems. The 16 quantities $g_{\mu\nu}$ are known as the metric tensor.[193] *The four quantities dx^μ are infinitesimal displacements in space and time.*

Postulates 15.2 and 15.3 are assumptions that we can only categorize as contingent at the present time because we do not have an a priori reason to motivate these assumptions.[194] Since we can conceive of other universes without these postulates they cannot be viewed as necessary.

---

[193] The repeated indices indicate a summation. In this case from 0 to 3 as shown in eq. 15.3

[194] There is an a postiori reason pointed out by Einstein and Poincaré that the equations of electromagnetism imply them. But since our goal is to construct electromagetism these postulates can only be asserted.

If we expand eq. 15.2 it is equivalent to

$$d\tau^2 = g_{00}dx^0dx^0 + g_{11}dx^1dx^1 + g_{22}dx^2dx^2 + g_{33}dx^3dx^3 \qquad (15.3)$$

which equals

$$d\tau^2 = c^2dt^2 - dx^2 - dy^2 - dz^2 \qquad (15.4)$$

using the familiar form of the time and rectangular space coordinates.[195]

15.3 The Lorentz Group

The metric tensor $g_{\mu\nu}$ for rectangular coordinates has the matrix form G = diag(1, –1, –1, –1):

$$G = \begin{bmatrix} 1 & 0 & 0 & 0 \\ 0 & -1 & 0 & 0 \\ 0 & 0 & -1 & 0 \\ 0 & 0 & 0 & -1 \end{bmatrix} \qquad (15.5)$$

The invariant interval under a transformation between rectangular coordinate systems (with "primed" and "unprimed" coordinates) has the form of eq. 15.4 for the unprimed coordinates and the same form for the primed coordinates:

$$d\tau^2 = c^2dt'^2 - dx'^2 - dy'^2 - dz'^2 \qquad (15.6)$$

That is the meaning of invariant in this context.

In matrix form we can define an "unprimed" coordinate column vector by

---

[195] $g_{00} = +c^2$, $g_{11} = g_{22} = g_{33} = -1$ in rectangular coordinate systems.

$$a = \begin{bmatrix} t \\ x \\ y \\ z \end{bmatrix} \qquad (15.7)$$

and its "primed" coordinate equivalent by

$$a' = \begin{bmatrix} t' \\ x' \\ y' \\ z' \end{bmatrix} \qquad (15.7)$$

If a and a' are the coordinates of the same point in the respective coordinate systems then they are related by a Lorentz transformation $\Lambda(\mathbf{v})$ with the form

$$a' = \Lambda(\mathbf{v})a \qquad (15.8)$$

where **v** is the relative velocity of the coordinate systems. The form of the transformation is constrained by postulates 15.2 and 15.3 to be[196]

$$\Lambda(\mathbf{v}) = \begin{bmatrix} \gamma & -\gamma v_x & -\gamma v_y & -\gamma v_z \\ -\gamma v_x & 1 + (\gamma - 1)v_x^2/v^2 & (\gamma - 1)v_x v_y/v^2 & (\gamma - 1)v_x v_z/v^2 \\ -\gamma v_y & (\gamma - 1)v_x v_y/v^2 & 1 + (\gamma - 1)v_y^2/v^2 & (\gamma - 1)v_y v_z/v^2 \\ -\gamma v_z & (\gamma - 1)v_x v_z/v^2 & (\gamma - 1)v_y v_z/v^2 & 1 + (\gamma - 1)v_z^2/v^2 \end{bmatrix} \qquad (15.9)$$

where $\gamma = (1 - v^2)^{-1/2}$, $\mathbf{v} = (v_x, v_y, v_z)$, $v = |\mathbf{v}|$ and we set $c = 1$ for convenience.[197] The set of all matrices of the form of $\Lambda(\mathbf{v})$ for $v < c$

[196] We shall consider only the proper, orthochronous Lorentz group at this point. We assume that the primed and unprimed coordinate systems have parallel axes. So there is no rotation of axes embodied in eq. 15.9.

[197] One can set c = 1 by an appropriate choice of time and spatial distance scales. The demonstration that $\Lambda(\mathbf{v})$ has the form given by eq. 15.9 can be found in many textbooks.

forms a matrix representation of the Lorentz group. Elements of the Lorentz group satisfy the relation

$$\Lambda(\mathbf{v})^T G \Lambda(\mathbf{v}) = G \tag{15.10}$$

which is regarded as the defining relation of the Lorentz group, where the superscript ^{T} specifies the transpose of the matrix.

The Lorentz group, with which we are familiar, relates the coordinates of an event in two coordinate systems that differ by a relative velocity whose magnitude is less than the speed of light. The inhomogenous Lorentz group includes coordinate displacements.[198]

The group elements of homogeneous Lorentz group can be expressed in terms of the generators **K** of boosts to coordinate systems moving at a constant velocity **v** and the generators **J** of purely spatial rotations by

$$\Lambda(\mathbf{v}, \boldsymbol{\theta}) = \exp[i\omega\hat{\mathbf{u}}\cdot\mathbf{K} + i\boldsymbol{\theta}\cdot\mathbf{J}] \tag{15.11}$$

where the vector $\boldsymbol{\theta}$ is a 3-vector specifying the rotation angles, and where $\mathbf{v} = \hat{\mathbf{u}} \tanh\omega$, $\hat{\mathbf{u}}\bullet\hat{\mathbf{u}} = 1$.

The boost transformation $\Lambda(\mathbf{v}) = \Lambda(\mathbf{v}, \mathbf{0})$ has the form

$$\Lambda(\mathbf{v}) = \exp[i\omega\hat{\mathbf{u}}\cdot\mathbf{K}] \tag{15.12}$$

Its matrix form is eq. 15.9. The matrix form can be expressed in terms of the unit normalized velocity vector $\mathbf{u} = (u_x, u_y, u_z)$ and ω as

$$\Lambda(\omega, \mathbf{u}) = \Lambda(\mathbf{v}) \tag{15.13}$$

$$= \begin{bmatrix} \cosh(\omega) & -\sinh(\omega)u_x & -\sinh(\omega)u_y & -\sinh(\omega)u_z \\ -\sinh(\omega)u_x & 1 + (\cosh(\omega) - 1)u_x^2 & (\cosh(\omega) - 1)u_x u_y & (\cosh(\omega) - 1)u_x u_z \\ -\sinh(\omega)u_y & (\cosh(\omega) - 1)u_x u_y & 1 + (\cosh(\omega) - 1)u_y^2 & (\cosh(\omega) - 1)u_y u_z \\ -\sinh(\omega)u_z & (\cosh(\omega) - 1)u_x u_z & (\cosh(\omega) - 1)u_y u_z & 1 + (\cosh(\omega) - 1)u_z^2 \end{bmatrix}$$

[198] See Weinberg (1995) for a discussion of the inhomogeneous Lorentz group.

This definition of the general form of proper, orthochronous, Lorentz boost matrices $\Lambda(\omega, \mathbf{u})$ will be used in subsequent sections to define faster-than-light boost transformations.

The vector form of a Lorentz boost transformation is

$$\mathbf{x}' = \mathbf{x} + (\gamma - 1)\mathbf{x}\cdot\mathbf{v}\,\mathbf{v}/v^2 - \gamma\mathbf{v}t \qquad (15.14)$$
$$t' = \gamma(t - \mathbf{v}\cdot\mathbf{x}/c^2)$$

where $\gamma = (1 - \beta^2)^{-1/2}$ with $\beta = v/c = v$ (since we set $c = 1$).

15.4 The Nature of $\Lambda(\omega, \mathbf{u})$ for Complex ω

We now turn to the case of complex ω where we will see faster-than-light Lorentz transformations will be defined using complex values of ω (section 15.6). Since for any complex value z

$$\cosh^2(z) - \sinh^2(z) = 1 \qquad (15.15)$$

it follows that for any complex value of ω:

$$\Lambda(\omega, \mathbf{u})^T G \Lambda(\omega, \mathbf{u}) = G \qquad (15.16)$$

eq. 15.16 implies $\Lambda(\omega, \mathbf{u})$ is a member of the Lorentz group, or of the complex Lorentz group.[199]

For certain values of the imaginary part of ω the matrix $\Lambda(\omega, \mathbf{u})$ has a particularly simple form, similar to that of $\Lambda(\omega, \mathbf{u})$ for real ω, but which generates boosts to relative velocities greater than the speed of light. Among these values are:

$$\omega = \omega_\pm = \omega \pm i\pi/2 \qquad (15.17)$$

Later we will see that the alternate choices $\omega_\pm$ correspond to specific physical situations relating to parity.

[199] The complex Lorentz group is defined as the group of all complex transformations that satisfy eq. 15.16.

15.5 Complex Lorentz Group

In the preceding section we saw that the parameter ω can be complex and the boost transformation will still satisfy the Lorentz condition eq. 15.10. More generally we can consider complex homogeneous Lorentz transformations $\Lambda(\mathbf{v}, \boldsymbol{\theta})$ which can be represented by eq. 15.11 with complex parameters ω, $\hat{\mathbf{u}}$, and $\boldsymbol{\theta}$ where $\hat{\mathbf{u}}$ and $\boldsymbol{\theta}$ are complex 3-vectors.

In general $\Lambda(\mathbf{v}, \boldsymbol{\theta})$ is a transformation between coordinate systems that have complex coordinates. One coordinate system is moving at a constant complex velocity with respect to the other. And the coordinate systems do not have parallel spatial axes in general.

Within the complex Lorentz group, denoted L(C),[200] *there are subsets of boosts with important physical roles in the derivation of form of The Standard Model. In particular we will see that certain classes of boosts generate faster-than-light transformations. These transformations can be divided into subclasses of "left-handed" and "right-handed" transformations based on the quantum field theories that they lead to. Further within each subclass there are subclasses of transformations that naturally lead to Dirac-like free field equations that can be described as lepton-like and quark-like.*

Thus these boosts are a key ingredient in understanding the form of The Standard Model.

15.6 Faster-than-Light Transformations

In this section we will substitute $\omega_\pm$ for ω in $\Lambda(\omega, \mathbf{u})$ and then show that we obtain two sets of possible transformations from sublight reference frames to faster-than-light reference frames. One set of transformations where $\omega_L = \omega + i\pi/2$ will be called *left-handed superluminal boosts*. They eventually lead to the "left-handed" Standard Model. We denote members of this set, $\Lambda_L(\omega, \mathbf{u})$, with the subscript "L".

The other set of boosts where $\omega_R = \omega - i\pi/2$ will be called *right-handed superluminal boosts*. They eventually lead to a right-handed,

[200] Streater (2000) points out that the complex Lorentz group is essential to the proof of the CPT theorem.

unphysical,[201] version of the Standard Model. We denote members of this set of boosts, $\Lambda_R(\omega, \mathbf{u})$, with the subscript "R".

Before considering faster-than-light boosts we note the relation between ω in a Lorentz boost $\Lambda(\omega, \mathbf{u})$, and the magnitude of the relative velocity v for $v < 1$, is

$$\mathbf{v} = \hat{\mathbf{u}} \tanh\omega \qquad\qquad \hat{\mathbf{u}}\cdot\hat{\mathbf{u}} = 1$$

$$\cosh(\omega) = \gamma = (1 - v^2)^{-1/2} \tag{15.18}$$
$$\sinh(\omega) = v\gamma = \beta\gamma$$

where $\beta = v = |\mathbf{v}|$.

15.7 Left-Handed Superluminal Transformations

Left-handed (proper orthochronous) superluminal boost transformations $\Lambda_L(\mathbf{v})$ have the same form as eq. 15.9 for ordinary (proper orthochronous) Lorentz boost transformations. However the magnitude of the relative velocity **v** is greater than the speed of light. Thus $\gamma = (1 - v^2)^{-1/2}$ is pure imaginary and $\Lambda_L(\mathbf{v})$ is complex.

$$\Lambda_L(\mathbf{v}) = \begin{bmatrix} \gamma & -\gamma v_x & -\gamma v_y & -\gamma v_z \\ -\gamma v_x & 1 + (\gamma - 1)v_x^2/v^2 & (\gamma - 1)v_x v_y/v^2 & (\gamma - 1)v_x v_z/v^2 \\ -\gamma v_y & (\gamma - 1)v_x v_y/v^2 & 1 + (\gamma - 1)v_y^2/v^2 & (\gamma - 1)v_y v_z/v^2 \\ -\gamma v_z & (\gamma - 1)v_x v_z/v^2 & (\gamma - 1)v_y v_z/v^2 & 1 + (\gamma - 1)v_z^2/v^2 \end{bmatrix} \tag{15.19}$$

This transformation raises several issues – the most prominent of which is the interpretation of the imaginary coordinates generated by the transformation. Imaginary coordinates would appear at first glance to be unphysical. However if we view the measurement of these quantities operationally: an observer measures distances with "rulers", and time with clocks, which both give real numeric values. Thus an observer *in any coordinate system* will always measure real numbers for time and

---

[201] Currently. If a right-handed counterpart to the current Standard Model surfaces at higher energies then the features emerging from right-handed superluminal boosts then become physically important.

space distances. However an observer *in another coordinate system* that is related to the first coordinate system by a superluminal transformation will view the coordinates in the first system as complex as eq. 15.19 indicates.

It is only when an observer's coordinate system is related to another coordinate system through a superluminal transformation that the *apparent* complex nature of coordinates becomes evident. Thus seemingly complex coordinates pose no new conceptual issues.

15.7.1 Cosh-Sinh Representation of Left-Handed Superluminal Boosts

We will now develop the representation of left-handed superluminal boost transformations in terms of cosh(ω) and sinh(ω) for later use in our discussion of tachyons. We find that we must use a complex $\omega_L \equiv \omega + i\pi/2$ to properly describe left-handed superluminal boosts. The relation between ω_L and v is different from eq. 15.18 for the case of left-handed superluminal boosts:

$$\begin{aligned} \cosh(\omega_L) &= i\,\sinh(\omega) = -\gamma = i\gamma_s \\ \sinh(\omega_L) &= i\,\cosh(\omega) = -\beta\gamma = i\beta\gamma_s \end{aligned} \qquad (15.20)$$

where $\beta = v > 1$, $\boldsymbol{\omega \geq 0}$, and

$$\gamma_s = (\beta^2 - 1)^{-1/2} \qquad (15.21)$$

Note

$$\begin{aligned} \sinh(\omega) &= \gamma_s \\ \cosh(\omega) &= \beta\gamma_s \end{aligned} \qquad (15.22)$$

Upon substituting ω_L for ω in eq. 15.13 we obtain another form for a left-handed superluminal transformation (equivalent to that of eq. 15.19):

$\Lambda_L(\omega, \mathbf{u}) = \Lambda(\omega + i\pi/2, \mathbf{u})$

$$= \begin{bmatrix} \cosh(\omega_L) & -\sinh(\omega_L)u_x & -\sinh(\omega_L)u_y & -\sinh(\omega_L)u_z \\ -\sinh(\omega_L)u_x & 1+(\cosh(\omega_L)-1)u_x^2 & (\cosh(\omega_L)-1)u_xu_y & (\cosh(\omega_L)-1)u_xu \\ -\sinh(\omega_L)u_y & (\cosh(\omega_L)-1)u_xu_y & 1+(\cosh(\omega_L)-1)u_y^2 & (\cosh(\omega_L)-1)u_yu_z \\ -\sinh(\omega_L)u_z & (\cosh(\omega_L)-1)u_xu_z & (\cosh(\omega_L)-1)u_yu_z & 1+(\cosh(\omega_L)-1)u_z^2 \end{bmatrix}$$

$$= \begin{bmatrix} i\gamma_s & -i\beta\gamma_s u_x & -i\beta\gamma_s u_y & -i\beta\gamma_s u_z \\ -i\beta\gamma_s u_x & 1 + (i\gamma_s - 1)u_x^2 & (i\gamma_s - 1)u_x u_y & (i\gamma_s - 1)u_x u_z \\ -i\beta\gamma_s u_y & (i\gamma_s - 1)u_x u_y & 1 + (i\gamma_s - 1)u_y^2 & (i\gamma_s - 1)u_y u_z \\ -i\beta\gamma_s u_z & (i\gamma_s - 1)u_x u_z & (i\gamma_s - 1)u_y u_z & 1 + (i\gamma_s - 1)u_z^2 \end{bmatrix} = \Lambda_L(\mathbf{v}) \qquad (15.23)$$

A simple case that illustrates a left-handed superluminal boost is to assume the relative velocity is in the x direction. Then eq. 15.23 becomes

$$\Lambda_L(\omega, \mathbf{u} = (1,0,0)) = \begin{bmatrix} i\gamma_s & -i\beta\gamma_s & 0 & 0 \\ -i\beta\gamma_s & i\gamma_s & 0 & 0 \\ 0 & 0 & 1 & 0 \\ 0 & 0 & 0 & 1 \end{bmatrix} \qquad (15.24)$$

implementing the coordinate transformation:

$$X' = \Lambda_L(\omega, \mathbf{u} = (1,0,0))X$$

or

$$\begin{aligned} t' &= i\gamma_s(t - \beta x) \\ x' &= i\gamma_s(x - \beta t) \\ y' &= y \\ z' &= z \end{aligned} \qquad (15.25)$$

The addition rule for the x-component of velocity can be computed for infinitesimal displacements in space and time:

$$v_x' = \Delta x' / \Delta t' = (\Delta x\, \gamma_s - \Delta t\, \beta\gamma_s)/(\Delta t\, \gamma_s - \Delta x\, \beta\gamma_s)$$

$$= (v_x - \beta)/(1 - \beta v_x) \qquad (15.26)$$

(where $\beta = w$ is the relative speed) in the limit $\Delta t \rightarrow 0$ where the x component of a particle's velocity in the unprimed frame is $v_x = \Delta x/\Delta t$. $\Delta t'$ is determined by

$$\Delta t' = i\Delta t\, \gamma_s(1 - \beta v_x) \qquad (15.27)$$

Note the velocity of light is the same in the primed and unprimed reference frames. (If $v_x = 1$ then $v_x' = 1$.) *Thus left-handed superluminal transformations preserve the constancy of the speed of light in all reference frames.* (Postulate 15.2)

Further note that increasing the value of ω in $\Lambda_L(\omega, \mathbf{u})$ corresponds to decreasing the magnitude of the relative velocity v since

$$v = \text{cotanh}(\omega) \qquad (15.28)$$

by eq. 15.22. Thus when v = 1 then ω = ∞, and when ω = 0 then v = ∞.

15.7.2 General Velocity Transformation Law – Left-Handed Superluminal Boosts

The general velocity transformation law for a particle moving with velocity **v** in the unprimed reference frame and velocity **v'** in the primed reference frame is

$$\mathbf{v}' = [\mathbf{v} + (\gamma - 1)\mathbf{w}\cdot\mathbf{v}\ \mathbf{w}/w^2 - \gamma\mathbf{w}]/[\ \gamma(1 - \mathbf{w}\cdot\mathbf{v})] \qquad (15.29)$$

where **w** is the relative velocity of the primed reference frame with respect to the unprimed reference frame, and $\gamma = (1 - w^2)^{-1/2}$. Eq. 15.29 is obtained by calculating the derivative d**x**'/dt' using eqs. 15.14. The relative velocity **w** can be greater or less than the speed of light. Eq. 15.29 implies

$$v'^2 = 1 + (v^2 - 1)(1 - w^2)/(1 - \mathbf{w}\cdot\mathbf{v})^2 \qquad (15.30)$$

The velocity transformation law can be used to determine the multiplication rules for subluminal and superluminal Lorentz transformations (next subsection).

15.7.3 Left-Handed Transformations Multiplication Rules

In this subsection we will determine the multiplication rules of left-handed subluminal and superluminal Lorentz boosts. To do this we will consider three reference frames: an unprimed frame, a "primed" frame moving with velocity **w** with respect to the unprimed frame, and a "double-primed" frame moving with velocity **v** with respect to the unprimed frame and velocity **v'** with respect to the primed frame. See Fig.15.2.

The velocity **v'** is related to **v** by eqs. 15.29 and 15.30. Think of the double-primed coordinate system as a particle or attached to a particle. In addition note that the transformation law from the unprimed to the double-primed reference frame must be the product of consecutive transformations (boosts) from the unprimed to the primed reference frames and then from the primed to the double-primed reference frames.

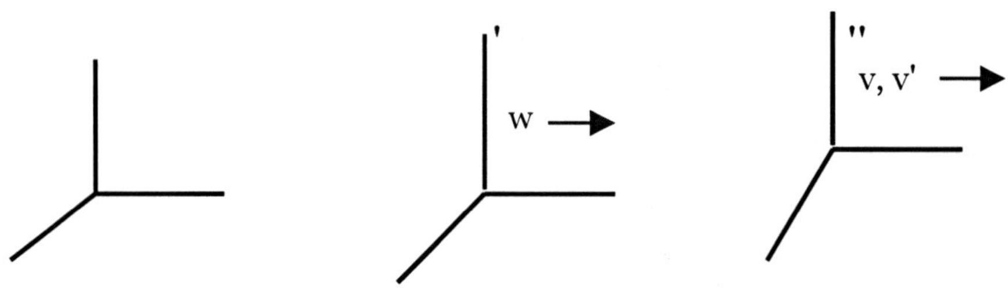

Figure 15.2. Three reference frames used to establish transformation multiplication rules.

Thus the transformations satisfy

$$\Lambda_?(\mathbf{v}) = \Lambda_?(\mathbf{v}')\Lambda_?(\mathbf{w}) \tag{15.31}$$

where the "?" subscripts indicate suluminal or superluminal transformations (boosts) depending on the magnitude of the relative velocity in the transformation's parentheses.

We now consider the cases using eq. 15.30:

1) If w > 1 and v' > 1

   then eq. 15.30 implies v < 1 and thus $\Lambda_?(\mathbf{v})$ is a subluminal transformation

$$\Lambda(\mathbf{v}) = \Lambda_L(\mathbf{v}')\Lambda_L(\mathbf{w}) \tag{15.32}$$

2) If w > 1, v' < 1

   then eq. 15.30 implies v > 1 and thus $\Lambda_?(\mathbf{v})$ is a superluminal transformation

$$\Lambda_L(\mathbf{v}) = \Lambda(\mathbf{v}')\Lambda_L(\mathbf{w}) \tag{15.33}$$

3) If w < 1, v' > 1

then eq. 15.30 implies v > 1 and thus $\Lambda_?(\mathbf{v})$ is a superluminal transformation

$$\Lambda_L(\mathbf{v}) = \Lambda_L(\mathbf{v'})\Lambda(\mathbf{w}) \tag{15.34}$$

4) If w < 1, v' < 1

then eq. 15.30 implies v < 1 and thus $\Lambda_?(\mathbf{v})$ is a Lorentz transformation

$$\Lambda(\mathbf{v}) = \Lambda(\mathbf{v'})\Lambda(\mathbf{w}) \tag{15.35}$$

where, in each above case, the transformation on the left side of the equation may be a boost or a combination of a boost and a spatial rotation. Thus we have obtained the multiplication rules for left-handed suluminal and superluminal Lorentz transformations.

15.7.4 Inverse of Left-Handed Transformations

The inverse of a Lorentz boost is

$$\Lambda^{-1}(\omega, \hat{\mathbf{u}}) = \exp[-i\omega\hat{\mathbf{u}}\cdot\mathbf{K}] \tag{15.36}$$

where $\boldsymbol{\omega} \geq \mathbf{0}$. Thus the inverse is generated by letting $\omega \rightarrow -\omega$. Note that since v = tanhω, the effect of $\omega \rightarrow -\omega$ is to let v → –v. In the case of superluminal left-handed boosts, since

$$\Lambda_L(\omega, \mathbf{u}) = \Lambda(\omega + i\pi/2, \mathbf{u}) = \exp[i(\omega + i\pi/2)\hat{\mathbf{u}}\cdot\mathbf{K}] \tag{15.37}$$

we find the inverse is

$$\Lambda_L^{-1}(\omega, \mathbf{u}) = \Lambda(-(\omega + i\pi/2), \mathbf{u}) = \exp[-i(\omega + i\pi/2)\hat{\mathbf{u}}\cdot\mathbf{K}] \tag{15.38}$$

where $\boldsymbol{\omega} \geq \mathbf{0}$. Since $\Lambda_L^{-1}(\omega, \mathbf{u})$ is not the hermitean conjugate of $\Lambda_L(\omega, \mathbf{u})$, superluminal boosts are not unitary. However unitarity is not required since even complex Lorentz group elements satisfy the defining relation of the Lorentz group (eq. 15.10). The effect of letting $\omega_L = \omega + i\pi/2 \rightarrow -(\omega + i\pi/2)$ is to let v → –v since

$$\tanh(\omega_L) = \beta = v \tag{15.39}$$

by eqs. 15.28.

15.8 Right-Handed Superluminal Transformations

When we transform between reference frames using a right-handed[202] superluminal boost the relation between ω and v also changes. The variable ω becomes $\omega_R = \omega - i\pi/2$ and

$$\cosh(\omega_R) = -i\sinh(\omega) = \gamma = -i\gamma_s \qquad (15.40)$$
$$\sinh(\omega_R) = -i\cosh(\omega) = \beta\gamma = -i\beta\gamma_s \qquad (15.41)$$

where $\beta = v > 1$ and $\boldsymbol{\omega \geq 0}$. Note (as before)

$$\sinh(\omega) = \gamma_s \qquad (15.42)$$
$$\cosh(\omega) = \beta\gamma_s$$

with

$$\gamma_s = (\beta^2 - 1)^{-1/2} \qquad (15.43)$$

Upon substituting ω_R for ω in eq. 15.13 we obtain the form of the right-handed superluminal boost:[203]

$$\Lambda_R(\omega, \mathbf{u}) = \Lambda(\omega - i\pi/2, \mathbf{u}) \qquad (15.44)$$

$$= \begin{bmatrix} -i\gamma_s & i\beta\gamma_s u_x & i\beta\gamma_s u_y & i\beta\gamma_s u_z \\ i\beta\gamma_s u_x & 1 + (-i\gamma_s - 1)u_x^2 & (-i\gamma_s - 1)u_x u_y & (-i\gamma_s - 1)u_x u_z \\ i\beta\gamma_s u_y & (-i\gamma_s - 1)u_x u_y & 1 + (-i\gamma_s - 1)u_y^2 & (-i\gamma_s - 1)u_y u_z \\ i\beta\gamma_s u_z & (-i\gamma_s - 1)u_x u_z & (-i\gamma_s - 1)u_y u_z & 1 + (-i\gamma_s - 1)u_z^2 \end{bmatrix}$$

A simple case that illustrates right-handed superluminal transformations is to assume a relative velocity in the x direction. Then eq. 15.44 becomes

---

202 We call these transformations right-handed because they lead eventually to an alternate Standard Model with right-handed SU(2) doublets and left-handed SU(2) singlets. This alternate right-handed Standard Model does not appear to correspond to experimental reality.

203 We note the singularities at $\beta = \pm 1$ or $\omega = \pm\infty$. As a result we have a branch cut in the complex ω-plane consisting of the entire real ω axis. Therefore three left-handed boosts are not equivalent to a right-handed boost but rather appear on another Riemann sheet.

$$\Lambda_R(\omega, \mathbf{u} = (1,0,0)) = \begin{bmatrix} -i\gamma_s & i\beta\gamma_s & 0 & 0 \\ i\beta\gamma_s & -i\gamma_s & 0 & 0 \\ 0 & 0 & 1 & 0 \\ 0 & 0 & 0 & 1 \end{bmatrix} \quad (15.45)$$

implementing the coordinate transformation:

$$X' = \Lambda_R(\omega, \mathbf{u})X$$

or

$$\begin{aligned} t' &= -i\gamma_s(t - \beta x) \\ x' &= -i\gamma_s(x - \beta t) \\ y' &= y \\ z' &= z \end{aligned} \quad (15.46)$$

Comparing eq. 15.45 with eq. 15.24 for a left-handed superluminal boost we see that

$$PT\Lambda_L(\omega, \mathbf{u} = (1,0,0)) = \begin{bmatrix} -i\gamma_s & i\beta\gamma_s & 0 & 0 \\ i\beta\gamma_s & -i\gamma_s & 0 & 0 \\ 0 & 0 & -1 & 0 \\ 0 & 0 & 0 & -1 \end{bmatrix}$$

where P is the parity operator and T is the time reversal operator. If we now apply a spatial rotation $\mathfrak{R}$ of π radians around the x axis then we obtain

$$\mathfrak{R}PT\Lambda_L(\omega, \mathbf{u} = (1,0,0))\mathfrak{R}^{-1} = \begin{bmatrix} -i\gamma_s & i\beta\gamma_s & 0 & 0 \\ i\beta\gamma_s & -i\gamma_s & 0 & 0 \\ 0 & 0 & 1 & 0 \\ 0 & 0 & 0 & 1 \end{bmatrix} \quad (15.47)$$

$$= \Lambda_R(\omega, \mathbf{u} = (1,0,0))$$

Since P and T commute with spatial rotations we find

$$\Lambda_R(\omega, \mathbf{u} = (1,0,0)) = PT\mathfrak{R}\Lambda_L(\omega, \mathbf{u} = (1,0,0))\mathfrak{R}^{-1} \quad (15.48)$$

or, more generally, performing additional spatial rotations:

$$\Lambda_R(\omega, \mathbf{u}) = PT\mathfrak{R}_u\mathfrak{R}\mathfrak{R}_w\Lambda_L(\omega, \mathbf{w})\mathfrak{R}_w^{-1}\mathfrak{R}^{-1}\mathfrak{R}_u^{-1} \tag{15.49}$$

or,

$$\Lambda_R(\omega, \mathbf{u}) = PT\mathfrak{R}_{tot}\Lambda_L(\omega, \mathbf{w})\mathfrak{R}_{tot}^{-1} \tag{15.50}$$

where **u** and **w** are unit vectors. Alternately,

$$\Lambda_L(\omega, \mathbf{w}) = PT\mathfrak{R}_{tot}^{-1}\Lambda_R(\omega, \mathbf{u})\mathfrak{R}_{tot} \tag{15.51}$$

or

$$\Lambda_L(\omega, \mathbf{w}) = PT\Lambda_R(\omega, \mathbf{u}') \tag{15.52}$$

for some unit vector **u'**.

Thus we have shown that PT can be used to relate left-handed and right-handed boosts in a one-to-one fashion. *The appearance of the parity operator P takes on great significance when we derive features of the Standard Model. The appearance of left-handed doublets and right-handed singlets stems directly from the implicit parity dependence of the left-handed sector of the superluminal part of the complex Lorentz group.*

For the right-handed boost the addition rule for the x-component of velocity can be computed for infinitesimal displacements in space and time:

$$\begin{aligned} v_x' = \Delta x'/\Delta t' &= (\Delta x\,\gamma_s - \Delta t\,\beta\gamma_s)/(\Delta t\,\gamma_s - \Delta x\,\beta\gamma_s) \\ &= (v_x - \beta)/(1 - \beta v_x) \end{aligned} \tag{15.53}$$

in the limit $\Delta t \rightarrow 0$ where the x component of a particle's velocity in the unprimed frame is $v_x = \Delta x/\Delta t$. Note if $v_x = 1$ then $v_x' = 1$. *Thus right-handed superluminal transformations also preserve the constancy of the speed of light in all reference frames.*

15.9 Inhomogeneous Left-Handed Lorentz Group Transformations

The *Left-Handed transformations of the Lorentz group* consist of the elements of the real Lorentz group plus left-handed superluminal boost transformations, and combinations of boosts and spatial rotations. Thus homogeneous left-handed superluminal transformations have the general form:

$$\Lambda_L(\mathbf{v}, \boldsymbol{\theta}) = \exp[i\omega_L \hat{\mathbf{u}}\cdot\mathbf{K} + i\boldsymbol{\theta}\cdot\mathbf{J}] \tag{15.54}$$

where $\omega_L' = \omega + i\pi/2$, $\boldsymbol{\theta}$ is the angular vector, and $\mathbf{J}$ is the angular momentum operator vector. Inhomogeneous left-handed superluminal transformations, which include displacements, can be expressed as

$$\Lambda_L(\mathbf{v}, \boldsymbol{\theta}, \mathbf{d}) = \exp[i\omega_L \hat{\mathbf{u}}\cdot\mathbf{K} + i\boldsymbol{\theta}\cdot\mathbf{J} - i\mathbf{d}\cdot\mathbf{P}] \tag{15.55}$$

where $\mathbf{P}$ is the momentum operator vector and $\mathbf{d}$ is a displacement vector.

We note

$$\det \Lambda_L(\omega, \mathbf{u}) = \pm 1 \tag{15.56}$$

The ordinary Lorentz group is divided into four disjoint subgroups that are often denoted:

$$\begin{aligned}
L_+^\uparrow&: \quad \det \Lambda(\omega, \mathbf{u}) = +1; \ \operatorname{sgn} \Lambda(\omega, \mathbf{u})^0{}_0 = +1 \\
L_-^\uparrow&: \quad \det \Lambda(\omega, \mathbf{u}) = -1; \ \operatorname{sgn} \Lambda(\omega, \mathbf{u})^0{}_0 = +1 \\
L_+^\downarrow&: \quad \det \Lambda(\omega, \mathbf{u}) = +1; \ \operatorname{sgn} \Lambda(\omega, \mathbf{u})^0{}_0 = -1 \\
L_-^\downarrow&: \quad \det \Lambda(\omega, \mathbf{u}) = -1; \ \operatorname{sgn} \Lambda(\omega, \mathbf{u})^0{}_0 = -1
\end{aligned} \tag{15.57}$$

where $\operatorname{sgn} \Lambda(\omega, \mathbf{u})^0{}_0$ is the sign of the 00 component of the $\Lambda(\omega, \mathbf{u})$ matrix. The various subgroups are related by the discrete transformations of parity P and time reversal T:

$$L_+^\uparrow \xrightarrow{P} L_-^\uparrow$$

$$L_+^\uparrow \xrightarrow{PT} L_+^\downarrow$$

$$L_+^\uparrow \xrightarrow{T} L_-^\downarrow$$

The left-handed superluminal transformations are disjoint in a somewhat different way. By eq. 15.56 the determinants are ±1. However the 0-0 matrix element of eq. 15.16 gives

$$\Lambda_L{}^0{}_0{}^2 - \Sigma_i\,(\Lambda_L{}^i{}_0)^2 = 1 \qquad (15.58)$$

The representation of superluminal boosts shows that each factor in eq. 15,58 is imaginary. Thus eq. 15.58 implies

$$\Sigma_i\,|\Lambda_L{}^i{}_0|^2 \geq 1 \qquad (15.59)$$

$$|\Lambda_L{}^0{}_0| \geq 0 \qquad (\text{not } \geq 1) \qquad (15.60)$$

where || indicates absolute value. Thus the magnitude of $\Lambda_L{}^0{}_0$ does not have a gap. Therefore left-handed superluminal transformations can be divided into two categories:

$$\begin{aligned} {}_L L_+&: \ \det \Lambda_L(\omega, \mathbf{u}) = +1 \\ {}_L L_-&: \ \det \Lambda_L(\omega, \mathbf{u}) = -1 \end{aligned} \qquad (15.61)$$

as one expects for complex Lorentz group transformations.[204]

Earlier we saw that under a PT transformation a left-handed superluminal transformation becomes a right handed superluminal transformation.

Again the various disjoint pieces are related by the discrete transformations of parity P and time reversal T:

$$\begin{aligned} {}_L L_+ &\xrightarrow{P} {}_L L_- \\ {}_L L_+ &\xrightarrow{T} {}_L L_- \end{aligned} \qquad (15.62)$$

[204] Streater (2000) p. 13.

15.10 Inhomogeneous Right-Handed Extended Lorentz Group

The inhomogeneous right-handed part of the complex Lorentz group[205] consists of the real Lorentz group plus right-handed superluminal transformations that have the form:

$$\Lambda_R(\mathbf{v}, \boldsymbol{\theta}, \mathbf{d}) = \exp[i\omega_R \hat{\mathbf{u}}\cdot\mathbf{K} + i\boldsymbol{\theta}\cdot\mathbf{J} - i\mathbf{d}\cdot\mathbf{P}] \qquad (15.63)$$

in general where $\omega_R = \omega - i\pi/2$.

15.11 General Forms of Superluminal Boosts

The group elements of homogeneous complex Lorentz group L(C) can be expressed in terms of the group generators as

$$\Lambda_C = \exp[i(\omega_r\hat{\mathbf{u}}_r + i\omega_i\hat{\mathbf{u}}_i)\cdot\mathbf{K} + i\boldsymbol{\theta}_c\cdot\mathbf{J}] \qquad (15.64)$$

where the vector $\boldsymbol{\theta}_c$ is a complex 3-vector, $\omega_r \geq 0$ and $\omega_i \geq 0$ are real numbers, and $\hat{\mathbf{u}}_r$ and $\hat{\mathbf{u}}_i$ are real normalized 3-vectors such that $\hat{\mathbf{u}}_r\cdot\hat{\mathbf{u}}_r = 1 = \hat{\mathbf{u}}_i\cdot\hat{\mathbf{u}}_i$. The generators of the homogeneous complex Lorentz group are **K**, and **J** just as for the homogeneous real Lorentz group.

We now focus on boosts because they will be crucial in the determination of the equations of motion of various types of spin ½ particles.

A boost has the form

$$\Lambda_C(\mathbf{v}_c) = \exp[i\omega\hat{\mathbf{w}}\cdot\mathbf{K}] \qquad (15.65)$$

where

$$\omega = (\omega_r^2 - \omega_i^2 + 2i\omega_r\omega_i\, \hat{\mathbf{u}}_r\cdot\hat{\mathbf{u}}_i)^{1/2} \qquad (15.66)$$

and

$$\hat{\mathbf{w}} = (\omega_r\hat{\mathbf{u}}_r + i\omega_i\hat{\mathbf{u}}_i)/\omega \qquad (15.67)$$

Since $\hat{\mathbf{u}}_r\cdot\hat{\mathbf{u}}_r = 1 = \hat{\mathbf{u}}_i\cdot\hat{\mathbf{u}}_i$

$$\hat{\mathbf{w}}\cdot\hat{\mathbf{w}} = 1 \qquad (15.68)$$

---

[205] Since γ_s has branch points at $v = \pm 1$ which corresponds to $\omega = \pm\infty$ for the left-handed and right-handed groups there is a cut along the real ω axis between $-\infty$ and $+\infty$. Therefore the product of three left-handed Lorentz transformations does not yield a right-handed transformation (as might be supposed from eqs. 2.43 and 2.51) but rather a left-handed transformation on the second sheet. A transformation with $\omega + 3i\pi/2$ is not equivalent to any with $\omega - i\pi/2$.

and the complex relative velocity is

$$\mathbf{v_c} = \hat{\mathbf{w}} \tanh(\omega) \tag{15.69}$$

Having placed boost transformations in the form of eq. 15.12 we can take advantage of the form of real proper orthchronous Lorentz boost transformations, eq. 15.13, and analytically continue to complex ω and complex unit vectors $\hat{\mathbf{w}}$ provided eq. 15.69 is satisfied. The resulting complex generalization will be the matrix form of proper boosts:

$$\Lambda_C(\mathbf{v_c}) = \exp[i\omega\hat{\mathbf{w}}\cdot\mathbf{K}] \equiv \Lambda_C(\omega, \hat{\mathbf{w}})$$

$$= \begin{bmatrix} \cosh(\omega) & -\sinh(\omega)\hat{w}_x & -\sinh(\omega)\hat{w}_y & -\sinh(\omega)\hat{w}_z \\ -\sinh(\omega)\hat{w}_x & 1+(\cosh(\omega)-1)\hat{w}_x^{\,2} & (\cosh(\omega)-1)\hat{w}_x\hat{w}_y & (\cosh(\omega)-1)\hat{w}_x\hat{w}_z \\ -\sinh(\omega)\hat{w}_y & (\cosh(\omega)-1)\hat{w}_x\hat{w}_y & 1+(\cosh(\omega)-1)\hat{w}_y^{\,2} & (\cosh(\omega)-1)\hat{w}_y\hat{w}_z \\ -\sinh(\omega)\hat{w}_z & (\cosh(\omega)-1)\hat{w}_x\hat{w}_z & (\cosh(\omega)-1)\hat{w}_y\hat{w}_z & 1+(\cosh(\omega)-1)\hat{w}_z^{\,2} \end{bmatrix} \tag{15.70}$$

Since analytic continuations are unique, the above form for $\Lambda_C(\mathbf{v_c})$ is well-defined and unique. It spans the complete set of proper complex Lorentz boosts.

We now will study six classes of boosts that have the property that they boost from a coordinate system with real time and space coordinates to a coordinate system with either a purely real or purely imaginary time, and real, imaginary or complex spatial coordinates. One pair of boost classes is left-handed. The other pair is right-handed. In each pair there is a class of boosts that lead to lepton-like free Dirac-like equations and a class that leads to quark-like free Dirac-like equations. The third pair of boosts leads to the conventional Dirac equation and a quark-like, Dirac-like equation. We will discuss these Dirac-like equations in detail later.

15.11.1 "Lepton-like" Left-Handed Boosts

If we let

$$\hat{\mathbf{u}}_\mathbf{i} = \hat{\mathbf{u}}_\mathbf{r} \equiv \hat{\mathbf{u}} \tag{15.71}$$

so that the vector $\hat{\mathbf{u}}_\mathbf{i}$ is parallel to $\hat{\mathbf{u}}_\mathbf{r}$, and let

$$\omega_i = \pi/2 \tag{15.72}$$

then $\Lambda_C(\mathbf{v_c})$ becomes a lepton-like left-handed boost:[206]

$$\Lambda_C = \exp[i(\omega_r + i\,\pi/2)\hat{\mathbf{u}}_r\cdot\mathbf{K}] \tag{15.73}$$

15.11.2 "Lepton-like" Right-Handed Boosts

If we let

$$\hat{\mathbf{u}}_i = -\hat{\mathbf{u}}_r \equiv -\hat{\mathbf{u}} \tag{15.74}$$

so that the vector $\hat{\mathbf{u}}_i$ is anti-parallel to $\hat{\mathbf{u}}_r$, and

$$\omega_i = -\pi/2 \tag{15.75}$$

then $\Lambda_C(\mathbf{v_c})$ becomes a right-handed boost:

$$\Lambda_C = \exp[i(\omega_r - i\,\pi/2)\hat{\mathbf{u}}_r\cdot\mathbf{K}] \tag{15.76}$$

15.11.3 "Quark-like" Left-Handed Boosts

If the real and imaginary relative vectors parts of $\hat{\mathbf{w}}$, namely $\hat{\mathbf{u}}_r$ and $\hat{\mathbf{u}}_i$, are perpendicular, $\hat{\mathbf{u}}_r\cdot\hat{\mathbf{u}}_i = 0$, then by eq. 15.66

$$\omega = (\omega_r^2 - \omega_i^2)^{1/2} \tag{15.77}$$

Thus ω is either pure real ($\omega_r \geq \omega_i$) or pure imaginary ($\omega_r < \omega_i$). We choose ω real, and reset

$$\omega = (\omega_r^2 - \omega_i^2)^{1/2} \rightarrow \omega' = (\omega_r^2 - \omega_i^2)^{1/2} + i\pi/2 = \omega + i\pi/2 \tag{15.78}$$

by adding $i\pi/2$ to the ω factor in eq. 15.65 since ω is a free parameter. Then the resulting Lorentz transformation then becomes a quark-like left-handed boost:[207]

$$\Lambda_C = \exp[i((\omega_r^2 - \omega_i^2)^{1/2} + i\pi/2)(\omega_r\hat{\mathbf{u}}_r + i\omega_i\hat{\mathbf{u}}_i)\cdot\mathbf{K}/\omega] \tag{15.79}$$

---

[206] We say "lepton-like" because we obtain a lepton-like Dirac-like equation using these boosts later.

[207] We say "quark-like" because we obtain a quark-like left-handed Dirac-like equation with complex spatial momentum terms using these boosts later.

15.11.4 "Quark-like" Right-Handed Boosts

If the real and imaginary relative vectors parts of $\hat{\mathbf{w}}$, namely $\hat{\mathbf{u}}_r$ and $\hat{\mathbf{u}}_i$, are perpendicular, $\hat{\mathbf{u}}_r \cdot \hat{\mathbf{u}}_i = 0$, then by eq. 15.66

$$\omega = (\omega_r^2 - \omega_i^2)^{1/2} \tag{15.80}$$

Thus ω again starts out either pure real ($\omega_r \geq \omega_i$) or pure imaginary ($\omega_r < \omega_i$). In this case we also choose ω real, and reset

$$\omega = (\omega_r^2 - \omega_i^2)^{1/2} \rightarrow \omega' = (\omega_r^2 - \omega_i^2)^{1/2} - i\pi/2 \tag{15.81}$$

by subtracting $i\pi/2$ from ω in eq. 15.65 since ω is a free parameter. Then the resulting Lorentz boost

$$\Lambda_C = \exp[i((\omega_r^2 - \omega_i^2)^{1/2} - i\pi/2)(\omega_r\hat{\mathbf{u}}_r + i\omega_i\hat{\mathbf{u}}_i)\cdot\mathbf{K}/\omega] \tag{15.82}$$

becomes a quark-like right-handed boost.[208]

15.11.5 "Quark-like" Boosts

If the real and imaginary relative vectors parts of $\hat{\mathbf{w}}$, namely $\hat{\mathbf{u}}_r$ and $\hat{\mathbf{u}}_i$, are perpendicular, $\hat{\mathbf{u}}_r \cdot \hat{\mathbf{u}}_i = 0$, then by eq. 15.66

$$\omega = (\omega_r^2 - \omega_i^2)^{1/2} \tag{15.83}$$

Thus ω again starts out either pure real ($\omega_r \geq \omega_i$) or pure imaginary ($\omega_r < \omega_i$). In this case choose ω_r real and use ω as defined by eq. 15.83.

Then the resulting Lorentz boost

$$\Lambda_C = \exp[i(\omega_r^2 - \omega_i^2)^{1/2}(\omega_r\hat{\mathbf{u}}_r + i\omega_i\hat{\mathbf{u}}_i)\cdot\mathbf{K}/\omega] \tag{15.84}$$

becomes a quark-like boost without handedness.[209]

---

[208] We say "quark-like" because we obtain a quark-like right-handed Dirac-like equation with complex spatial momentum terms using these boosts later.
[209] We say "quark-like" because we obtain a quark-like Dirac-like equation with complex spatial momentum terms using these boosts later.

15.11.6 "Dirac-like" Boosts

If we let

$$\hat{\mathbf{u}}_\mathbf{i} = \hat{\mathbf{u}}_\mathbf{r} \equiv \hat{\mathbf{u}} \tag{15.85}$$

so that the vector $\hat{\mathbf{u}}_\mathbf{i}$ is parallel to $\hat{\mathbf{u}}_\mathbf{r}$, and let

$$\omega_\mathbf{i} = 0 \tag{15.86}$$

then $\Lambda_C(\mathbf{v_c})$ becomes a Dirac-like boost:[210]

$$\Lambda = \exp[i\omega_\mathbf{r}\hat{\mathbf{u}}_\mathbf{r}\cdot\mathbf{K}] \tag{15.87}$$

This boost can be used to generate the free Dirac equation.

[210] We say "Dirac-like" because we obtain a Dirac equation using this boost later.

16. From a 16-Dimensional Complex General Relativity to the Complex Lorentz Group

In this chapter we develop a contingent space-time formalism with 16 complex dimensions that leads to a derivation of The Standard Model. This approach seems reasonable and enables us to develop a complete framework for the derivation.

However we could have started from a 4-dimensional complex space-time based on GL(4) combined with a GL(4) group space (giving GL(4)⊗GL(4)) that leads to the derivation of The Standard Model. The reader who wishes to skip to the beginning postulates for the derivation based on GL(4)⊗GL(4) can skip to section 16.3.

16.1 Complex, Sixteen Dimensional General Relativity (Optional)

The current climate of physics envisions a unified theory of the Standard Model and Quantum Gravity. Therefore it makes sense to define a General Relativistic framework for The Standard Model derivation realizing that a flat space-time approximation will be used in the derivation, and so the General Relativistic framework is not of direct significance in The Standard Model's derivation. Nevertheless, it will become evident that the flat space-time that ultimately results from this approach directly implies crucial Standard Model features such as Parity Violation, the form of the ElectroWeak sector, and color SU(3).

Postulate 16.1. The universe has sixteen complex dimensions and is described by a curved, 16-dimensional complex extension of 4-dimensional General Relativity. Complex, 16-dimensional General Relativity is invariant under complex coordinate transformations of the form:

$$X^{\mu} = f^{\mu}(X'^{0}, X'^{1}, ..., X'^{15}) \tag{16.1}$$

where X'^{μ} denotes an alternate complex coordinate system that is related to X^{μ} by continuous, holomorphic functions f^{μ}.

Postulate 16.2. We define an invariant interval[211] *with a metric $g_{\mu\nu}$ where*

$$d\tau^2 = g_{\mu\nu}dX^{\mu}dX^{\nu} \tag{16.2}$$

Note one important consequence of eq. 16.2 is that $g_{\mu\nu}$ is symmetric, $g_{\mu\nu} = g_{\nu\mu}$, in complex 16-dimensional space-time.[212] Under complex coordinate transformations (eq. 16.1) the complex interval $d\tau^2$ is invariant. The transformed vectors dX^{μ} are given by

$$dX^{\mu} = dX'^{\nu}\, \partial f^{\mu}(X'^{0}, X'^{1}, X'^{2}, X'^{3})/\partial X'^{\nu} \tag{16.3}$$

or, more simply,

$$dX^{\mu} = dX'^{\nu}\, \partial X^{\mu}/\partial X'^{\nu} \tag{16.4}$$

by the rules of partial differentiation. (In this book we define holomorphic in a conventional manner to mean analytic in *some* domain. Thus we will say $g_{\mu\nu}$, as we define it, is an holomorphic function of X^{μ}. Nevertheless $g_{\mu\nu}$ can have singularities, branch points, etc.) Since the components of each X^{μ} vector are complex variables, we can define the partial derivative of any function, or tensor, with respect to a coordinate X^{ν} to be:

$$\partial f(X)/\partial X^{\nu} = \lim_{w^{\nu} \to 0} [f(X^{\nu} + w^{\nu}) - f(X^{\nu})]/w^{\nu} \tag{16.5}$$

[211] Note that we use complex variable coordinates (and not their complex conjugates) throughout the development. This enables $g_{\mu\nu}$ (and other tensors such as the Riemann-Christoffel curvature tensor) to be a holomorphic (analytic) function of the coordinates – a feature that survives a change of coordinate system. Thus we obtain the benefits of analyticity. See Blaha (2004) for details of four complex dimensional General Relativity, particularly chapter 3.

[212] Another important consequence of eq. 16.2 is the matrix form of the Lorentz condition, $\Lambda_{16}^{T}G_{16}\Lambda_{16} = G_{16}$, uses the transpose just like the 4-dimensional Lorentz condition. In contrast the often used complex space-time interval $d\tau^2 = g_{\mu\nu}dX^{\mu}dX^{\nu *}$ does not imply $g_{\mu\nu}$ is symmetric and would have to use a different "Lorentz condition" $\Lambda_{16}^{\dagger}G_{16}\Lambda_{16} = G_{16}$ where $\Lambda_{16}^{\dagger}$ is the hermitian conjugate of Λ_{16}.

with no implied summation over ν and with the other coordinates X^{μ} held fixed when $\mu \neq \nu$. The only non-zero component of the vector w^{μ} is the νth component w^{ν} which is an arbitrary complex quantity that approaches zero in the limit. From a geometric viewpoint, the point $X^{\nu} + w^{\nu}$ can approach X^{ν} along any arbitrary curve in the complex X^{ν} plane in eq. 16.5. The limit in eq. 16.5 must exist and must be independent of the path along which $X^{\nu} + w^{\nu}$ approaches X^{ν}. The quantity $f(X^{\mu})$ is *differentiable* at the point X^{μ} if all first order partial derivatives with respect to all components of X^{μ} exist at the point X^{μ}.

Since the definition of a complex partial derivative is formally identical to that of a partial derivative with respect to a real variable we can make the following observation.

Observation: *All the formal rules of complex partial differentiation are the same as those of real partial differentiation. Thus the differentiation of sums, products, quotients and so on are formally the same as in the real variable case. In particular, a complex partial derivative will coincide with the corresponding real partial derivative on the "real axis" (i.e. for all X^{ν} real).*

Thus we are saved the labor of recalculating the equations of Riemannian geometry for the complex case although we now have to contend with sixteen dimensions.

Under complex coordinate transformations the metric tensor transforms as a covariant second rank tensor:

$$g_{\mu\nu} = g'_{\alpha\beta}\,\partial X'^{\alpha}/\partial X^{\mu}\;\partial X'^{\beta}/\partial X^{\nu} \tag{16.6}$$

$d\tau^2$ is invariant under complex coordinate transformations:

$$d\tau^2 = g_{\mu\nu}dX^{\mu}dX^{\nu} = d\tau'^2 = g'_{\mu\nu}dX'^{\mu}dX'^{\nu} \tag{16.7}$$

The inverse of $g_{\mu\nu}$, denoted $g^{\mu\nu}$, satisfies

$$g_{\alpha\nu}\,g^{\nu\beta} = \delta_{\alpha}^{\ \beta} \tag{16.8}$$

where $\delta_\alpha^{\ \beta}$ is the Kronecker delta function ($\delta_\alpha^{\ \beta} = 1$ if $\alpha = \beta$ and is zero otherwise), and transforms as a contravariant tensor under a complex coordinate transformation:

$$g'^{\mu\nu} = g^{\alpha\beta}\, \partial X'^{\mu}/\partial X^{\alpha}\ \partial X'^{\nu}/\partial X^{\beta} \tag{16.9}$$

The Principle of Equivalence in sixteen complex dimensions is somewhat problematic. It requires a definition of a local, flat space-time metric that is the analog of the 4-dimensional real space-time metric of Einstein's General Relativity Theory. The motion of a particle, moving under the influence of gravitational forces only, must be described as straight line motion in a free falling coordinate system x^μ where

$$d^2x^\mu/d\tau^2 = 0 \tag{16.10}$$

with

$$d\tau^2 = \eta_{16\mu\nu}dx^\mu dx^\nu \tag{16.11}$$

It is reasonable to assume the metric tensor $\eta_{16\mu\nu}$ in rectangular coordinates consists of diagonal entries with the values ±1. In light of the reduction of the 16-dimensional Lorentz group L_{16} eventually to the exterior product of two Lorentz groups in four dimensions $L_4 \otimes L_4$ in section 3.3, and the known flat space-time metric $\eta_{4\mu\nu}$ of our 4-dimensional space-time, we define the 16-dimensional flat space-time metric to be the outer product of the 4-dimensional metric $\eta_{4\mu\nu}$ with itself. (Note: $\eta_{400} = 1$; $\eta_{4ij} = -\delta_{ij}$ for i, j = 1, 2, 3; and $\eta_{4\mu\nu} = 0$ if $\mu \neq \nu$.) Thus,

Postulate 16.3. We assume a 16-dimensional Equivalence Principle: At any point one can transform to the coordinates of a locally flat space-time where

$$g'^{\mu\nu} = \eta_{16}^{\ \alpha\beta}\, \partial X'^{\mu}/\partial X^{\alpha}\ \partial X'^{\nu}/\partial X^{\beta} \tag{16.9a}$$

with

$$\begin{aligned}\eta_{16\mu\nu} &= diag(1, -1, -1, -1, -1, 1, 1, 1, -1, 1, 1, 1, -1, 1, 1, 1) \\ &\equiv \eta_{16}^{\ \mu\nu}\end{aligned} \tag{16.12}$$

The development of complex, 16-dimensional General Relativity proceeds from this point along the lines of the development of complex 4-dimensional General Relativity in Blaha (2004). We will not pursue it here because of the similarity and also because our goal is the derivation of the flat space-time Standard Model (for which this development is irrelevant.)

16.2 Flat Complex 16-Dimensional Space-time; GL(16), U(16), and L(16) (Optional)

The general linear group GL(16) in sixteen complex dimensions consists of all non-singular linear coordinate transformations with complex coefficients. The $\gamma\lambda$-matrix elements of the 256 generators $M_{\alpha\beta}$ of this group have the form:

$$(M_{\alpha\beta})_{\gamma\lambda} = \delta_{\alpha\gamma}\delta_{\beta\lambda} \qquad (16.13)$$

where $\delta_{\alpha\beta} = 1$ if $a = \beta$ and is zero otherwise, and where $a, \beta, \gamma, \lambda = 0, \ldots,$ 15. The generators satisfy the commutation relation

$$[M_{\alpha\beta}, M_{\kappa\lambda}] = \delta_{\beta\kappa}M_{\alpha\lambda} - \delta_{\alpha\lambda}M_{\kappa\beta} \qquad (16.14)$$

GL(16) group transfomations can be expressed in the form:

$$W(a) = \exp(ia^{\alpha\beta}M_{\alpha\beta}) = \exp(i\eta_{16}{}^{\alpha\gamma}\eta_{16}{}^{\beta\delta}a_{\alpha\beta}M_{\gamma\delta}) \qquad (16.15)$$

where the constants $a^{\alpha\beta}$ are complex numbers.

16.2.1 Relation of GL(16) to U(16)

The unitary group in 16 dimensions U(16) is based on the unitary condition $U^{\dagger}U = I$. Its 256 generators can be defined in terms of the GL(16) generators by the symmetric matrices:

$$U^{s}{}_{\alpha\beta} = M_{\alpha\beta} + M_{\alpha\beta} \qquad (16.16a)$$

and by the anti-symmetric matrices:

$$U^{a}{}_{\alpha\beta} = i(M_{\alpha\beta} - M_{\alpha\beta}) \qquad (16.16b)$$

for $\alpha \neq \beta$, and, for diagonal elements,

$$U_{\alpha\alpha} = M_{\alpha\alpha} \tag{16.17}$$

Linear combinations of the matrices, eqs. 16.16 – 16.17, with real coefficients generate all elements of the Lie algebra of U(16) while linear combinations of the matrices, eqs. 16.16 – 16.17, with *complex* coefficients generate all elements of the Lie algebra of GL(16). Thus the groups U(16) and GL(16) are intimately related.

The general form of a unitary transformation is

$$U(a) = \exp(ia^k U_k) \tag{16.18}$$

where the constants a^k are real and k is a numbering of the generators defined by eqs. 16.16 – 16.17. If the a^k are allowed to be complex, then the set of U(a) are the elements of the GL(16) group.

16.2.2 The Complex, 16-Dimensional Lorentz Group

The coordinates of inertial reference frames in sixteen complex dimensions are related by transformations of the complex, 16-dimensional inhomogeneous Lorentz group, denoted L(16). To determine the form of the L(16) generators we consider infinitesimal "rotations" and translations. We denote translations by d^μ and homogeneous "rotations" by $\Lambda^\alpha{}_\beta$. The identity rotation is $\Lambda_{16}{}^\alpha{}_\beta = \delta^\alpha{}_\beta$. An infinitesmal rotation can be expresssed in the form

$$\Lambda_{16}{}^\alpha{}_\beta = \delta^\alpha{}_\beta + \omega^\alpha{}_\beta \tag{16.19}$$

The defining feature of Lorentz transformations is

$$\Lambda_{16}{}^T G_{16}\,\Lambda_{16} = G_{16} \tag{16.20}$$

where Λ_{16} is a Lorentz transformation expressed in matrix form, $\Lambda_{16}{}^T$ is the transpose of Λ_{16}, and G_{16} is the metric $\eta_{16\mu\nu}$ expressed as a (diagonal) matrix (eq. 16.12). In the case of infinitesimal rotations eq. 16.20 becomes

$$(\delta^\alpha{}_\mu + \omega^\alpha{}_\mu)\eta_{16\alpha\beta}\,(\delta^\beta{}_\nu + \omega^\beta{}_\nu) = \eta_{16\mu\nu} \tag{16.21}$$

Using $\eta_{16\alpha\beta}$ to lower indices and $\eta_{16}{}^{\alpha\beta}$ to raise indices we have

$$\omega_{\beta\mu} = \eta_{16\alpha\beta}\omega^{\alpha}{}_{\mu} \tag{16.22}$$

so that eq. 16.20 implies the antisymmetry of $\omega_{\mu\beta}$

$$\omega_{\mu\beta} + \omega_{\beta\mu} = 0 \tag{16.23}$$

to leading order in ω.

The infinitesimal, unitary, linear Hilbert space transformation corresponding to eq. 16.19 to which we add a translation by d_μ is

$$U(\omega, d) = I + \tfrac{1}{2}\, i\, \omega_{\alpha\beta}J^{\alpha\beta} - id_\mu P^\mu \tag{16.24}$$

where the unitarity of U requires the 120 operators $J^{\alpha\beta}$ and 16 operators P^μ to be hermitian, and eq. 16.23 requires $J^{\alpha\beta}$ to be antisymmetric in α and β. After some further conventional considerations the following complex, 16-dimensional Poincaré group generator algebra commutation relations result with the metric $\eta_{16\mu\nu}$ and its inverse $\eta_{16}{}^{\mu\nu}$.

$$\begin{aligned}
&[J^{\alpha\beta}, J^{\kappa\lambda}] = i(\eta_{16}{}^{\alpha\kappa}J^{\beta\lambda} + \eta_{16}{}^{\lambda\alpha}J^{\kappa\beta} - \eta_{16}{}^{\beta\kappa}J^{\alpha\lambda} - \eta_{16}{}^{\lambda\beta}J^{\kappa\alpha})\\
&[P^\alpha, J^{\kappa\lambda}] = i(\eta_{16}{}^{\alpha\lambda}P^\kappa - \eta_{16}{}^{\alpha\kappa}P^\lambda)\\
&[P^\alpha, P^\kappa] = 0
\end{aligned} \tag{16.25}$$

The momentum operators are

$$\mathbf{P} = (P^0, P^1, \ldots , P^{15}) \tag{16.26}$$

The boost 15-vector is

$$\mathbf{K} = (\, J^{10}, J^{20}, \ldots , J^{150}) \tag{16.27}$$

The angular momentum 15-vector is

$$\mathbf{J} = (\, J^{1415}, J^{151}, J^{12}, J^{23}, J^{34}, \ldots , J^{1314}) \tag{16.28}$$

The form of an inhomogeneous complex 16-dimensional Lorentz, Hilbert space transformation is

$$U(\mathbf{v}, \boldsymbol{\theta}, \mathbf{d}) = \exp[i\mathbf{a}\cdot\mathbf{K} + i\boldsymbol{\theta}\cdot\mathbf{J} - i\mathbf{d}\cdot\mathbf{P}] \tag{16.29}$$

where **a**, **θ** and **d** are complex.

We leave the development of the commutation relations of the Lorentz transformation generators as an exercise for the reader.

The remainder of the development of Special and General Relativity in 16 complex dimensions is a generalization of the development of complex, 4-dimensional General and Special Relativity of Blaha (2004) to which the reader is referred.

16.3 GL(4)⊗GL(4) Subgroup (Required)

The postulates, and discussions, of sections 16.1 and 16.2 are not necessary to our derivation of the Standard Model. However they seem to furnish a natural setting within which the derivation follows smoothly.

The following postulate appears to be the proper starting point for the postulates required for the derivation.

Postulate 16.4. We assume that GL(16) is simply broken to GL(4)⊗GL'(4) through some mechanism and this group is the basis of the form of the Standard Model.[213]

Variations[214] on this postulate are possible that would also lead to the Standard Model.

The mechanism for the breakdown to GL(4)⊗GL'(4) remains to be determined. Knowledge of its nature is not required for our derivation. So we leave its determination to the future.

Taking the simplest approach to the breakdown of GL(16) to GL(4)⊗GL'(4) we assume:

Postulate 16.5. After breakdown, a complex 16-dimensional vector can be represented as the outer product of two complex 4-dimensional vectors

---

[213] The prime character will be used to distinguish the GL(4) factors. The GL(4) factor's coordinate space representation collapses to the Minkowski space-time with which we are familiar. GL'(4) denotes another GL(4) group. GL'(4)'s coordinate space representation collapses to a point completely leaving only the fermion generations and mixing matrices as its residue.

[214] For example, alternate postulate: GL(16) is broken to GL(4)⊗U(4). This postulate seems less satisfactory to the author since the groups in the product are different, and the author feels that there ought to be an initial symmetry between the group leading to our space-time and the group leading to the generations.

$$x^{\mu} \rightarrow y^{i} y'^{j} \qquad (16.30)$$

where μ ranges from 0 to 15, and i, j range from 0 through 3. The 4-dimensional metrics are assumed to be the outer product of the Lorentz metric

$$\eta_{16\mu\nu} \rightarrow diag(1, -1, -1, -1) \otimes diag(1, -1, -1, -1) \qquad (16.31)$$

$$= (\eta_4) \otimes (\eta'_4)$$

using a matrix notation.

The generators and group elements of each GL(4) group can be expressed in the form of eqs. 16.13 – 16.16 using 4×4 matrices; and can be related to U(4) with expressions similar to eqs. 16.16 – 16.18 with 4×4 matrix rows and columns.

16.4 The Complex, 4-Dimensional Lorentz Group, L(C)

In this section we will consider a subgroup of GL(4) that is a form of complex, 4-dimensional Lorentz group that is conventionally labeled L(C). We will define a 4×4 matrix representation for GL(4)'s sixteen generators:

$$(M_{\alpha\beta})_{\gamma\lambda} = \delta_{\alpha\gamma}\delta_{\beta\lambda} \qquad (16.32)$$

where $\delta_{\alpha\beta} = 1$ if $\alpha = \beta$ and is zero otherwise, and where $\alpha, \beta, \gamma, \lambda = 0, \ldots,$ 3. The generators satisfy the commutation relation (as before)

$$[M_{\alpha\beta}, M_{\kappa\lambda}] = \delta_{\beta\kappa}M_{\alpha\lambda} - \delta_{\alpha\lambda}M_{\kappa\beta} \qquad (16.33)$$

We define the standard boost generator vector **K**'s components with

$$K^{i} = -i(M^{0i} + M^{i0}) \qquad (16.34)$$

for i = 1, 2, 3 where $\eta_4^{\mu\nu}$ is used to raise indices in the conventional way. The rotation generator vector **J**'s components are

$$J^{i} = -i\epsilon^{ijk}M_{jk} \qquad (16.35)$$

where i, j, k = 1,2,3.

These generators satisfy the familiar Lorentz generator commutation relations

$$\begin{aligned} [K_i, K_j] &= -i\varepsilon_{ijk}J_k \\ [J_i, J_j] &= i\varepsilon_{ijk}J_k \\ [J_i, K_j] &= i\varepsilon_{ijk}K_k \end{aligned} \tag{16.36}$$

The L(C) group transformations in 4-dimensional complex coordinate space are

$$\Lambda_C = \exp[i\mathbf{a}\cdot\mathbf{K} + i\boldsymbol{\theta}_c\cdot\mathbf{J}] \tag{16.37}$$

where **a** and $\boldsymbol{\theta}_c$ are complex vectors. Upon adding complex translations **d** we obtain complex Poincaré group transformations:

$$P_C = \exp[i\mathbf{a}\cdot\mathbf{K} + i\boldsymbol{\theta}_c\cdot\mathbf{J} + i\mathbf{d}\cdot\mathbf{P}] \tag{16.38}$$

The fact that L(C) group transformations have a complex parameter **a** will be crucial in the next chapter when we define the four species of fermions: charged leptons, neutral leptons, up-type quarks and down-type quarks.

16.5 New Notation for GL(4) Generators Bringing Out the Subgroups: L(C), U(3) and U(1)

It is possible to define the 16 generators of GL(4) using a notation based on Pauli matrices. This approach brings out the subgroups contained within GL(4) in a direct way. The Pauli matrices σ_i are

$$\sigma_1 = \begin{bmatrix} 0 & 1 \\ 1 & 0 \end{bmatrix} \qquad \sigma_2 = \begin{bmatrix} 0 & -i \\ -i & 0 \end{bmatrix} \qquad \sigma_3 = \begin{bmatrix} 1 & 0 \\ 0 & -1 \end{bmatrix} \tag{16.39}$$

The outer product of two Pauli matrices can be used to produce 4×4 matrices which we will express using a functional notation:

$$\sigma_i\sigma''_k = \sigma_i(\sigma''_k) \tag{16.40}$$

where the double prime on the Pauli matrix differentiates between the two Pauli matrices in the outer product. (Note the Pauli matrices are <u>not</u>

matrix multiplied in eq. 16.40. Rather they represent the matrix elements $[\sigma_i]_{ab}[\sigma'_k]_{cd}$ where a, b, c, and d equal 1 or 2.) The functional form in eq. 16.40 implements the concept that σ''_k appears in each of the four matrix elements of σ_i multiplied by the value of that σ_i matrix element. For example

$$\sigma_2\sigma''_3 = \sigma_2(\sigma''_3)$$

$$= \begin{bmatrix} 0 & 0 & -i & 0 \\ 0 & 0 & 0 & i \\ i & 0 & 0 & 0 \\ 0 & -i & 0 & 0 \end{bmatrix} \qquad (16.41)$$

We also define $\sigma_0 = I$, the two-dimensional identity matrix, and use the notation $\sigma_0(\sigma''_i) = \sigma_0\sigma''_i$. For example,

$$\sigma_0(\sigma''_3) = \sigma_0\sigma''_3$$

$$= \begin{bmatrix} 1 & 0 & 0 & 0 \\ 0 & -1 & 0 & 0 \\ 0 & 0 & 1 & 0 \\ 0 & 0 & 0 & -1 \end{bmatrix} \qquad (16.42)$$

Similarly we define $\sigma_0(\sigma''_0) = \sigma_0\sigma''_0 = 4 \times 4$ identity matrix, and $\sigma_i(\sigma''_0) = \sigma_i\sigma''_0$.

The $\sigma_a(\sigma''_b)$ representation for a, b = 0, 1, 2, 3 can be expressed in terms of Dirac matrices as

$$\begin{aligned} \sigma_0(\sigma''_k) &= \tfrac{1}{2}\varepsilon_{ijk}\sigma^{ij} = \tfrac{1}{4}\, i\varepsilon_{ijk}[\gamma^i,\gamma^j] \\ \sigma_1(\sigma''_k) &= \gamma^0\gamma^k \\ \sigma_2(\sigma''_k) &= -i\gamma^k \\ \sigma_3(\sigma''_k) &= \tfrac{1}{2}\varepsilon_{ijk}\gamma^0\sigma^{ij} \\ \sigma_1(\sigma''_0) &= \gamma^5 \\ \sigma_2(\sigma''_0) &= -i\gamma^5\gamma^0 \\ \sigma_3(\sigma''_0) &= \gamma^0 \end{aligned} \qquad (16.43)$$

for k = 1, 2, 3. Furthermore the $\sigma_i(\sigma''_k)$ matrices are hermitean and their square is the 4×4 identity matrix:

$$[\sigma_i(\sigma''_k)]^\dagger = \sigma_i(\sigma''_k) \qquad (16.44)$$
$$[\sigma_i(\sigma''_k)]^2 = \sigma_0(\sigma''_0)$$

for i, k = 0,1,2,3.

We now define ten symmetric matrices

$$\Omega_{+cd} = ½[\sigma_c(\sigma''_d) + \sigma_d(\sigma''_c)] \qquad (16.45)$$

(with $c \le d$) in terms of $\sigma_i(\sigma''_k)$ (eq. 16.40). It also will be important to define the six antisymmetric matrices (with $c \le d$)

$$\Omega_{-cd} = ½[\sigma_c(\sigma''_d) - \sigma_d(\sigma''_c)] \qquad (16.46)$$

We now show that linear combinations of the matrices defined by eqs. 16.45 and 16.46 form representations of the generators of the Lorentz group, the U(3) group and a U(1) group. There are sixteen matrices with six matrices serving as Lorentz group generators, one matrix generator of U(1), and nine matrices serving as U(3) generators.

16.5.1 L(C) Lorentz Group Generators

Six matrices defined by eqs. 16.45 – 16.46 can be used to define boost and rotation generator matrices of the Lorentz group. We define the Lorentz boost generators (in our notation):

$$K_i = i\Omega_{-0i} \qquad (16.47)$$

and the three rotation generators

$$J_i = \Omega_{+0i} \qquad (16.48)$$

(Note the hermitean conjugate relations $K_i^\dagger = -K_i$, and $J_i^\dagger = J_i$ are the same as those of the corresponding Lorentz generators in our notation.)

Their commutation relations are

$$[K_i, K_j] = -i\varepsilon_{ijk}J_k$$
$$[J_i, J_j] = i\varepsilon_{ijk}J_k \qquad (16.49)$$
$$[J_i, K_j] = i\varepsilon_{ijk}K_k$$

16.5.2 U(1) Generator

The matrix $\sigma_0(\sigma'_0) = \Omega_{+00}$ serves as the generator of a U(1) symmetry.

16.5.3 U(3) Generators

The nine matrices Ω_{-ik} and Ω_{+ik} (for i, k = 1, 2, 3), together with linear combinations of Ω_{+0k} and Ω_{-0k}, form a 4 × 4 representation of the U(3) generators with the 4th row and 4th column consisting of 0's.

16.5.4 Gell-Mann SU(3) Generators

$$\begin{aligned}
F_1 &= \tfrac{1}{2}(\Omega_{+13} - \Omega_{-13} + \Omega_{+01} + \Omega_{-01}) \\
F_2 &= \tfrac{1}{2}(\Omega_{+23} - \Omega_{-23} + \Omega_{+02} + \Omega_{-02}) \\
F_3 &= \tfrac{1}{2}(\Omega_{+33} + \Omega_{+03} + \Omega_{-03}) \\
F_4 &= \tfrac{1}{2}(\Omega_{+10} + \Omega_{-10} + \Omega_{+13} + \Omega_{-13}) \\
F_5 &= \tfrac{1}{2}(\Omega_{+20} + \Omega_{-20} + \Omega_{+23} + \Omega_{-23}) \\
F_6 &= \tfrac{1}{2}(\Omega_{+11} + \Omega_{+22}) \\
F_7 &= \tfrac{1}{2}(\Omega_{+21} + \Omega_{-21} - \Omega_{+12} - \Omega_{-12}) \\
F_8 &= (2\sqrt{3})^{-1}(-\Omega_{+00} + 2\Omega_{+33} + \Omega_{+03} - 5\Omega_{-03})
\end{aligned} \tag{16.50}$$

They satisfy the standard SU(3) commutation relations:

$$[F_i, F_j] = i\, f_{ijk} F_k \tag{16.51}$$

where f_{ijk} are the totally antisymmetric SU(3) structure constants for i, j, k = 1, 2, …,8. Note: the Lorentz generators do not commute with the F_i.

$$\begin{aligned}
[F_i, K_j] &\neq 0 \\
[F_i, J_j] &\neq 0
\end{aligned} \tag{16.52}$$

16.5.5 Additional U(1) Generator needed for U(3)

$$F_9 = \tfrac{1}{4}(3\Omega_{+00} - \Omega_{+33} + 2\Omega_{+03}) \tag{16.53}$$

This generator consists of a 3×3 identity matrix with the element in the fourth row and column having the value 0.

We note also

$$[F_i, F_9] = 0 \qquad (16.54)$$
$$[F_9, K_j] \neq 0$$
$$[F_9, J_j] \neq 0$$

for j = 1, 2, 3 and i = 1, 2, …, 8.

16.6 The GL(4) Subgroup Transformations have Complex Coefficients

The group elements of the L(C), U(3) and U(1) algebras defined in section 16.5 are transformations with complex coefficients since these are subgroups of GL(4) whose transformations have the general form of eq. 16.15 where $a^{\alpha\beta}$ are complex numbers.

Thus the L(C) subgroup of GL(4) has the form of eq. 16.37 with complex constant vectors **a** and $\boldsymbol{\theta}_c$. We now postulate

Postulate 16.6. The GL(4) group has a familiar coordinate space representation except that it is a complex space-time with a complex Lorentz group denoted L(C) whose elements Λ_C satisfy $\Lambda_C{}^T G_4 \Lambda_C = G_4$. L(C) is in turn broken to the real Lorentz group L of space-time that we physically observe.

The U(3) algebra of generators defined in section 16.5 is used with nine complex coefficients a_i to define elements of GL(3)

$$V_C(a) = \exp[ia_k F_k] \qquad (16.55)$$

(Note we don't distinguish between a_k and a^k in this discussion.) Alternately these same generators can be used to define elements of the unitary group U(3) if the coefficients a_i are real:

$$U_3(a) = \exp[ia_k F_k] \qquad (16.56)$$

Similarly the U(1) generator Ω_{+00} defines the U(1) group element

$$U_1(b) = \exp[ib\Omega_{+00}] \qquad (16.57)$$

if the constant b is real.

Since

$$[U_3(a), U_1(b)] = 0 \qquad (16.58)$$

for all real a and b, these elements can be used to form a U(3)⊗U(1) group consisting of the elements

$$U(a,b) = U_3(a)U_1(b) \qquad (16.59)$$

with the constants a_i and b real.

16.7 How GL(4) and GL'(4) Combine to Yield Four Fermion Species and Fermion Generations

Since the elements of the L(C) algebra and the U(3)⊗U(1) algebra do not commute, and since one cannot combine internal symmetries with space-time symmetries without violating "no go' theorems, we will use the L(C) boost transformations of GL(4) to generate the four fermion species: charged leptons, neutral leptons, up-type quarks and down-type quarks. We will use the U(3) subgroup (or U(4) subgroup) of the other GL(4) (denoted GL'(4)) to create the fermion generations.

Thus we postulate the source of the fermion generations with

Postulate 16.7. The GL'(4) group has a space-time that is reduced to a point. However it has a U(3) subgroup that gives the 3 known fermion generations, OR it may have 4 generations based on GL'(4)'s U(4) subgroup.

The pattern of breakdowns of these GL(4) groups is indicated in Fig. 16.1.

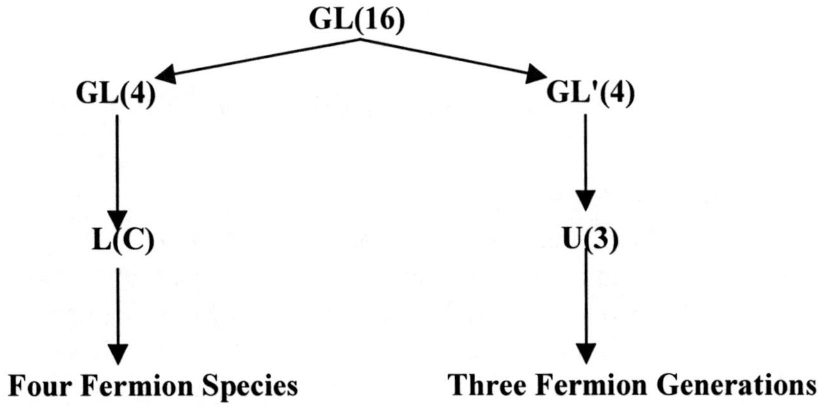

Figure 16.1. The hierarchy of groups that appear in the breakdown from GL(16) and lead to the known 3 generation fermion spectrum.

By having two GL(4) groups which separately generate the four species and the fermion generations we avoid issues associated with the fact that the L(C) and U(3) (or U(4)) subgroups of GL(4) do not commute with each other.

17. Fermion Particle Species

17.1 Four Fermion Species

One of the main features of the Standard Model is the presence of exactly four fermion species: charged lepton, neutral lepton, up-type quark, and down-type quark. The distinguishing features of these fermion species is their appearance in ElectroWeak doublets, their parity violating form of ElectroWeak interactions, and the presence of Strong interactions for quarks – the main qualitative difference between quarks and leptons.

In this chapter we will show how these features follow from the L(C) and U(3) subgroups of GL(4) and, as such, can justly be called geometrical features.[215]

We identify the free Dirac equation as the free dynamical equation of charged leptons such as electrons; a free tachyon equation as the free dynamical equation of neutral leptons (neutrinos); a free Dirac equation with complex 3-momenta as the free dynamical equation of up-type quarks; and a tachyon Dirac equation with complex 3-momenta as the free dynamical equation of down-type quarks.[216] These equations will be derived from L(C) boosts applied to the Dirac equation of a particle at rest.

17.2 Deriving the Free Dynamical Equations of Motion for Fermions

The starting point for our derivation of the four species' dynamical equations is the homogeneous L(C) subgroup of GL(4). Its group elements can be expressed as

$$\Lambda_C = \exp[i\mathbf{a}\cdot\mathbf{K} + i\boldsymbol{\theta}_c\cdot\mathbf{J}] \qquad (17.1)$$

[215] The detailed discussion and proofs of the statements and equations in this chapter and the following chapters may be found in Blaha (2007b).

[216] An alternate fomulation wherein quarks have real energies and 3-momenta will also be considered later.

$$= \exp[i(\omega_r\hat{\mathbf{u}}_r + i\omega_i\hat{\mathbf{u}}_i)\cdot\mathbf{K} + i\boldsymbol{\theta}_c\cdot\mathbf{J}] \tag{17.2}$$

where the vectors **a** and $\boldsymbol{\theta}_c$ are complex 3-vectors. Eq. 17.2 expresses **a** in terms of complex constants: real unit vectors with real constants $\omega_r \geq 0$ and $\omega_i \geq 0$; and with $\hat{\mathbf{u}}_r$ and $\hat{\mathbf{u}}_i$ being real normalized 3-vectors such that $\hat{\mathbf{u}}_r\cdot\hat{\mathbf{u}}_r = 1 = \hat{\mathbf{u}}_i\cdot\hat{\mathbf{u}}_i$. The generators of the homogeneous group are **K** and **J** same as the real Lorentz group generators. Their matrix form is given by eqs. 16.34 and 16.35.

The general form of a pure L(C) boost can be expressed in terms of a complex relative velocity $\mathbf{v_c}$ between reference frames as

$$\Lambda_C(\mathbf{v}_c) \equiv \Lambda_C(\omega, \mathbf{v}_c) = \exp[i\omega\hat{\mathbf{w}}\cdot\mathbf{K}] \tag{17.3}$$

where

$$\omega = (\omega_r^2 - \omega_i^2 + 2i\omega_r\omega_i\,\hat{\mathbf{u}}_r\cdot\hat{\mathbf{u}}_i)^{1/2} \tag{17.4}$$

and

$$\hat{\mathbf{w}} = (\omega_r\hat{\mathbf{u}}_r + i\omega_i\hat{\mathbf{u}}_i)/\omega \tag{17.5}$$

Since $\hat{\mathbf{u}}_r\cdot\hat{\mathbf{u}}_r = 1 = \hat{\mathbf{u}}_i\cdot\hat{\mathbf{u}}_i$

$$\hat{\mathbf{w}}\cdot\hat{\mathbf{w}} = 1 \tag{17.6}$$

and the complex relative velocity is seen to be

$$\mathbf{v}_c = \hat{\mathbf{w}}\tanh(\omega) \tag{17.7}$$

The matrix form of proper (det $\Lambda_C = 1$) L(C) coordinate boosts is

$$\Lambda_C(\mathbf{v}_c) = \begin{bmatrix} \cosh(\omega) & -\sinh(\omega)\hat{w}_x & -\sinh(\omega)\hat{w}_y & -\sinh(\omega)\hat{w}_z \\ -\sinh(\omega)\hat{w}_x & 1 + (\cosh(\omega) - 1)\hat{w}_x^2 & (\cosh(\omega) - 1)\hat{w}_x\hat{w}_y & (\cosh(\omega) - 1)\hat{w}_x\hat{w}_z \\ -\sinh(\omega)\hat{w}_y & (\cosh(\omega) - 1)\hat{w}_x\hat{w}_y & 1 + (\cosh(\omega) - 1)\hat{w}_y^2 & (\cosh(\omega) - 1)\hat{w}_y\hat{w}_z \\ -\sinh(\omega)\hat{w}_z & (\cosh(\omega) - 1)\hat{w}_x\hat{w}_z & (\cosh(\omega) - 1)\hat{w}_y\hat{w}_z & 1 + (\cosh(\omega) - 1)\hat{w}_z^2 \end{bmatrix} \tag{17.8}$$

The free dynamical equations of the four fermion species will be generated by L(C) boosts of the free Dirac equation of a particle at rest with the *requirement that the time variable is real in the resulting field equations*. The procedure can most easily be performed in momentum space and the coordinate space version of the generated equation determined from the momentum space version. We therefore postulate:

Postulate 17.1. The dynamical equations of the allowed types of free fermions can be generated by L(C) boosts from the Dirac equation for a fermion at rest subject to the requirement that there be only one real time variable in each resulting dynamical differential equation.

Postulate 17.2. Each free fermion field has the simplest possible lagrangian from which the canonically conjugate momentum, the energy-momentum tensor, the equal-time (or equal light-front) anti-commutation relations, and other physically important quantities can be derived.

17.2.1 Dirac Equation – Free Charged Lepton Dynamical Equation

In this section we describe a well-known method of obtaining the Dirac equation by conventional Lorentz boosts[217] of the spinor wave function of a particle at rest. This section amplifies the discussion of section 13.6 (eq. 13.14 in particular) where we showed that the Operator Logic spinor formalism leads to the Dirac equation if we introduce coordinates and perform Lorentz boosts. *Thus the most fundamental aspects of Physics can be viewed as Operator Logic statements combined with space-time coordinates and Lorentz transformations. We will see that the four fermion species naturally emerge from this basis and that The Standard Model follows upon the further introduction of local vector boson interactions.*

A generic positive energy plane wave solution of the Dirac equation for a particle at rest with rest energy m is

$$\psi(x) = e^{-imt}w(0) \tag{17.9}$$

with w(0) a four component spinor column vector. It satisfies the momentum space Dirac equation for a particle at rest:[218]

---

[217] A Lorentz boost is a special case of an L_C boost with $\omega_i = 0$ and thus a real $\omega = \omega_r$ and a real $\hat{\mathbf{w}}$.

[218] Compare to eq. 13.14.

$$(m\gamma^0 - m)e^{-imt}w(0) = 0 \qquad (17.10)$$

The 4 x 4 matrix form of a Lorentz transformation with relative velocity **v** of the Dirac matrices is

$$S^{-1}(\Lambda(\mathbf{v}))\gamma^{\nu}S(\Lambda(\mathbf{v})) = \Lambda^{\nu}{}_{\mu}(\mathbf{v})\gamma^{\mu} \qquad (17.11)$$

where $S(\Lambda(\mathbf{v}))$ is

$$S(\Lambda(\mathbf{v})) = \exp(-i\omega\sigma_{0i}v_i/(2|\mathbf{v}|)) = \exp(-\omega\gamma^0\boldsymbol{\gamma}\cdot\mathbf{v}/(2|\mathbf{v}|))$$

$$= \cosh(\omega/2)I + \sinh(\omega/2)\gamma^0\boldsymbol{\gamma}\cdot\mathbf{p}/|\mathbf{p}| \qquad (17.12)$$

with real $\omega = \text{arctanh}(|\mathbf{v}|)$ and real **v** where $\cosh(\omega/2) = [(E+m)/(2m)]^{\frac{1}{2}}$ and $\sinh(\omega/2) = |\mathbf{p}|[2m(E+m)]^{-\frac{1}{2}}$. Also

$$S^{-1}(\Lambda(\mathbf{v})) = \gamma^0 S^{\dagger}(\Lambda(\mathbf{v}))\gamma^0 = \exp(\omega\gamma^0\boldsymbol{\gamma}\cdot\mathbf{v}/(2|\mathbf{v}|))$$

$$= \cosh(\omega/2)I - \sinh(\omega/2)\gamma^0\boldsymbol{\gamma}\cdot\mathbf{p}/|\mathbf{p}| \qquad (17.13)$$

If we now apply $S(\Lambda(\mathbf{v}))$ to the momentum space Dirac equation of a particle at rest (eq. 17.10) we find

$$\begin{aligned} 0 &= S(\Lambda(\mathbf{v}))(m\gamma^0 - m)e^{-imt}w(0) \\ &= [mS(\Lambda(\mathbf{v}))\gamma^0 S^{-1}(\Lambda(\mathbf{v})) - m]S(\Lambda(\mathbf{v}))w(0) \end{aligned}$$

A straightforward evaluation shows

$$mS(\Lambda(v))\gamma^0 S^{-1}(\Lambda(v)) = g_{\mu\nu}p^{\mu}\gamma^{\nu} = \not{p} \qquad (17.14)$$

where $p^0 = (p^2 + m^2)^{\frac{1}{2}}$, $\mathbf{p} = \gamma m\mathbf{v}$, and $p = |\mathbf{p}|$. In addition we define

$$S(\Lambda(v))w(0) = w(p) \qquad (17.15)$$

a positive energy Dirac spinor. Therefore the Dirac equation in momentum space has the familiar form:

$$(\not{p} - m)e^{-ip\cdot x}w(p) = 0 \qquad (17.16)$$

where the exponential factor, mt, is also boosted to p·x. Eq. 17.16 implies the free, coordinate space Dirac equation:

$$(i\gamma^{\mu}\partial/\partial x^{\mu} - m)\psi(x) = 0 \qquad (17.17)$$

We identify this equation as the dynamical equation of a free charged lepton.

17.2.2 L(C) Complex Boost for a Dirac Spinor

The form of the L(C) spinor boost transformation corresponding to the coordinate transformation eq. 17.8 is:

$$\begin{aligned} S_C(\omega, \mathbf{v_c}) \equiv S_C &= \exp(-i\omega\sigma_{0k}\hat{w}_k/2) = \exp(-\omega\gamma^0\boldsymbol{\gamma}\cdot\hat{\mathbf{w}}/2) \\ &= \cosh(\omega/2)I + \sinh(\omega/2)\gamma^0\boldsymbol{\gamma}\cdot\hat{\mathbf{w}} \end{aligned} \qquad (17.18)$$

with $\mathbf{v_c}$ and $\hat{\mathbf{w}}$ defined by eqs. 17.7 and 17.5 respectively. The inverse transformation is

$$\begin{aligned} S_C^{-1}(\omega, \mathbf{v_c}) &= \gamma^2\gamma^0 K^{-1} S_C^{\dagger} K \gamma^0\gamma^2 = \gamma^2\gamma^0 S_C^{\,T}\gamma^0\gamma^2 = \exp(\omega\gamma^0\boldsymbol{\gamma}\cdot\hat{\mathbf{w}}/2) \\ &= \cosh(\omega/2)I - \sinh(\omega/2)\gamma^0\boldsymbol{\gamma}\cdot\hat{\mathbf{w}} \end{aligned} \qquad (17.19)$$

where the superscript T denotes the transpose and K is the complex conjugation operator (that also appears in the time-reversal operator). Note that S_C is not unitary just as $S(\Lambda(\mathbf{v}))$ (eq. 17.12) is not unitary.

We now apply an L(C) spinor boost (eq. 17.18) to the Dirac equation for a particle at rest (eqs. 13.14 (Operator Logic statement) and 17.10) in this more general case of complex ω and $\hat{\mathbf{w}}$.

$$\begin{aligned} 0 &= S_C(\omega, \mathbf{v_c}))(m\gamma^0 - m)e^{-imt}w(0) \\ &= [mS_C\gamma^0 S_C^{-1} - m]e^{-imt}S_C w(0) \end{aligned} \qquad (17.20)$$

where $S_C = S_C(\omega, \mathbf{v_c})$. After some algebra we find

$$mS_C\gamma^0 S_C^{-1} = m[\cosh(\omega)\gamma^0 - \sinh(\omega)\boldsymbol{\gamma}\cdot\hat{\mathbf{w}}] \qquad (17.21)$$

We will use these complex boosts to generate the other fermion species' Dirac-like equations.

17.2.3 Tachyon Dirac Equation – Free Neutral Lepton Dynamical Equation

The development of the L(C) spinor boost transformation (eqs. 17.18 – 17.21) leads to two possible forms of the tachyon Dirac equation. One form will lead to a free lagrangian theory with physical left-handed neutrinos. The other form leads to a free lagrangian theory with physical right-handed neutrinos.

17.2.3.1 Form Leading to a Left-Handed Neutrino Theory

If the real and imaginary relative vectors parts of $\hat{\mathbf{w}}$, namely $\hat{\mathbf{u}}_r$ and $\hat{\mathbf{u}}_i$, are parallel, then $\hat{\mathbf{u}}_r \cdot \hat{\mathbf{u}}_i = 1$ and

$$\omega = \omega_r + i\omega_i \qquad (17.22)$$

Eqs. 17.18 and 17.19 then imply

$$mS_C\gamma^0 S_C^{-1} = m[\cosh(\omega_r)\cos(\omega_i) + i\sinh(\omega_r)\sin(\omega_i)]\gamma^0 -$$

$$- m[\sinh(\omega_r)\cos(\omega_i) + i\cosh(\omega_r)\sin(\omega_i)]\boldsymbol{\gamma}\cdot\hat{\mathbf{u}}_r \qquad (17.23)$$

or

$$mS_C\gamma^0 S_C^{-1} = \cos(\omega_i)\gamma\cdot p_r + i\sin(\omega_i)\gamma\cdot p_i \qquad (17.24)$$

where

$$p_r^0 = m\cosh(\omega_r) \qquad p_i^0 = m\sinh(\omega_r) \qquad (17.25)$$

and

$$\mathbf{p}_r = m\hat{\mathbf{u}}_r\sinh(\omega_r) \qquad \mathbf{p}_i = m\hat{\mathbf{u}}_r\cosh(\omega_r) \qquad (17.26)$$

If $\omega_i = 0$, then we recover the momentum space Dirac equation eq. 17.16 (or Operator Logic statement eq. 13.14). If $\omega_i = \pi/2$, then we obtain the left-handed momentum space tachyon equation:

$$mS_C\gamma^0 S_C^{-1} = i\gamma\cdot p_i \qquad (17.27)$$

and the tachyon energy and momentum expressions

$$\mathbf{p} = m\mathbf{v}\gamma_s \qquad E = m\gamma_s \qquad (17.28)$$

where $\sinh(\omega) = \gamma_s = (\beta^2 - 1)^{-1/2}$ with $\beta = v/c > 1$. Also

$$S_C w(0) = w_C(p) \tag{17.29}$$

is a tachyon spinor. (See Blaha (2007b).)

The momentum space tachyonic Dirac equation is

$$(i\not{p} - m)e^{ip\cdot x}w_T(p) = 0 \tag{17.30}$$

where p·x = Et – **p·x** after performing a corresponding L(C) boost in the exponential factor. Thus the positive energy wave is transformed into a negative energy wave by the superluminal boost transformation.

If we apply $i\not{p}$ to we find the tachyon mass condition is satisfied

$$-E^2 + \mathbf{p}^2 = m^2 \tag{17.31}$$

Transforming back to coordinate space we obtain the "left-handed" *tachyonic Dirac equation*:

$$(\gamma^\mu \partial/\partial x^\mu - m)\psi_T(x) = 0 \tag{17.32}$$

17.2.3.2: Form Leading to a Right-handed Neutrino Theory

If the real and imaginary relative vectors parts of $\hat{\mathbf{w}}$, $\hat{\mathbf{u}}_\mathbf{r}$ and $\hat{\mathbf{u}}_\mathbf{i}$, are anti-parallel $\hat{\mathbf{u}}_\mathbf{r} = -\hat{\mathbf{u}}_\mathbf{i}$, then $\hat{\mathbf{u}}_\mathbf{r}\cdot\hat{\mathbf{u}}_\mathbf{i} = -1$ and

$$\omega = \omega_r - i\omega_i \tag{17.33}$$

Then

$$mS_C\gamma^0 S_C^{-1} = m[\cosh(\omega_r)\cos(\omega_i) - i\sinh(\omega_r)\sin(\omega_i)]\gamma^0 -$$

$$- m[\sinh(\omega_r)\cos(\omega_i) - i\cosh(\omega_r)\sin(\omega_i)]\boldsymbol{\gamma}\cdot\hat{\mathbf{u}}_\mathbf{r} \tag{17.34}$$

or

$$mS_C\gamma^0 S_C^{-1} = \cos(\omega_i)\gamma\cdot p_r - i\sin(\omega_i)\gamma\cdot p_i \tag{17.35}$$

where

$$p_r^0 = m\cosh(\omega_r) \qquad p_i^0 = m\sinh(\omega_r) \tag{17.36}$$

and

$$\mathbf{p_r} = m\hat{\mathbf{u}}_\mathbf{r}\sinh(\omega_r) \qquad \mathbf{p_i} = m\hat{\mathbf{u}}_\mathbf{r}\cosh(\omega_r) \tag{17.37}$$

If $\omega_i = \pi/2$, then we obtain the right-handed momentum space tachyon equation.[219]

$$(-\gamma^\mu\partial/\partial x^\mu - m)\psi_T(x) = 0 \qquad (17.38)$$

The sign difference between eqs. 17.32 and 17.38 is significant taking account of the required sign of the mass term in the lagrangian due to positivity requirements of the associated hamiltonian. It leads to the difference between a theory with physical interacting, left-handed neutrinos and a theory with physical interacting, right-handed neutrinos as we shall see.

17.2.4 Quark Dirac Equation – Free Up-type Quark Dynamical Equation

There are two other cases where we can obtain fermion dynamical equations with a *real* time variable and real energy. In one case we set $\hat{\mathbf{u}}_\mathbf{r}\cdot\hat{\mathbf{u}}_\mathbf{i} = 0$ and have a real ω.

If the real and imaginary relative vectors parts of $\hat{\mathbf{w}}$, namely $\hat{\mathbf{u}}_\mathbf{r}$ and $\hat{\mathbf{u}}_\mathbf{i}$, are perpendicular, $\hat{\mathbf{u}}_\mathbf{r}\cdot\hat{\mathbf{u}}_\mathbf{i} = 0$, then by eq. 17.4

$$\omega = (\omega_r^2 - \omega_i^2)^{1/2} \qquad (17.39)$$

Thus ω is either pure real ($\omega_r \geq \omega_i$) or pure imaginary ($\omega_r < \omega_i$).

The momentum space equation generated by the corresponding L(C) spinor boost (eqs. 17.12 and 17.13) is

$$\{m \cosh(\omega)\gamma^0 - m \sinh(\omega)\boldsymbol{\gamma}\cdot(\omega_r\hat{\mathbf{u}}_\mathbf{r} + i\omega_i\hat{\mathbf{u}}_\mathbf{i})/\omega - m\}e^{-ip.x}w_c(p) = 0 \qquad (17.40)$$

Defining the momentum 4-vector

$$p = (p^0, \mathbf{p}) \qquad (17.41)$$

where

$$p^0 = m \cosh(\omega) \qquad \mathbf{p} = \mathbf{p}_r + i\mathbf{p}_i \qquad (17.42)$$

with

---

[219] We note that $\gamma_s = (\beta^2 - 1)^{-1/2}$, if expressed in terms of ω, has a branch cut extending from <-∞, +∞> in the complex ω plane so values of ω with positive imaginary parts are physically different from values of ω with negative imaginary parts.

$$\mathbf{p_r} = m\omega_r\hat{\mathbf{u}}_r \sinh(\omega)/\omega \qquad \mathbf{p_i} = m\omega_i\hat{\mathbf{u}}_i \sinh(\omega)/\omega \tag{17.43}$$

$$\mathbf{p_r}\cdot\mathbf{p_i} = 0 \tag{17.44}$$

then we obtain a positive energy Dirac-like equation

$$[p\cdot\gamma - m]e^{-ip\cdot x}w_c(p) = 0$$

or

$$[p^0\gamma^0 - (\mathbf{p_r} + i\mathbf{p_i})\cdot\boldsymbol{\gamma} - m]e^{-ip\cdot x}w_c(p) = 0 \tag{17.45}$$

with a complex 3-momentum **p** and the 4-momentum mass shell condition:

$$p^2 = p^{0\,2} - \mathbf{p_r}\cdot\mathbf{p_r} + \mathbf{p_i}\cdot\mathbf{p_i} = m^2 \tag{17.46}$$

Note

$$|\mathbf{v_c}| = |\mathbf{p}|/p^0 = [(\mathbf{p_r} + i\mathbf{p_i})\cdot(\mathbf{p_r} + i\mathbf{p_i})]^{1/2}/p^0 = \tanh(\omega) \tag{17.47}$$

and so the Lorentz factor

$$\gamma = \cosh(\omega) \tag{17.48}$$

Eq. 17.45 is the momentum space equivalent of the wave equation

$$[i\gamma^0\partial/\partial t + i\boldsymbol{\gamma}\cdot(\nabla_r + i\nabla_i) - m]\psi_{Cu}(t, \mathbf{x_r}, \mathbf{x_i}) = 0 \tag{17.49}$$

where $\mathbf{x} = \mathbf{x_r} - i\mathbf{x_i}$, and where the grad operators ∇_r and ∇_i are with respect to $\mathbf{x_r}$ and $\mathbf{x_i}$ respectively. Since $\hat{\mathbf{u}}_r\cdot\hat{\mathbf{u}}_i = 0$, which in turn implies eq. 17.44, we see that there is a subsidiary condition on the wave function

$$\nabla_r\cdot\nabla_i\, \psi_{Cu}(t, \mathbf{x_r}, \mathbf{x_i}) = 0 \tag{17.50}$$

We will call the particles satisfying eqs. 17.49 and 17.50 *complexons.*

We note that eq. 17.49 is covariant under the real Lorentz group and eq. 17.50 can be easily put into covariant form since the difference of these 4-vectors squared (which is a real Lorentz group invariant):

$$[\gamma^0\partial/\partial t + i\boldsymbol{\gamma}\cdot(\nabla_r + i\nabla_i)]^2 - [\gamma^0\partial/\partial t + i\boldsymbol{\gamma}\cdot(\nabla_r - i\nabla_i)]^2 = 4\nabla_r\cdot\nabla_i.$$

We identify eq. 17.49 as the dynamical equation of a free, up-type quark.[220] Later we will show that the fourier solution of the equation has a SU(3) global symmetry.

17.2.5 Left-handed Quark Dirac Equation – Free Down-type Quark Dynamical Equation

In this case we set $\hat{\mathbf{u}}_\mathbf{r}\cdot\hat{\mathbf{u}}_\mathbf{i} = 0$. Then by eq. 17.4

$$\omega = (\omega_\mathbf{r}^2 - \omega_\mathbf{i}^2)^{1/2}$$

Thus ω again starts out either pure real ($\omega_\mathbf{r} \geq \omega_\mathbf{i}$) or pure imaginary ($\omega_\mathbf{r} < \omega_\mathbf{i}$). In this case we also choose ω real, and reset

$$\omega = (\omega_\mathbf{r}^2 - \omega_\mathbf{i}^2)^{1/2} \rightarrow \omega' = (\omega_\mathbf{r}^2 - \omega_\mathbf{i}^2)^{1/2} + i\pi/2 = \omega + i\pi/2$$

by adding $i\pi/2$ to ω in eq. 15.65 since ω is a free parameter and proceed as we did in the prior tachyon case.[221]. Then the resulting Lorentz boost

$$\Lambda_C = \exp[i((\omega_\mathbf{r}^2 - \omega_\mathbf{i}^2)^{1/2} + i\pi/2)(\omega_\mathbf{r}\hat{\mathbf{u}}_\mathbf{r} + i\omega_\mathbf{i}\hat{\mathbf{u}}_\mathbf{i})\cdot\mathbf{K}/\omega] \qquad (17.51)$$

becomes a quark-like left-handed boost. The resulting tachyon dynamical equation is

$$[\gamma^0\partial/\partial t + \gamma\cdot(\nabla_\mathbf{r} + i\nabla_\mathbf{i}) - m]\psi_{Cd}(x) = 0 \qquad (17.52)$$

with the constraint equation

$$\nabla_\mathbf{r}\cdot\nabla_\mathbf{i}\ \psi_{Cd}(t, \mathbf{x}_\mathbf{r}, \mathbf{x}_\mathbf{i}) = 0 \qquad (17.53)$$

We will call the particles satisfying eqs. 17.52 and 17.53 *tachyonic complexons.*

---

220 As we point out later in the chapter on experimental questions we have chosen to consider the possibility that the free dynamical equations for up-type and down-type quarks are the same as those for charged leptons and neutral leptons respectively, and the possibility that quarks are complexons. The global SU(3) symmetry implicit in complexons strongly suggested that quarks be identified with complexons.

221 Here again the choice of ω in eq. 17.51 leads to a left-handed quark ElectroWeak sector while the choice $\omega' = \omega - i\pi/2$ leads to a right-handed quark ElectroWeak sector.

We identify eq. 17.52 as the dynamical equation of a free, down-type, left-handed quark. Later we will show that the fourier solution of the equation also has an SU(3) global symmetry.

Thus we have seen how to generate thedynamical equations of the four fermion species from eq. 13.14 – a generic Operator Logic statement – which is equivalent to the dynamical equation of a fermion at rest. This transition is the key to the transition from Plato's Realm of Ideas to the Realm of Reality.

17.3 Free Lagrangians of the Four Fermion Species

In defining the lagragians for the four species that lead to their dynamical equations in the usual canonical way, we require the conventional quantum field theory feature that the hamiltonian derived from the lagrangian is hermitean. We will develop a separate lagrangian for each species and combine them later in the derivation of the form of the Standard Model. The interested readers can derive these lagragians for themselves or see their derivation in Blaha (2007b).

Since the Standard Model is formulated as a quantum field theory and calculations are often based on its path integral formulation it appears reasonable to make the following essentially mathematical postulates[222] (assumptions):

Postulate 17.3. The Standard Model is a quantum field theory with a lagrangian and a path integral formulation. The mathematics of these constructs is assumed to be correct.

Postulate 17.4. The free field approximation to the Standard Model is a canonical quantum field theory but not necessarily quantized on light-like surfaces.

17.3.1 Free Charged Lepton Lagrangian

The Dirac equation lagrangian is

$$\mathcal{L} = \bar{\psi}(i\gamma^{\mu}\partial/\partial x^{\mu} - m)\psi(x) \tag{17.54}$$

where

$$\bar{\psi} = \psi^{\dagger}\gamma^{0}$$

[222] Postulates will be numbered and appear in bold type as seen above.

and $\psi^\dagger$ is the hermitean conjugate of ψ.

17.3.2 Free Neutral Lepton Lagrangian

$$\mathcal{L}_T = \psi_T^{\,S}(\gamma^\mu\partial/\partial x^\mu - m)\psi_T(x) \qquad (17.55)$$

where

$$\psi_T^{\,S} = \psi_T^{\,\dagger}\, i\gamma^0\gamma^5 \qquad (17.56)$$

The peculiar form of the tachyon lagrangian is necessitated by the hermiticity of the hamiltonian calculated from it.

17.3.3 Free Up-type Quark Lagrangian

The lagrangian density of a free up-type complexon quark is

$$\mathcal{L}_{Cu} = \bar{\psi}_{Cu}(i\gamma^\mu D_\mu - m)\psi_{Cu}(x) \qquad (17.57)$$

where $\bar{\psi}_{Cu} = \psi_{Cu}^{\,\dagger}\gamma^0$ and

$$\psi_{Cu}^{\,\dagger} = [\psi_{Cu}(\mathbf{x_r}, \mathbf{x_i})]^\dagger \,|_{\mathbf{x_i} = -\mathbf{x_i}} \qquad (17.58)$$

$$D_0 = \partial/\partial x^0$$
$$D_k = \partial/\partial x_r^{\,k} + i\,\partial/\partial x_i^{\,k} \qquad (17.59)$$

for k = 1, 2, 3. The action

$$I = \int d^7x\,\mathcal{L}_{Cu} \qquad (17.60)$$

is invariant under real Lorentz transformations involving (x^0, $\mathbf{x_r}$) with $\mathbf{x_i}$ held constant. It is easy to show that this action is also real.

17.3.4 Free Down-type Quark Lagrangian

The simplest, physically acceptable, free, spin ½ tachyon lagrangian density[223] for ψ_{Cd} is:

$$\mathcal{L}_{Cd} = \psi_{Cd}^{\,C}(x)(\gamma^0\partial/\partial t + \boldsymbol{\gamma}\cdot(\nabla_r + i\nabla_i) - m)\psi_{Cd}(x) \qquad (17.61)$$

---

[223] Leading to a left-handed quark ElectroWeak sector.

where

$$\psi_{Cd}{}^{C}(x) = [\psi_{Cd}(x)]^{\dagger}|_{\mathbf{x_i} = -\mathbf{x_i}}\, i\gamma^0\gamma^5 \tag{17.62}$$

In words, eq. 17.62 states take the hermitean conjugate of $\psi_{Cd}(x)$; change $\mathbf{x_i}$ to $-\mathbf{x_i}$; and then post-multiply by the indicated factors.

The free tachyon complexon, action

$$I = \int d^7x \mathcal{L}_{Cd} \tag{17.63}$$

is invariant under real Lorentz transformations involving $(x^0, \mathbf{x_r})$ with $\mathbf{x_i}$ held constant. And the action can be shown to be real.

17.4 Canonical Commutation Relations of the Four Fermion Species

17.4.1 Free Charged Lepton Commutation Relations

The canonically conjugate momentum determined by the lagrangian density eq. 17.54 is

$$\pi_a = \partial\mathcal{L}/\partial\dot{\psi}_a \equiv \partial\mathcal{L}/\partial(\partial\psi_a/\partial t) = i\psi^{\dagger}{}_a \tag{17.64}$$

The resulting non-zero, canonical, equal time, anti-commutation relations are

$$\{\pi_a(x), \psi_b(x')\} = i\,\delta_{ab}\,\delta^3(x - x')$$

or

$$\{\psi^{\dagger}{}_a(x), \psi_b(x')\} = \delta_{ab}\,\delta^3(x - x') \tag{17.65}$$

17.4.2 Free Neutral Lepton Commutation Relations

We identify neutral leptons with tachyons. Having defined a suitable tachyon lagrangian we can now proceed to its canonical quantization. The conjugate momentum can be calculated from the lagrangian density eq. 17.55 is

$$\pi_{Ta} = \partial\mathcal{L}_T/\partial\dot{\psi}_{Ta} \equiv \partial\mathcal{L}_T/\partial(\partial\psi_{Ta}/\partial t) = -i(\psi_T{}^{\dagger}\gamma^5)_a \tag{17.66}$$

The resulting non-zero, canonical anti-commutation relations are

$$\{\pi_{T_a}(x), \psi_{Tb}(x')\} = i\,\delta_{ab}\,\delta^3(x - x')$$

or

$$\{\psi_{T\,a}^{\dagger}(x), \psi_{Tb}(x')\} = -\,[\gamma^5]_{ab}\,\delta^3(x - x') \tag{17.67}$$

The presence of the γ^5 in eq. 17.67 indicates that parity differentiates between the left-handed and right-handed field anti-commutation relations. We define left-handed and right-handed fields using a transformed set of Dirac matrices:

$$\gamma^0 = \begin{bmatrix} 0 & -I \\ -I & 0 \end{bmatrix} \quad \gamma^i = \begin{bmatrix} 0 & \sigma_i \\ -\sigma_i & 0 \end{bmatrix} \quad \gamma^5 = \begin{bmatrix} I & 0 \\ 0 & -I \end{bmatrix} \tag{17.68}$$

which are obtained from the usual Dirac matrices by applying the unitary transformation $U = 2^{-\frac{1}{2}}(I + \gamma^5\gamma^0)$. Note I is the 4×4 identity matrix. The γ^5 chirality operator's eigenvalues define handedness: +1 corresponds to right-handed; and –1 corresponds to left-handed:

$$\gamma^5\psi_{TL} = -\psi_{TL} \qquad\qquad \gamma^5\psi_{TR} = \psi_{TR} \tag{17.69a}$$

The projection operators:

$$\begin{aligned} C^{\pm} &= \tfrac{1}{2}(I \pm \gamma^5) \\ C^{+} + C^{-} &= I \\ C^{\pm\,2} &= C^{\pm} \\ C^{+}C^{-} &= 0 \end{aligned} \tag{17.69b}$$

can be used to define left and right handed tachyon fields

$$\begin{aligned} \psi_{TL} &= C^{-}\psi_{T} \\ \psi_{TR} &= C^{+}\psi_{T} \end{aligned} \tag{17.70}$$

Eq. 17.67 implies the anti-commutation relations

$$\{\psi_{TLa}{}^{\dagger}(x), \psi_{TLb}(x')\} = C^{-}{}_{ab}\delta^3(x - x') \tag{17.71}$$

$$\{\psi_{TRa}{}^{\dagger}(x), \psi_{TRb}(x')\} = -\, C^{+}{}_{ab}\delta^3(x - x') \tag{17.72}$$

$$\{\psi_{TLa}{}^{\dagger}(x), \psi_{TRb}(x')\} = \{\psi_{TRa}{}^{\dagger}(x), \psi_{TLb}(x')\} = 0 \tag{17.73}$$

The lagrangian density of eq. 17.55 decomposes into left-handed and right-handed parts (modulo the mass term):

$$\begin{aligned}\mathcal{L}_T = \psi_{TL}{}^{\dagger}\gamma^0 i\gamma^\mu\partial_\mu\psi_{TL} - \psi_{TR}{}^{\dagger}\gamma^0 i\gamma^\mu\partial_\mu\psi_{TR} - \\ - im[\psi_{TR}{}^{\dagger}\gamma^0\psi_{TL} - \psi_{TL}{}^{\dagger}\gamma^0\psi_{TR}]\end{aligned} \tag{17.74}$$

Noting the sign difference between the left-handed and right-handed kinetic terms in eq. 17.74 we see that the right-handed kinetic term (having the wrong sign) leads to an unphysical quantum field theory. This difference is the source of parity violation in the Standard Model, and the source of its left-handedness, as we will show in a later section.

17.4.3 Free Up-type Quark Commutation Relations

There are two choices for the up-type free quark (and corrrespondingly two choices in the interacting quark case). The first choice is to assume a free up-type quark theory similar to the theory of the free charged lepton considered earlier in this section based on the lagrangian eq. 17.54. Pursuing this approach would lead to the conventional Standard Model quark up-type sector.[224]

However it is clear there is a profound difference between quarks and leptons in nature. Quarks are always bound; leptons are not necessarily bound. Quarks experience the strong interaction; leptons do not. Consequently it is of interest to consider a second case embodied in the third type of fermion as specified in eqs. 17.57 – 17.60. In this case we shall see that the field solutions contain an embedded global SU(3) symmetry. This symmetry leads us to postulate that quarks are in the $\underline{3}$ representation of an SU(3) symmetry that we identify with color.

In this second case the conjugate momentum is

$$\pi_{Cua} = \partial\mathcal{L}/\partial\dot{\psi}_{Cua} \equiv \partial\mathcal{L}/\partial(\partial\psi_{Cua}/\partial x^0) = i\psi_{Cu}{}^{\dagger}{}_a \tag{17.75}$$

[224] We consider this possibility later and call the resulting theory the Standard Standard Model.

where a is a spinor index. The non-zero anti-commutation relation, *at first glance*, appears to be

$$\{\psi_{Cu}{}^{\dagger}{}_{a}(x), \psi_{Cub}(y)\} = \delta_{ab}\, \delta^3(x_r - y_r)\delta^3(x_i - y_i) \qquad (17.76)$$

However the constraint eq. 17.50 *requires* the anti-commutator to be

$$\{\psi_{Cu}{}^{\dagger}{}_{a}(x), \psi_{Cub}(y)\} = -\,\delta_{ab}\delta'(\nabla_r\cdot\nabla_i/m^2)[\delta^3(x_r - y_r)\delta^3(x_i - y_i)] \qquad (17.77)$$

where all ∇_r and ∇_i are ∇ derivatives with respect to x, and where $\delta'(\nabla_r\cdot\nabla_i)$ is the derivative of a delta function with the argument being differential operators such as those in eq. 17.50. The minus sign is due to the presence of a *derivative* of a delta-function and is not an issue.

17.4.4 Free Down-type Quark Commutation Relations

The conjugate momentum can be calculated from the lagrangian density eq. 17.61:

$$\pi_{Cda} = \partial\mathcal{L}_{Cd}/\partial\dot{\psi}_{Cda} \equiv \partial\mathcal{L}_{Cd}/\partial(\partial\psi_{Cda}/\partial t) = -i([\psi_{Cd}(x)]^{\dagger}|_{\mathbf{x_i} = -\mathbf{x_i}}\gamma^5)_a \qquad (17.78)$$

The resulting non-zero, canonical anti-commutation relations are presumably

$$\{\pi_{Cda}(x), \psi_{Cdb}(y)\} = i\,\delta_{ab}\, \delta^3(x_r - y_r)\delta^3(x_i - y_i) \qquad (17.79)$$

based on locality in both the real and imaginary coordinates. However, or taking account of the constraint eq. 17.53 it must be modified to

$$\{\psi_{Cd}{}^{\dagger}{}_{a}(x)|_{\mathbf{x_i} = -\mathbf{x_i}}, \psi_{Cdb}(y)\} = [\gamma^5]_{ab}\delta'(\nabla_r\cdot\nabla_i/m^2)[\delta^6(x - y)] \qquad (17.80)$$

As in the case of the neutral lepton commutation relations the presence of a γ^5 in eq. 17.80 will differentiate between left and right-handed field anti-commutators:

$$\{\psi_{CdL\alpha}{}^{\dagger}(x)|_{\mathbf{x_i} = -\mathbf{x_i}}, \psi_{CdLb}(y)\} = -C^{-}{}_{ab}\delta'(\nabla_r\cdot\nabla_i/m^2)[\delta^6(x - y)] \qquad (17.81)$$

$$\{\psi_{CdR\alpha}{}^{\dagger}(x)|_{\mathbf{x_i} = -\mathbf{x_i}}, \psi_{CdRb}(y)\} = C^{+}{}_{ab}\delta'(\nabla_r\cdot\nabla_i/m^2)[\delta^6(x - y)] \qquad (17.82)$$

$$\{\psi_{CdL\alpha}{}^{\dagger}(x)|_{\mathbf{x_i} = -\mathbf{x_i}}, \psi_{CdRb}(y)\} = \{\psi_{CdR\alpha}{}^{\dagger}(x)|_{\mathbf{x_i} = -\mathbf{x_i}}, \psi_{CdLb}(x')\} = 0 \qquad (17.83)$$

The left-handed fields again have a physically acceptable sign and lead to the left-handedness of the Standard Model quark sector.

17.5 Fourier Expansions of Free Fermion Fields

In this section we will simply list the fourier expansions, and some salient features, of the four species of fermion fields. The interested reader can read about the details of the fourier expansions in Blaha (2007b). We will express the fourier expansions in terms of light front variables[225] for the Dirac and tachyon species although they are required only for a correct canonical *tachyon* quantization.[226] (If equal-time canonical quantization is attempted, then non-localities,[227] and other problems, appear.) The light front variables are:

$$x^{\pm} = (x^0 \pm x^3)/\surd 2 \qquad (17.84)$$
$$\partial/\partial x^{\pm} \equiv \partial^{\mp} \equiv (\partial/\partial x^0 \pm \partial/\partial x^3)/\surd 2$$

with the "transverse" coordinate variables, x^1 and x^2, unchanged.

---

[225] L. Susskind, Phys. Rev. **165**, 1535 (1968); K. Bardakci and M. B. Halpern Phys. Rev. **176**, 1686 (1968), S. Weinberg, Phys. Rev. **150**, 1313 (1966); J. Kogut and D. Soper, Phys. Rev. **D1**, 2901 (1970); J. D. Bjorken, J. Kogut, and D. Soper, Phys. Rev. **D3**, 1382 (1971); R. A. Neville and F. Rohrlich, Nuov. Cim. **A1**, 625 (1971); F. Rohrlich, Acta Phys Austr. Suppl. **8**, 277 (1971); S-J Chang, R. Root, and T-M Yan, Phys. Rev. **D7**, 1133 (1973); S-J Chang, and T-M Yan, Phys. Rev. **D7**, 1147 (1973); T-M Yan, Phys. Rev. **D7**, 1761 (1973); T-M Yan, Phys. Rev. **D7**, 1780 (1973); C. Thorn, Phys. Rev. **D19**, 639 (1979); and references therein.

[226] This was first demonstrated in Blaha (2006) who showed tachyon fields must be separated into left-handed and right-handed parts, and then second quantized using light-front coordinates, to obtain local, equal light-front anti-commutators.

[227] See G. Feinberg, Phys. Rev. **159**, 1089 (1967) for example.

17.5.1 Free Charged Lepton Field Fourier Expansion

The fourier expansion of the free charged lepton field is the conventional Dirac field expansion. In light-front coordinates the free, "+" wave function Fourier expansion[228] of a Dirac field is:

$$\psi^+(x) = \sum_{\pm s} \int d^2p dp^+ N^+(p)\theta(p^+)[b^+(p, s)u^+(p, s)e^{-ip\cdot x} + \\ + d^{+\dagger}(p, s)v^+(p, s)e^{+ip\cdot x}] \quad (17.85)$$

where

$$N^+(p) = [m/((2\pi)^3 p^+)]^{1/2} \quad (17.86)$$

with $u^+(p, s)$ and $v^+(p, s)$ being projections of the conventional spinors u and v; and with the non-zero creation and annihilation operator anti-commutators:

$$\{b^+(q,s), b^{+\dagger}(p,s')\} = \delta_{ss'}\delta^2(\mathbf{q} - \mathbf{p})\delta(q^+ - p^+)$$
$$\{d^+(q,s), d^{+\dagger}(p,s')\} = \delta_{ss'}\delta^2(\mathbf{q} - \mathbf{p})\delta(q^+ - p^+) \quad (17.87)$$

17.5.2 Free Neutral Lepton Field Fourier Expansion

The free neutral lepton field fourier expansion is somewhat more involved since it is a tachyon field. The free, "+" light-front, *left-handed* tachyon wave function Fourier expansion is:

$$\psi_{TL}{}^+(x) = \sum_{\pm s} \int d^2p dp^+ N_{TL}{}^+(p)\theta(p^+)[b_{TL}{}^+(p, s)u_{TL}{}^+(p, s)e^{-ip\cdot x} + \\ + d_{TL}{}^{+\dagger}(p, s)v_{TL}{}^+(p, s)e^{+ip\cdot x}] \quad (17.88)$$

with

$$N_{TL}{}^+(p) = [2m|\mathbf{p}|/((2\pi)^3(p^+(p^+ - p^-) + p_\perp{}^2))]^{1/2} \quad (17.89)$$

and where the non-zero anti-commutators of the Fourier coefficient operators are

$$\{b_{TL}{}^+(q,s), b_{TL}{}^{+\dagger}(p,s')\} = \delta_{ss'}\delta^2(\mathbf{q} - \mathbf{p})\delta(q^+ - p^+)$$
$$\{d_{TL}{}^+(q,s), d_{TL}{}^{+\dagger}(p,s')\} = \delta_{ss'}\delta^2(\mathbf{q} - \mathbf{p})\delta(q^+ - p^+) \quad (17.90)$$

---

[228] See S-J Chang and T- M. Yan Phys. Rev. D7, (1973) for a detailed presentation on light-front (infinite momentum frame) quantization of Dirac fields as well as Blaha (2007b).

The spinors are

$$u_{TL}^{+}(p, s) = C^{-} R^{+} S_C(\omega, \mathbf{v_c}) w^1(0)$$
$$u_{TL}^{+}(p, -s) = C^{-} R^{+} S_C(\omega, \mathbf{v_c}) w^2(0)$$
$$v_{TL}^{+}(p, s) = C^{-} R^{+} S_C(\omega, \mathbf{v_c}) w^3(0)$$
$$v_{TL}^{+}(p, -s) = C^{-} R^{+} S_C(\omega, \mathbf{v_c}) w^4(0)$$

where the superscript "T" indicates the transpose, $R^{\pm} = ½(I \pm \gamma^0\gamma^3)$ are ± field light-front projection operators[229], $S_C(\omega, \mathbf{v_c})$ is an L(C) spinor boost, and the column 4-vector components $[w^i(0)]^k = \delta^{ik}$.

The case of *right-handed* free neutral lepton fields (tachyons) is similar to the left-handed case with only two differences: a minus sign in the creation and annihilation operator anti-commutation relations, and the use of right-handed projection operators. The right-handed tachyon wave function light-front Fourier expansion has the form:

$$\psi_{TR}^{+}(x) = \sum_{\pm s} \int d^2p\,dp^{+} N_{TR}^{+}(p)\theta(p^{+})[b_{TR}^{+}(p, s)u_{TR}^{+}(p, s)e^{-ip\cdot x} + \\ + d_{TR}^{+\dagger}(p, s)v_{TR}^{+}(p, s)e^{+ip\cdot x}]$$

with the non-zero anti-commutators of the Fourier coefficient operators:

$$\{b_{TR}^{+}(q,s), b_{TR}^{+\dagger}(p,s')\} = -\delta_{ss'}\delta^2(\mathbf{q} - \mathbf{p})\delta(q^{+} - p^{+})$$
$$\{d_{TR}^{+}(q,s), d_{TR}^{+\dagger}(p,s')\} = -\delta_{ss'}\delta^2(\mathbf{q} - \mathbf{p})\delta(q^{+} - p^{+})$$

The right-handed spinors are

$$u_{TR}^{+}(p, s) = C^{+} R^{+} S_C(\omega, \mathbf{v_c}) w^1(0)$$
$$u_{TR}^{+}(p, -s) = C^{+} R^{+} S_C(\omega, \mathbf{v_c}) w^2(0)$$
$$v_{TR}^{+}(p, s) = C^{+} R^{+} S_C(\omega, \mathbf{v_c}) w^3(0)$$
$$v_{TR}^{+}(p, -s) = C^{+} R^{+} S_C(\omega, \mathbf{v_c}) w^4(0)$$

17.5.3 Free Up-Type Complexon Quark Field Fourier Expansion

We believe quarks are complexons although this is an experimental question that remains to be resolved. Otherwise the quark

---

[229] Blaha (2007b) p. 48.

field fourier expansions would be the same as the lepton fourier expansions described in the previous two sub-sections.

We will express the fourier expansion of the up-type quark field in conventional coordinates for the sake of illustration and because they are more familiar to most physicists. The form of the up-type complexon quark field, ignoring generations and color indices temporarily, and taking account of the subsidiary condition is[230, 231]

$$\psi_{Cu}(x_r, x_i) = \sum_{\pm s} \int d^3p_r d^3p_i\, N_C(p)\delta(\mathbf{p_r}\cdot\mathbf{p_i}/m^2)[b_{Cu}(p,s)u_{Cu}(p, s)e^{-i(p\cdot x + p^*\cdot x^*)/2} + \\ + d_{Cu}^\dagger(p,s)v_{Cu}(p, s)e^{+i(p\cdot x + p^*\cdot x^*)/2}] \quad (17.91)$$

where $\mathbf{p} = \mathbf{p_r} + i\mathbf{p_i}$, $\mathbf{x} = \mathbf{x_r} - i\mathbf{x_i}$, $p\cdot x = p^0x^0 - \mathbf{p}\cdot\mathbf{x}$, and where we use

$$(p\cdot x + p^*\cdot x^*)/2 = p^0x^0 - \mathbf{p_r}\cdot\mathbf{x_r} - \mathbf{p_i}\cdot\mathbf{x_i} \quad (17.92)$$

in the exponentials in order to avoid divergences that would appear in the calculation of the equal-time commutator, the Feynman propagator and other quantities of interest after second quantization. Note that

$$(\nabla_\mathbf{r} + i\nabla_\mathbf{i})e^{-i(p\cdot x + p^*\cdot x^*)/2} = i(\mathbf{p_r} + i\mathbf{p_i})e^{-i(p\cdot x + p^*\cdot x^*)/2} \quad (17.93)$$

and

$$(\nabla_\mathbf{r} + i\nabla_\mathbf{i})e^{-ip^*\cdot x^*} = 0 \quad (17.94)$$

for all p.

Further we note

$$N_C(p) = [2m/((2\pi)^6p^0)]^{1/2} \quad (17.95)$$

and

$$u_{Cu}(p, s) = S_C(\omega, \mathbf{v_c})w^1(0)$$

---

[230] Note that when $|\mathbf{p_i}| \geq |\mathbf{p_r}|$ (for imaginary $\omega = (\omega_r^2 - \omega_i^2)^{1/2}$) the 3-momentum becomes imaginary $\mathbf{p}\cdot\mathbf{p} < 0$. However, since we will be identifying confined quarks with this type of particle – much modified by a confining color quark interaction – the issue of an imaginary 3-momentum in the hypothetical free quark case becomes moot. We note the energy gap between positive and negative energy states disappears so E = 0 is possible. Thus real Lorentz transformations can mix positive and negative energy states. The solution is to do all calculations in the light-front frame as we do for tachyons. Then the mixing issue is resolved. In the present case we second quantize on the "time-front" for illustrative purposes.

[231] We scale $\mathbf{p_r}\cdot\mathbf{p_i}$ with m^2 in the delta function for convenience. In the case of a zero mass particle some other scale could be used.

$$u_{Cu}(p, -s) = S_C(\omega, \mathbf{v_c})w^2(0)$$
$$v_{Cu}(p, s) = S_C(\omega, \mathbf{v_c})w^3(0)$$
$$v_{Cu}(p, -s) = S_C(\omega, \mathbf{v_c})w^4(0) \quad (17.96)$$

The momentum 4-vector $p = (p^0, \mathbf{p})$ is related to the other quantities by

$$p^0 = m\cosh(\omega) \qquad \mathbf{p} = \mathbf{p_r} + i\mathbf{p_i}$$
$$\mathbf{p_r} = m\omega_r\hat{\mathbf{u}}_r \sinh(\omega)/\omega \qquad \mathbf{p_i} = m\omega_i\hat{\mathbf{u}}_i \sinh(\omega)/\omega$$
$$\omega = (\omega_r^2 - \omega_i^2)^{1/2}$$
$$\mathbf{p_r}\cdot\mathbf{p_i} = 0$$
$$\hat{\mathbf{w}} = (\omega_r\hat{\mathbf{u}}_r + i\omega_i\hat{\mathbf{u}}_i)/\omega$$
$$\hat{\mathbf{w}}\cdot\hat{\mathbf{w}} = 1$$
$$\mathbf{v_c} = \hat{\mathbf{w}}\tanh(\omega) \quad (17.97)$$

The non-zero anti-commutators of the Fourier coefficient operators are

$$\{b_{Cu}(p,s), b_{Cu}^\dagger(p'^*,s')\} = \delta_{ss'}\delta^3(\mathbf{p_r} - \mathbf{p'}_{r'})\delta^3(\mathbf{p_i} + \mathbf{p'}_{i'})$$
$$\{d_{Cu}(p,s), d_{Cu}^\dagger(p'^*,s')\} = \delta_{ss'}\delta^3(\mathbf{p_r} - \mathbf{p'}_{r'})\delta^3(\mathbf{p_i} + \mathbf{p'}_{i'}) \quad (17.98)$$

17.5.4 Free Down-Type Complexon Quark Field Fourier Expansion

It is necessary that we perform the fourier expansion of the down-type quark field in light-front coordinates in order to obtain a local canonical quantization.

17.5.4.1 Left-Handed Down-Type Complexon Quark Field Fourier Expansion

The independent, left-handed down-type complexon quark field Fourier expansion (free, "+" light-front, left-handed tachyonic complexon), ignoring fermion generations and color indices temporarily, and taking account of the subsidiary condition, is

$$\psi_{CdL}{}^{+}(x_r, x_i) = \sum_{\pm s}\int d^2p_r dp^+ d^3p_i\, N_{CdL}{}^{+}(p)\theta(p^+)\cdot$$
$$\cdot\delta((p_i{}^3(p^+ - p^-)/\surd 2 + \mathbf{p}_{r\perp}\cdot\mathbf{p}_{i\perp})/m^2)\cdot$$
$$\cdot[b_{CdL}{}^{+}(p, s)u_{CdL}{}^{+}(p, s)e^{-i(p\cdot x + p^*\cdot x^*)/2} +$$
$$+ d_{CdL}{}^{+\dagger}(p, s)v_{CdL}{}^{+}(p, s)e^{+i(p\cdot x + p^*\cdot x^*)/2}] \quad (17.99)$$

where

$$N_{CdL}^{+}(p) = (2\pi)^{-3}(2m/p^{+})^{1/2}$$

and

$$\begin{aligned} u_{CdL}^{+}(p, s) &= C^{-} R^{+} S_C(\omega',\mathbf{v_c})w^1(0) \\ u_{CdL}^{+}(p, -s) &= C^{-} R^{+} S_C(\omega',\mathbf{v_c})w^2(0) \\ v_{CdL}^{+}(p, s) &= C^{-} R^{+} S_C(\omega',\mathbf{v_c})w^3(0) \\ v_{CdL}^{+}(p, -s) &= C^{-} R^{+} S_C(\omega',\mathbf{v_c})w^4(0) \end{aligned} \tag{17.100}$$

where

$$\omega' = \omega + i\pi/2 \tag{17.101}$$

The momentum 4-vector $p = (p^0, \mathbf{p})$ is related to the other quantities by eq. 17.97 with ω replaced by $\omega' = \omega + i\pi/2$ in all the relations of eq. 17.97.

The non-zero anti-commutators of the Fourier coefficient operators are:

$$\begin{aligned} \{b_{CdL}(p,s), b_{CdL}^{\dagger}(p'^*,s')\} &= 2^{-1/2}\delta_{ss'}\delta(p^+ - p'^+)\delta^2(\mathbf{p_r} - \mathbf{p'_{r'}})\delta^3(\mathbf{p_i} + \mathbf{p'_{i'}}) \\ \{d_{CdL}(p,s), d_{CdL}^{\dagger}(p'^*,s')\} &= 2^{-1/2}\delta_{ss'}\delta(p^+ - p'^+)\delta^2(\mathbf{p_r} - \mathbf{p'_{r'}})\delta^3(\mathbf{p_i} + \mathbf{p'_{i'}}) \end{aligned} \tag{17.102}$$

17.5.4.2 Right-Handed Down-Type Complexon Quark Field Fourier Expansion

The independent, right-handed down-type complexon quark field Fourier expansion (free, "+" light-front, right-handed tachyonic complexon), ignoring fermion generations and color indices temporarily, and taking account of the subsidiary condition, is

$$\begin{aligned} \psi_{CdR}^{+}(x_r, x_i) = \sum_{\pm s} \int d^2p_r dp^+ d^3p_i\, N_{CdR}^{+}(p)\theta(p^+)\cdot \\ \cdot\delta((p_i^3(p^+ - p^-)/\sqrt{2} + \mathbf{p}_{r\perp}\cdot\mathbf{p}_{i\perp})/m^2)\cdot \\ \cdot[b_{CdR}^{+}(p, s)u_{CdR}^{+}(p, s)e^{-i(p\cdot x + p^*\cdot x^*)/2} + \\ + d_{CdR}^{+\dagger}(p, s)v_{CdR}^{+}(p, s)e^{+i(p\cdot x + p^*\cdot x^*)/2}] \end{aligned} \tag{17.103}$$

where

$$N_{CdR}^{+}(p) = (2\pi)^{-3}(2m/p^{+})^{1/2}$$

and

$$u_{CdR}{}^{+}(p, s) = C^{+}\, R^{+}\, S_C(\omega',\mathbf{v_c})w^1(0)$$
$$u_{CdR}{}^{+}(p, -s) = C^{+}\, R^{+}\, S_C(\omega',\mathbf{v_c})w^2(0)$$
$$v_{CdR}{}^{+}(p, s) = C^{+}\, R^{+}\, S_C(\omega',\mathbf{v_c})w^3(0)$$
$$v_{CdR}{}^{+}(p, -s) = C^{+}\, R^{+}\, S_C(\omega',\mathbf{v_c})w^4(0) \qquad (17.104)$$

where

$$\omega' = \omega + i\pi/2 \qquad (17.101)$$

The momentum 4-vector $p = (p^0, \mathbf{p})$ is related to the other quantities by eq. 17.97 with ω replaced by $\omega' = \omega + i\pi/2$ in all the relations of eq. 17.97.[232]

The non-zero anti-commutators of the Fourier coefficient operators are:

$$\{b_{CdR}(p,s), b_{CdR}{}^{\dagger}(p'^*,s')\} = -2^{-1/2}\,\delta_{ss'}\delta(p^+ - p'^+)\delta^2(\mathbf{p_r} - \mathbf{p'_{r'}})\delta^3(\mathbf{p_i} + \mathbf{p'_{i'}})$$
$$\{d_{CdR}(p,s), d_{CdR}{}^{\dagger}(p'^*,s')\} = -2^{-1/2}\delta_{ss'}\,\delta(p^+ - p'^+)\delta^2(\mathbf{p_r} - \mathbf{p'_{r'}})\delta^3(\mathbf{p_i} + \mathbf{p'_{i'}})$$
$$(17.105)$$

17.6 A Global SU(3) Symmetry of Complexon Quarks

We will now consider a global SU(3) covariance implicit in eqs. 17.49, 17.50, 17.52, and 17.53. The defining property of the group SU(3) is that it preserves the invariance of inner products of complex 3-vectors of the form:

$$u^*\cdot v = u^{1*}v^1 + u^{2*}v^2 + u^{3*}v^3 \qquad (17.106)$$

If we examine the dynamical equation eq. 17.49 (the case of eq. 17.52 differs only in details) we see that the differential operator is covariant under a global SU(3) transformation U of the complex spatial 3-coordinates:

$$[i\gamma^0\partial/\partial t + i\mathbf{D_c}^*\cdot\gamma - m] = [i\gamma^0\partial/\partial t + i\mathbf{D_c}'^*\cdot\gamma' - m] \qquad (17.107)$$

where

$$\mathbf{D_c}^* = \nabla_c = \nabla_r + i\nabla_i$$

and

---

232 Please note that this right-handed field together with the preceding left-handed field combine to give the left-handed form of the ElectroWeak sector. The choice $\omega' = \omega - i\pi/2$ would lead to a right-handed ElectroWeak theory and is not physically relevant.

$$\gamma^a = U^{ab}\gamma'^b \qquad (17.108a)$$
$$D_c{*}^a = D_c'{*}^b U^{ab}{*} \qquad (17.108b)$$

where $U^\dagger = U^{-1}$. We now wish to exhibit the covariance of eq. 17.49. Since we can view the three spatial γ-matrices as SU(3) 3-vectors, we can express eq. 17.108 as the result of an SU(3) rotation V of the γ-matrices (on the spinor indices)

$$V\gamma^a V^{-1} = U^{ab}\gamma'^b \qquad (17.109)$$

where V is a 4×4 reducible representation[233] of SU(3), namely, 3 + 1. In a manner similar to the covariance proof of the Dirac equation[234] we see that eq. 17.49 is covariant under SU(3) transformations:

$$V[i\gamma^0\partial/\partial t + i\boldsymbol{\gamma}\cdot\mathbf{D}_c{*} - m]V^{-1}V\psi_C(t, \mathbf{x}_r, \mathbf{x}_i) = 0$$

or

$$[i\gamma^{0\prime}\partial/\partial t' + i\mathbf{D}_c'{*}\cdot\boldsymbol{\gamma}' - m]V\psi_C(t, \mathbf{x}_r, \mathbf{x}_i) = 0 \qquad (17.110)$$

(Note $\gamma^{0\prime} = V\gamma^0V^{-1}$ and t' = t.) The SU(3) transformed wave function $\psi_C'(t, \mathbf{x}')$ is

$$\psi_C'(t', \mathbf{x}') = V\psi_C(t, \mathbf{x}) = V\psi_C(t', U\mathbf{x}') \qquad (17.111)$$

Thus the complexon Dirac equation is covariant under global SU(3).

The subsidiary condition, eq. 17.50, is also covariant under an SU(3) rotation:

$$\nabla_r'{*}\cdot\nabla_i'\psi_C'(t, \mathbf{x}') = \nabla_r\cdot\nabla_i\, V\psi_C(t, \mathbf{x}) = V\nabla_r{*}\cdot\nabla_i\, \psi_C(t, \mathbf{x}) = 0 \qquad (17.112)$$

We now examine the transformation of the wave function eq. 17.111 under the SU(3) transformation U. If we define

$$q^{*\mu} = (q^0, \mathbf{q}^*) = (p^0, \mathbf{p}_r + i\mathbf{p}_i) = (p^0, \mathbf{p}) = p^\mu \qquad (17.113)$$

---

233 Eqs. 16.50, and 16.53 furnish a reducible 4×4 matrix representation described later. The generators of this reducible representation are ${}_4F_i = F_i + \Omega_{+00} - F_9$ with F_i being the Gell-Mann SU(3) generators for i = 1, 2, …, 8.

234 For example see Bjorken (1964) pp. 18 – 20.

then $\psi_C(t, \mathbf{x})$ will be seen to be covariant form under a SU(3) transformation:

$$\psi_C(t, x) = \sum_{\pm s} \int d^3q_r d^3q_i \, N_C(p^0)\delta(\mathbf{q}_r^*\cdot\mathbf{q}_i/m^2)[b_C(q^*,s)u_C(q^*,s)e^{-i(q^*\cdot x + q\cdot x^*)/2} + \\ + d_C^\dagger(q^*,s)v_C(q^*,s)e^{+i(q^*\cdot x + q\cdot x^*)/2}] \quad (17.114)$$

Note both terms in each exponential are separately invariant under global SU(3). ($\mathbf{q}_r^* = \mathbf{q}_r$ since $\mathbf{q}_r$ is real.)

Eq. 17.111 implies that the spinors appearing in eq. 17.114 are covariant under SU(3) transformations

$$u_C'(q'^*,s') = Vu_C(q^*,s) \quad (17.115)$$
$$v_C'(q'^*,s') = Vv_C(q^*,s) \quad (17.116)$$

The fourier coefficients, if second quantized in a complex spatial coordinate generalization of the usual way, also have covariant anti-commutation relations under an SU(3) transformation:

$$\{b_C(q,s), b_C^\dagger(q'^*,s')\} = \delta_{ss'}\delta^3(q_r - q'_{r'})\delta^3(q_i - q'_{i'}) \quad (17.117)$$

Under an SU(3) transformation, z = Uq and z' = Uq', the right side of eq. 17.117 transforms to

$$\delta^3(q_r - q'_{r'})\delta^3(q_i - q'_{i'}) \to \delta^3(z_r - z'_{r'})\delta^3(z_i - z'_{i'})/|\partial(q)/\partial(z)| \\ = \delta^3(z_r - z'_{r'})\delta^3(z_i - z'_{i'}) \quad (17.118)$$

where

$$|\partial(q)/\partial(z)| = |\partial(q_r^1,q_r^2,q_r^3,q_i^1, q_i^2, q_i^3)/\partial(z_r^1,z_r^2,z_r^3,z_i^1, z_i^2, z_i^3)| = 1 \quad (17.119)$$

is the Jacobian of the transformation U. The fourier coefficients transform trivially under SU(3):

$$b_C(q^*,s) \to b_C(z^*,s) \quad (17.120)$$

Since the integrand transforms as

$$\int d^3q_r d^3q_i \rightarrow \int d^3z_r d^3z_i \,|\partial(q)/\partial(z)| = \int d^3z_r d^3z_i \qquad (17.121)$$

we see that the wave function $\psi_C(t, \mathbf{x})$ transforms covariantly.

17.6.1 Global SU(3) Generators

The generators of the global SU(3) symmetry under discussion have a 4×4 matrix reducible representation ($\underline{3}$ + $\underline{1}$) based on eqs. 16.50, and 16.53. The generators of this reducible representation are

$$_4F_i = F_i + \Omega_{+00} - F_9 \qquad (17.122)$$

with F_i being the Gell-Mann SU(3) generators for i = 1, 2, …, 8. The form of these matrices is

$$_4F_i = \left[\begin{array}{c|c} F_i & \begin{matrix} 0 \\ 0 \\ 0 \end{matrix} \\ \hline \begin{matrix} 0 & 0 & 0 \end{matrix} & 1 \end{array}\right] \qquad (17.123)$$

Projection operators can be defined to project out the $\underline{3}$ and the $\underline{1}$ representations of complexon spinor fields:

$$P_1 = \Omega_{+00} - F_9$$

and (17.124)

$$P_3 = F_9$$

where Ω_{+00} is the identity matrix. Thus the $\underline{3}$ complexon field is

$$\psi_{C3}(t, \mathbf{x_r}, \mathbf{x_i}) = P_3\psi_C(t, \mathbf{x_r}, \mathbf{x_i}) \qquad (17.125)$$

while the $\underline{1}$ complexon field is

$$\psi_{C1}(t, \mathbf{x_r}, \mathbf{x_i}) = P_1\psi_C(t, \mathbf{x_r}, \mathbf{x_i}) \qquad (17.126)$$

Since P_1 and P_3 do not commute with Lorentz transformations, a Lorentz transformation mixes ψ_{C1} and ψ_{C3}.[235] Since P_1 and P_3 do not commute with $\gamma 5$, left-handed and right-handed complexons would also be mixed by these projection operators.

However the appearance of an SU(3) invariance in complexon dynamics is suggestive of the color local SU(3) of the Strong interactions. **In chapter 19 we show that the global SU(3) symmetry described in this subsection can be interpreted as related to the local color SU(3) symmetry of quarks. This point gives us a strong theoretical inclination to regard color SU(3) as having a geometrical origin that is ultimately due to GL(4).**[236]

---

235 At this point it is worth noting that the construction of complexon fields based on a boost from a particle rest state guarantees that a reference frame exists in which any complexon particle has a single real time variable. Similarly a reference frame exists for a set of complexon particles (that is within a Lorentz of the center of momentum frame) with a single real time variable. The time variables of the individual complexon particles in the set are complex in general but are functions of the center of momentum real time variable. So there is only one real time variable for each complexon in the set although the time variable of an individual particle may be a complex function of the real center of momentum time variable.

236 Some may feel that associating color SU(3) with this essentially GL(4) space-time SU(3) leads to issues similar to those that led to the demise of SU(6) in the 1960's. (See Coleman, S., Phys. Rev. **138** B1262 (1965) for a discussion of these issues.) However, we evade these issues as will be seen later.

18. ElectroWeak Sector of the Standard Model

As we progress in our derivation of the Standard Model we will begin to see that many Standard Model features follow from the boost properties of the L(C) group and the requirement that the free dynamical equation of each of the fermion species results from a boost from a free fermion at rest.

Thus we will show that extended space-time geometrical considerations are directly the source of the form of the Standard Model. We believe that all Standard Model features (those known and those to be discovered in future experiments) will be found to originate in the geometry of extended space-time.

This chapter begins the connection of the form of the Standard with dynamics based on the L(C) group. In this chapter we will construct a lepton doublet consisting of a Dirac field and a tachyon field, develop the dynamical equations of the doublet as well as its transformation under the L(C) and the real Lorentz groups, and then show how parity violation and the other features of the dynamical part of the ElectroWeak sector follow. Then we develop the quark ElectroWeak sector of the Standard Model for both complexon quarks and "normal" quarks.

18.1 L(C) Transformations of Dirac and Tachyon Equations

An L(C) boost of a Dirac field can transform it into a spin ½ tachyon field, and vice versa:

$$\begin{aligned} S_C(\omega, \mathbf{v_c})\psi(x) &\rightarrow \psi_T'(x') \\ S_C(\omega, \mathbf{v_c})\psi_T(x) &\rightarrow \psi'(x') \end{aligned} \qquad (18.1a)$$

if $\omega = \omega_r + i\pi/2$ and $\mathbf{v_c}$ is real. The dynamical equations' operators then transform as

$$S_C(\omega, \mathbf{v_c})(\gamma^\mu\partial/\partial x^\mu - m)S_C^{-1}(\omega, \mathbf{v_c}) = (i\gamma^\mu\partial/\partial x'^\mu - m) \quad (18.1b)$$
$$S_C(\omega, \mathbf{v_c})\gamma^5(i\gamma^\mu\partial/\partial x^\mu - m)\gamma^5 S_C^{-1}(\omega, \mathbf{v_c}) = (\gamma^\mu\partial/\partial x'^\mu - m)$$

where

$$x'^\mu = \Lambda_C{}^\mu{}_\nu(\omega, \mathbf{v_c})x^\nu \quad (18.1c)$$
$$\partial/\partial x'^\mu = \Lambda_C{}^\nu{}_\mu(\omega, \mathbf{v_c})\partial/\partial x^\nu$$

Eq. 18.1c can be written in matrix form as

$$X' = \Lambda_C(\mathbf{v_c})X$$

Eqs. 18.1a – 18.1c imply

$$S_C(\omega, \mathbf{v_c})(\gamma^\mu\partial/\partial x^\mu - m)\psi_T(x) = (i\gamma^\mu\partial/\partial x'^\mu - m)S_C(\omega, \mathbf{v_c})\psi_T(x)$$
$$= (i\gamma^\mu\partial/\partial x'^\mu - m)\psi'(x') \quad (18.1d)$$

and

$$S_C(\omega, \mathbf{v_c})\gamma^5(i\gamma^\mu\partial/\partial x^\mu - m)\psi(x) = (\gamma^\mu\partial/\partial x'^\mu - m)S_C(\omega, \mathbf{v_c})\gamma^5\psi(x)$$
$$= (\gamma^\mu\partial/\partial x'^\mu - m)\psi_T'(x') \quad (18.1e)$$

where

$$\psi'(x') = S_C(\omega, \mathbf{v_c})\psi_T(x) \quad (18.1f)$$

and

$$\psi_T'(x') = S_C(\omega, \mathbf{v_c})\gamma^5\psi(x) \quad (18.1g)$$

We conclude that both the Dirac equation and the tachyon equation are not covariant under an L(C) boost.

18.2 Extension of Dirac & Tachyon Equations to a Doublet Format

We will now consider the issue of generalizing the Dirac equation to include the tachyon equation so that the generalized equation is covariant under both real Lorentz transformations and L(C) transformations.

The direct method to obtain a generalized equation that is covariant under both types of transformations is to define an 8 × 8 matrix generalization with a doublet field consisting of a Dirac field and a tachyon field. Let

$$đ(x) = \begin{bmatrix} (\gamma^\mu\partial/\partial x^\mu - m) & 0 \\ 0 & (i\gamma^\mu\partial/\partial x^\mu - m) \end{bmatrix} \tag{18.2}$$

be an 8 × 8 matrix differential operator with the 4 × 4 matrix elements shown, and let

$$\Psi(x) = \begin{bmatrix} \psi_T(x) \\ \psi(x) \end{bmatrix} \tag{18.3}$$

be an 8 row field composed of a Dirac field and a tachyon field. Then the extended L(C) covariant equation is

$$đ(x)\Psi(x) = 0 \tag{18.4}$$

We now define the 8 × 8 L(C) transformation

$$S_{L8}(\Lambda_C(\omega, \mathbf{v_c})) = \begin{bmatrix} 0 & S_C(\omega, \mathbf{v_c})\gamma^5 \\ S_C(\omega, \mathbf{v_c}) & 0 \end{bmatrix} \tag{18.5}$$

with inverse transformation

$$S_{L8}^{-1}(\Lambda_C(\omega, \mathbf{v_c})) = \begin{bmatrix} 0 & S_C^{-1}(\omega, \mathbf{v_c}) \\ \gamma^5 S_C^{-1}(\omega, \mathbf{v_c}) & 0 \end{bmatrix} \tag{18.6}$$

Applying S_{L8} to eq. 18.4 yields

$$0 = S_{L8}(\Lambda_C(\omega, \mathbf{v_c}))đ(x)\Psi(x) = đ(x')\Psi'(x') \tag{18.7}$$

where

$$\Psi'(x') = \begin{bmatrix} S_C\gamma^5\psi(x) \\ S_C\psi_T(x) \end{bmatrix} = \begin{bmatrix} \psi_T'(x') \\ \psi'(x') \end{bmatrix} \tag{18.8}$$

This exercise leads us to the postulate:

Postulate 18.1. A Dirac field and a tachyon field form a doublet whose dynamical equation is covariant under the 8 × 8 L(C) transformation eq. 18.5.

Note: L(C) covariance requires the tachyon and the Dirac particle must have the same absolute value for the bare mass.

It is easy to show that the extended Dirac equation eq. 18.4 is also covariant under conventional Lorentz transformations in the 8 × 8 representation:

$$S_8(\Lambda(v)) = \begin{bmatrix} S(\Lambda(v)) & 0 \\ 0 & S(\Lambda(v)) \end{bmatrix} \tag{18.9}$$

with inverse

$$S_8^{-1}(\Lambda(v)) = \begin{bmatrix} S^{-1}(\Lambda(v)) & 0 \\ 0 & S^{-1}(\Lambda(v)) \end{bmatrix} \tag{18.10}$$

and also the alternate Lorentz transformation:

$$S_{8A}(\Lambda(v)) = \begin{bmatrix} 0 & S(\Lambda(v)) \\ S(\Lambda(v)) & 0 \end{bmatrix} \tag{18.11}$$

with inverse

$$S_{8A}^{-1}(\Lambda(v)) = \begin{bmatrix} 0 & S^{-1}(\Lambda(v)) \\ S^{-1}(\Lambda(v)) & 0 \end{bmatrix} \tag{18.12}$$

Under a conventional Lorentz transformation we find

$$0 = S_8(\Lambda(v))đ(x)\Psi(x) = đ(x')\Psi'(x') \qquad (18.13)$$
$$0 = S_{8A}(\Lambda(v))đ(x)\Psi(x) = đ(x')\Psi'(x')$$

The lagrangian density that corresponds to our 8-dimensional construction is

$$\mathcal{L}_8 = \overline{\Psi}(x)đ(x)\Psi(x) \qquad (18.14)$$

where

$$\overline{\Psi}(x) = \Psi^\dagger\Gamma^0 \qquad (18.15)$$

and

$$\Gamma^0 = \begin{bmatrix} i\gamma^0\gamma^5 & 0 \\ 0 & \gamma^0 \end{bmatrix} \qquad (18.16)$$

The action

$$I = \int d^4x\mathcal{L}_8 \qquad (18.17)$$

is invariant under the 8×8 Lorentz transformations S_8 and S_{8A}.

The Hamiltonian density for the 8×8 formulation is

$$\mathcal{H}_8(x) = \begin{bmatrix} i\psi_T^\dagger\gamma^5(\boldsymbol{\alpha}\cdot\nabla + \beta m)\psi_T & 0 \\ 0 & \psi^\dagger(-i\boldsymbol{\alpha}\cdot\nabla + \beta m)\psi \end{bmatrix} \qquad (18.18)$$

18.3 Non-Invariance of the Extended Equation's Action under an L(C) Transformation

The action eq. 18.17 is *not* invariant under 8×8 L(C) transformations. The fundamental cause of this non-invariance is the three dimensional nature of space. In the case of Dirac particles one can define a Lorentz invariant action because time is one-dimensional. Thus one can use $\psi^\dagger\gamma^0 = \bar{\psi}$ to form the Dirac field lagrangian and action. A key factor in the Lorentz invariance is the relation between the inverse and hermitean conjugate of the spinor boost operator

$$\gamma^0 S^{-1}\gamma^0 = S^\dagger \qquad (18.19)$$

In the case of the tachyon lagrangian and action, L(C) invariance is not possible because the tachyonic equivalent to eq. 18.19 is[237]

$$S_L^{-1}(\Lambda_C(\omega, \mathbf{v_c}))\boldsymbol{\gamma}\cdot \mathbf{v_c}/|\mathbf{v_c}| = S_L^{-1}(\Lambda_C(\omega, \mathbf{v_c}))\boldsymbol{\gamma}\cdot\mathbf{p}/|\mathbf{p}|$$
$$= i\gamma^0 S_L^\dagger(\Lambda_C(\omega, \mathbf{v_c})) \qquad (18.20)$$

where $\mathbf{p} = m\gamma_s\mathbf{v_c}$. The appearance of $\boldsymbol{\gamma}\cdot\mathbf{p}/|\mathbf{p}|$ in eq. 18.20 precludes the invariance of the free tachyon action under L(C).

We will now show the effect of an L(C) boost transformation (eqs. 18.5 and 18.6) on the lagrangian density eq. 18.14. The two non-zero parts of the lagrangian density $\mathcal{L}_8$ (eq. 18.14) are

$$\mathcal{L}_1 = \psi_T^\dagger i\gamma^0\gamma^5(\gamma^\mu\partial/\partial x^\mu - m)\psi_T(x) \qquad (18.21)$$
$$\mathcal{L}_2 = \psi^\dagger\gamma^0(i\gamma^\mu\partial/\partial x^\mu - m)\psi(x) \qquad (18.22)$$

The effect of the transformation eqs. 18.5 and 18.6 on these terms is

$$\mathcal{L}_1' = \psi_T^\dagger i\gamma^0\gamma^5 S_L^{-1}S_L(\gamma^\mu\partial/\partial x^\mu - m)\, S_L^{-1}S_L\psi_T(x)$$
$$= \psi_T^\dagger i\gamma^0\gamma^5 S_L^{-1}(i\gamma^\mu\partial/\partial x'^\mu - m)S_L\psi_T(x)$$
$$= -\psi_T^\dagger S_L^\dagger\gamma^5(\boldsymbol{\gamma}\cdot\mathbf{p}/|\mathbf{p}|)(i\gamma^\mu\partial/\partial x'^\mu - m)S_L\psi_T(x)$$
$$= \psi'^\dagger(x')(\boldsymbol{\gamma}\cdot\mathbf{p}/|\mathbf{p}|)\gamma^5(i\gamma^\mu\partial/\partial x'^\mu - m)\psi'(x') \qquad (18.23)$$

using eqs. 18.20, 18.1f and 18.1g; and

$$\mathcal{L}_2' = \psi^\dagger\gamma^0\gamma^5 S_L^{-1}S_L\gamma^5(i\gamma^\mu\partial/\partial x^\mu - m)\gamma^5 S_L^{-1}S_L\gamma^5\psi(x)$$
$$= \psi^\dagger\gamma^0\gamma^5 S_L^{-1}(\gamma^\mu\partial/\partial x'^\mu - m)S_L\gamma^5\psi(x)$$
$$= i\psi^\dagger\gamma^5 S_L^\dagger(\boldsymbol{\gamma}\cdot\mathbf{p}/|\mathbf{p}|)(\gamma^\mu\partial/\partial x'^\mu - m)S_L\gamma^5\psi(x)$$
$$= i\psi_T'^\dagger(x')(\boldsymbol{\gamma}\cdot\mathbf{p}/|\mathbf{p}|)(\gamma^\mu\partial/\partial x'^\mu - m)\psi_T'(x') \qquad (18.24)$$

where $\psi'(x')$ is a solution of the Dirac equation obtained by a superluminal L(C) boost (by $\mathbf{v_c} = \mathbf{p}/(\gamma_s m)$) of the tachyon field and where

---

[237] This relation is derivable from eq. 17.19.

$\psi_T'(x')$ is a solution of the tachyon equation obtained by the same boost of the Dirac field. Eqs. 18.23 and 18.24 clearly show that $\mathcal{L}_8$ is *not* invariant under L(C) transformations in general.

Consequently the action of eq. 18.17 is only invariant under inhomogeneous Lorentz transformations. *This state of affairs is actually an advantage when we derive features of the Standard Model because it will be seen to prevent any interplay between unbroken ElectroWeak SU(2) rotations and L(C) transformations.*

Postulate 18.2. A lepton doublet in any generation consists of a charged Dirac field and a neutral tachyon field (the neutrino). By convention the neutrino field is the topmost component in a lepton doublet. The equations of motion of the lepton doublet are eqs. 18.4 with $\Psi(x)$ = $\Psi_{\ell e}(x)$ (eq. 18.25 below).

Postulate 18.3. A quark doublet in any generation consists of an up-type quark Dirac field and a down-type quark tachyon field. By convention the up-type quark field is the topmost component in a quark doublet. (The quark fields may be "normal" or complexon – an experimental issue.)

18.3.1 The Lepton Doublet

Thus the electron generation doublet is:

$$\Psi_{\ell e}(x) = \begin{bmatrix} \psi_{T\nu} \\ \psi_e \end{bmatrix} = \begin{bmatrix} \nu_e \\ e \end{bmatrix} \tag{18.25}$$

With this interpretation we can introduce SU(2)⊗U(1) gauge interactions and develop leptonic ElectroWeak theory naturally.

We have identified two of the four types of spin ½ fermions as leptons. The remaining two types of spin ½ fermions – complexons – ψ_C

and ψ_{CT} seem to naturally correspond with quarks since their equations of motion and wave functions have a natural global SU(3) symmetry.[238]

We therefore associate a color SU(3) symmetry with these two types of spin ½ complexons. The Electroweak doublet of first generation quarks is then

$$\Psi_{q1}{}^{a}(x) = \begin{bmatrix} \psi_C{}^{a} \\ \psi_{CT}{}^{a} \end{bmatrix} = \begin{bmatrix} u^a \\ d^a \end{bmatrix} \qquad (18.26)$$

where u is the up quark and d is the down quark[239] and a is a color index with a = 1, 2, or 3.

The rationale for constructing quark doublets is the same as the leptonic case: We wish to define a generalization of the "Dirac-like" equations of motion that is covariant under general L(C) boosts.

18.4 L(C) Covariance of Complexon (Quark) Doublet Dynamical Equations

An L(C) spinor boost of a complexon can change it into a tachyonic complexon and vice versa.[240] We will denote an L(C) spinor boost that generates an ordinary complexon (from a free quark at rest) as S_{CLu} (eqs. 17.39-17.50) and an L(C) spinor boost that generates a tachyon complexon (from a free quark at rest) as S_{CLd} (eqs. 17.51-

[238] The global SU(3) symmetry of complexons makes their identification with quarks reasonable. However, the complexon theory that we develop and use for quark dynamics in the Standard Model is <u>not</u> required. So we will consider both cases: non-complexon quarks and complexon quarks.

[239] The lepton situation is clear in the sense that charged leptons cannot be tachyons since their masses are known. Thus tachyonic neutrinos are the only currently allowed possibility. The quark situation is somewhat unclear. We have provisionally chosen the "down" type of quark (d, s, and b) as tachyonic. The association of bound states of these quarks such as the K^0 and B^0 systems which are known to have CP violation, and the CP violation engendered by tachyons, encourages this interpretation.

In addition, $W^{\pm}$ charge asymmetry in $p\bar{p}$ collisions indicate the d sea in a proton is greater than the u sea (K. Abe et al, PRL **74**, 850 (1995)) as does the asymmetry of Drell-Yan production in deep inelastic scattering on p and n targets (A. Baldit et al, Phys. Lett. **B332**, 244 (1994)). These results are to be expected since there is no mass gap for a d tachyon sea while there is a mass gap for a u Dirac particle sea. Complexon quarks may explicate the discrepancies between theory and experiment in the spin structure functions of the parton model for nucleons.

[240] The considerations of this section (and chapter) would still apply if quarks were not complexons but rather Dirac fields and tachyon fields.

17.53).[241] We will define the up-type complexon quark field as $\psi_{Cu}(x)$ and the down-type (tachyon) quark field as $\psi_{Cd}(x)$. S_{CLd} is a superluminal boost that transforms a Dirac-like quark into a tachyon quark and vice versa:

$$S_{CLd}\psi_{Cu}(x) \rightarrow \psi_{Cd}'(x') \qquad (18.27)$$
$$S_{CLd}\psi_{Cd}(x) \rightarrow \psi_{Cu}'(x')$$

Similarly the differential operator used in the equations of motion can also be transformed.

$$S_{CLd}(\gamma^\mu D_\mu - m)S_{CLd}^{-1} = (i\gamma^\mu D'_\mu - m)$$
$$S_{CLd}\gamma^5(i\gamma^\mu D_\mu - m)\gamma^5 S_{CLd}^{-1} = (\gamma^\mu D'_\mu - m) \qquad (18.28)$$

where D_μ is given by eq. 17.59 and where

$$x'^\mu = \Lambda_{CLd}{}^\mu{}_\nu(\omega, \mathbf{v_c})x^\nu \qquad (18.29)$$
$$D'_\mu = \Lambda_{CLd}{}^\nu{}_\mu(\omega, \mathbf{v_c})D_\nu$$

In matrix form

$$X' = \Lambda_{CLd}(\omega, \mathbf{v_c})X \qquad (18.30)$$

Eqs. 18.27 – 18.29 imply

$$S_{CLd}(\gamma^\mu D_\mu - m)\psi_{Cd}(x) = (i\gamma^\mu D'_\mu - m)S_{CLd}\psi_{Cd}(x)$$
$$= (i\gamma^\mu D'_\mu - m)\psi_{Cu}'(x') \qquad (18.31)$$

and

$$S_{CLd}\gamma^5(i\gamma^\mu D_\mu - m)\psi_{Cu}(x) = (\gamma^\mu D'_\mu - m)S_{CLd}\gamma^5\psi_{Cu}(x)$$
$$= (\gamma^\mu D'_\mu - m)\psi_{Cd}'(x') \qquad (18.32)$$

where

$$\psi_{Cu}'(x') = S_{CLd}\psi_{Cd}(x) \qquad (18.33)$$

and

$$\psi_{Cd}'(x') = S_{CLd}\gamma^5\psi_{Cu}(x) \qquad (18.34)$$

---

[241] The reader might think that the generation of both "normal" fermions and tachyonic fermions from a Dirac particle at rest makes them equivalent. This would be true if L_C covariance were not broken to Lorentz covariance as it is. The L_C breakdown makes "normal" fermions different from tachyonic fermions.

Thus neither free complexon dynamical equation is L(C) covariant.

18.4.1 Doublet Dynamical Equation for Complexon Quarks

We will now generalize the complexon quark dynamical equations as we did for lepton dynamical equations so that the generalized equation is covariant under both Lorentz transformations and L(C) boosts. The approach is the same except that the spatial derivatives have an imaginary part and the quark tachyon field is now the low component of the doublet.

So we define an 8 × 8 matrix generalization:

$$đ_C(x) = \begin{bmatrix} (i\gamma^\mu D_\mu - m) & 0 \\ 0 & (\gamma^\mu D_\mu - m) \end{bmatrix} \tag{18.35}$$

with D_μ given by eq. 17.59 and

$$\Psi_C(x) = \begin{bmatrix} \psi_{Cu}(x) \\ \psi_{Cd}(x) \end{bmatrix} \tag{18.36}$$

where $\Psi_C(x)$ is an 8 component field composed of a up-type quark complexon field and a down-type (tachyonic) complexon field.

Then the generalized complexon equation is

$$đ_C(x)\Psi_C(x) = 0 \tag{18.37}$$

We now define the 8 × 8 L(C) spinor boost transformation

$$S_{CL8} = \begin{bmatrix} 0 & S_{CLd} \\ S_{CLd}\gamma^5 & 0 \end{bmatrix} \tag{18.38}$$

with inverse transformation

$$S_{CL8}^{-1} = \begin{bmatrix} 0 & \gamma^5 S_{CLd}^{-1} \\ S_{CLd}^{-1} & 0 \end{bmatrix} \tag{18.39}$$

Applying S_{CL8} to eq. 18.37 yields

$$0 = S_{CL8}đ_C(x)\Psi_C(x) = đ_C(x')\Psi_C'(x') \tag{18.40}$$

where

$$\Psi_C'(x') = \begin{bmatrix} S_{CL8}\psi_{Cd}(x) \\ S_{CL8}\gamma^5\psi_{Cu}(x) \end{bmatrix} = \begin{bmatrix} \psi_{Cu}'(x') \\ \psi_{Cd}'(x') \end{bmatrix} \tag{18.41}$$

Thus the generalized complexon equation is covariant under L(C) boosts. Covariance requires the complexon, and tachyonic complexon, must have the same absolute value for the mass.

It is easy to show that the generalized complexon equation is also covariant under conventional real Lorentz transformations represented as 4×4 diagonal blocks in an 8×8 matrix representation. (The demonstration is analogous to eqs. 18.9 – 18.13.)

The lagrangian density that corresponds to our 8-dimensional construction is

$$\mathcal{L}_{C8} = \overline{\Psi}_C(x)đ_C(x)\Psi_C(x) \tag{18.42}$$

where

$$\overline{\Psi}_C(x) = \Psi_C^{\dagger}|_{\mathbf{x_i} = -\mathbf{x_i}}\,\Gamma_C^{0} \tag{18.43}$$

and

$$\Gamma_C^{0} = \begin{bmatrix} \gamma^0 & 0 \\ 0 & i\gamma^0\gamma^5 \end{bmatrix} \tag{18.44}$$

The action

$$I = \int d^4x\,\mathcal{L}_{C8} \tag{18.45}$$

is invariant under Lorentz transformations S_8 and S_{8A} (eqs. 18.9 – 18.12).

The Hamiltonian density for the 8-dimensional theory is

$$\mathcal{H}_{C8}(x) = \begin{bmatrix} \psi_{Cu}{}^{\dagger}(-i\boldsymbol{\alpha}\cdot\nabla_C + \beta m)\psi_{Cu} & 0 \\ 0 & i\psi_{Cd}{}^{\dagger}\gamma^5(\boldsymbol{\alpha}\cdot\nabla_C + \beta m)\psi_{Cd} \end{bmatrix} \tag{18.46}$$

where

$$\nabla_C = \mathbf{D} \tag{18.47}$$

is the spatial vector part of D^{μ}.

18.4.2 Non-Invariance of the Generalized Free Quark Complexon Action under an L(C) Boost

The action 18.45 is not invariant under L(C) boosts. The reason is similar to that of section 18.3 for the "leptonic" type of particle: there is no simple relation between the hermitean conjugate of an L(C) spinor boost and its inverse that is similar to eq. 18.19 for the Dirac boost case.

Consequently the action of eq. 18.45 is only invariant under inhomogeneous real Lorentz transformations. *This state of affairs is again an advantage when we derive features of the Standard Model because it prevents any interplay between unbroken ElectroWeak SU(2) rotations and* L(C) *transformations in the quark complexon sector.*

18.5 Postulates of the Derivation of the Leptonic ElectroWeak Sector

In this section we will derive the leptonic sector of the Standard Model based on the L(C) covariance of the doublet equations of motion. *The naturalness of the derivation, and its close connection to the L(C) group, strongly suggest the Standard Model, which was grown by theorists from experiment, has an undeniable truth that will likely survive the passage of time. The basis of the derivation in a more fundamental theoretical framework raises the hope that we have found a newer, deeper level of understanding of elementary particle dynamics.*

We will add certain new postulates that provide a basis for the derivation of the form of the leptonic sector of the Standard Model:

Postulate 18.4. The form of the equations of motion of the unbroken leptonic sector of the Standard Model are determined by covariance under the L(C) group, and local gauge symmetries. Electron number is not gauged.

Postulate 18.5. In each generation leptonic particles consist of spin ½ Dirac particles (electron-like) with charge -1 and tachyons (neutrinos) of charge zero as well as their anti-particles.[242]

Postulate 18.6. A neutrino is a tachyon with a non-zero bare mass that may be changed by symmetry breaking effects.

Postulate 18.7. The interacting extensions of doublet equations of motion have interactions consisting of local Yang-Mills fields. Gauge fields are massless "before" spontaneous symmetry breaking and are thus conventional local Yang-Mills gauge fields without a tachyon equivalent.

Postulate 18.8. L(C) group covariance, and gauge symmetries of the dynamical equations of motion, are spontaneously broken through the appearance of additional fermion and vector boson mass terms generated by a mechanism such as the Higgs mechanism or the Dimensional Mass Mechanism (sections 21.3 and 21.4 below).

18.6 Derivation of the Leptonic ElectroWeak Sector of the Standard Model

The steps of the derivation are:

A. L(C) group covariance of the free dynamical equations of motion requires that leptons be described by a generalization of the Dirac equation to an 8×8 matrix form in section 18.2 and eq. 18.25 based on a doublet consisting of a charged Dirac particle and a tachyon neutrino.

---

[242] If neutrinos are Majorana particles then the derivation must be modified.

B. The bare masses of these doublet particles have the same numeric value (before symmetry breaking).

C. The leptonic sector free field lagrangian (without gauge fields introduced yet) is explicitly

$$\mathcal{L}_{\text{freelep}} = \Psi^\dagger(x) \begin{bmatrix} \gamma^0\gamma^5 i(\gamma^\mu\partial/\partial x^\mu - m) & 0 \\ 0 & \gamma^0(i\gamma^\mu\partial/\partial x^\mu - m) \end{bmatrix} \Psi(x) \tag{18.48}$$

Focussing on the derivative term we see that it can be put in the form

$$\Psi^\dagger(x)\gamma^0 \begin{bmatrix} \gamma^5 & 0 \\ 0 & I_4 \end{bmatrix} i\gamma^\mu\partial/\partial x^\mu\, \Psi(x) = \Psi^\dagger(x)\gamma^0[C^+ I - C^- \sigma_3] i\gamma^\mu\partial/\partial x^\mu \Psi(x) \tag{18.49}$$

where I_4 is a 4×4 identity matrix, where I (an identity matrix) and σ_3 are 2 × 2 matrices, and where C^+ and C^- are defined in eq. 17.69. The Pauli matrices σ_i are defined by eq. 16.39.

Expression 18.49 can be re-expressed in terms of left-handed and right-handed fields as

$$\Psi_L{}^\dagger(x)\gamma^0 i\gamma^\mu\partial/\partial x^\mu\Psi_L(x) - \Psi_R{}^\dagger(x)\gamma^0 i\gamma^\mu\partial/\partial x^\mu\sigma_3\Psi_R(x) \tag{18.50}$$

D. At this point we are in a position to introduce couplings to gauge fields. In view of the doublet nature of the fields $\Psi_L(x)$ and $\Psi_R(x)$ it would appear, at first glance, that the symmetry group of the gauge fields would be $SU(2)_L \otimes SU(2)_R$. However the right-handed tachyon field in expression 18.50 has the wrong sign in the lagrangian, as has been noted in the previous discussion of the free tachyon lagrangian and anti-commutator. Consequently the right-handed tachyon field *cannot* have trilinear or higher order couplings. If it did have such interactions then it would rapidly degrade to lower and lower energy by the emission of particles since right-handed leptonic tachyons can exist in principle as free particles (modulo possible Higgs interaction terms). (In this regard the situation is similar to that of time-like photons, except that the set of tachyon physical states cannot be defined in a manner analogous to Gupta-Bleuler electrodynamics where the timelike and longitudinal photons "cancel" each other so that only transverse photons have physical effects.) *Thus there can be no right-handed trilinear or higher leptonic tachyon interactions.*[243] The doublet nature of the left-handed sector implies at least local SU(2) symmetries implemented with a covariant derivative.

The restricted nature of the right-handed leptonic sector indicates that *only* the Dirac particle in the "right-handed doublet" can have an interaction. Also the appearance of σ_3 in the right-handed term in expression 18.50 breaks SU(2) invariance <u>if</u> the left-handed covariant derivative (eq. 18.51 below) were substituted for $\partial/\partial x^\mu$ in the right-handed term. Thus a U(1) local gauge field interaction, restricted to the Dirac field member of the right-handed doublet and coupling to both members of the left-handed doublet, is the only allowed possibility. Without the U(1) interaction, the "right-handed doublet" would have no trilinear or higher order elementary particle interactions and would be physically irrelevant (except gravitationally) in the unbroken gauge theory before spontaneous breakdown.

Putting these symmetries together we obtain a left-handed covariant derivative implementing local SU(2)⊗U(1) invariance:

$$D_{L\mu} = \partial/\partial x^\mu + \tfrac{1}{2} i g_2 \boldsymbol{\sigma}\cdot\mathbf{W}_\mu + \tfrac{1}{2} i g' B_\mu \qquad (18.51)$$

and a right-handed covariant derivative[244]

$$\begin{aligned} D_{R\mu} &= \partial/\partial x^\mu \sigma_3 + \tfrac{1}{2} i g' B_\mu |Q| \sigma_3 \\ &= \partial/\partial x^\mu \sigma_3 + \tfrac{1}{2} i g' B_\mu |Q| \qquad (18.52) \end{aligned}$$

---

[243] Right-handed neutrinos must interact with gravitons due to their mass-energy and the universality of the gravitational interaction. The extreme weakness of the gravitational interaction mitigates this effect.

[244] The coupling constants are defined by $e = -g'\cos\theta_W = g_2 \sin\theta_W$.

where Q is the charge operator using the relation $|Q|\sigma_3 = |Q|$ for leptons. We use the absolute value of Q in order to achieve consistency in form with the right-handed quark sector described in the next chapter. As a result expression 18.50 becomes

$$\Psi_L^\dagger(x)\gamma^0 i\gamma^\mu D_{L\mu}\Psi_L(x) - \Psi_R^\dagger(x)\gamma^0 i\gamma^\mu D_{R\mu}\Psi_R(x) \qquad (18.53)$$

Thus the leptonic sector of the lagrangian[245] (modulo mass/Higgs terms or the Dimensional Mass Mechanism (sections 21.3 and 21.4 below)) is

$$\begin{aligned}\mathcal{L}_{\text{lep1}} &= \Psi_L^\dagger\gamma^0 i\gamma^\mu D_{L\mu}\Psi_L - \Psi_R^\dagger\gamma^0 i\gamma^\mu D_{R\mu}\Psi_R \qquad (18.54)\\ &= \Psi_L^\dagger\gamma^0 i\gamma^\mu D_{L\mu}\Psi_L + \overline{\psi}_{eR}\gamma^0 i\gamma^\mu D_{R\mu}\psi_{eR} - \overline{\psi}_{\upsilon R}\gamma^0 i\gamma^\mu \partial/\partial x^\mu\psi_{\upsilon R}\\ &= \Psi_L^\dagger\gamma^0 i\gamma^\mu D_{L\mu}\Psi_L + \overline{\psi}_{eR}\gamma^0 i\gamma^\mu(\partial/\partial x^\mu + \tfrac{1}{2}ig'B_\mu)\psi_{eR} -\\ &\quad - \overline{\psi}_{\upsilon R}\gamma^0 i\gamma^\mu \partial/\partial x^\mu\psi_{\upsilon R}\end{aligned}$$

where we identify the tachyon as a neutrino and the Dirac particle as its charged lepton partner. Our leptonic sector lagrangian is now the usual leptonic sector ElectroWeak lagrangian with a tachyon neutrino.

E. Local gauge invariance prior to symmetry breaking: The gauge field sector has the usual local Yang-Mills lagrangian terms and the B field has lagrangian terms similar to that of the QED lagrangian.

F. Spontaneous symmetry breaking of gauge symmetry, and of L(C) group covariance, via the Higgs mechanism (or an alternative mechanism such as the Dimensional Mass Mechanism (sections 21.3 and 21.4)) can be implemented in such a way as to give the electron its known mass as well as the massive vector bosons. Since spontaneous symmetry breaking breaks L(C) covariance to Lorentz covariance it is a moot point whether the Higgs sector (if there is one) exhibits a similar covariance.

---

245 Note again that the gauge fields do not have a tachyon equivalent since they are initially massless prior to spontaneous symmetry breaking.

This concludes the derivation of leptonic sector of the Standard Model. We have shown that the form of the leptonic sector of the Standard Model is fundamental in nature and based on L(C) group covariance of the equations of motion (or L(C) covariance) and the postulate that leptons can be L(C) boosted from a rest state subject to the restriction of one real time. *Thus the leptonic ElectroWeak sector is directly based on geometry.*

It is then not correct to view the ElectroWeak sector of the Standard Model as the result of the breakdown of a larger internal symmetry group to SU(2)⊗U(1). On the contrary, the SU(2)⊗U(1) interaction form is a consequence of the L(C) covariance of the free dynamical equations.

18.7 Postulates of the Quark ElectroWeak Sector

The derivation of the form of the quark ElectroWeak sector of the Standard Model is very similar to the preceding derivation of the form of the leptonic ElectroWeak sector – but with some important points of difference. The primary difference is that we can identify quarks with complexons[246] or with a "normal" doublet. We consider both cases in this derivation. (As we will see it is plausible to assume complexon quarks in view of the appearance of global SU(3) symmetry in complexon field dynamics. Since complexons have very different spin characteristics compared to Dirac fields, complexon quarks may also resolve difficulties in parton spin studies.)

The quark ElectroWeak sector postulate is:

Postulate 18.9. The equations of motion of free quark doublets in the unbroken form of the ElectroWeak sector of the Standard Model are determined by covariance under the complex Lorentz group L(C), and local gauge symmetries. Baryon number is not gauged.

---

[246] This choice seemed to be the appropriate choice in view of the global SU(3) invariance of complexons. However we could have opted for a Dirac particle and ordinary tachyon combined in a doublet and developed a quark ElectroWeak sector in exactly the same form as the conventional Standard Model ElectroWeak sector.

18.8 Derivation of the Quark ElectroWeak Sector of the Standard Model

The derivation:

A. L(C) covariance of the dynamical equations of motion requires that spin ½ particles be described by equations[247] generalized to an 8×8 matrix form (eqs 18.35 – 18.37) based on a doublet consisting of a Dirac-type particle and a tachyon. (They may or may not be complexons.) In in the case of quarks the Dirac-type particle is the top component in the doublet and the tachyon is the bottom component.

B. Thus the 8×8 quark matrix formalism is:

$$đ_q(x)\Psi_C(x) = 0 \qquad (18.37q)$$

where[248]

$$đ_q(x) = \begin{bmatrix} (i\gamma^\mu D_{q\mu} - m_0) & 0 \\ 0 & (\gamma^\mu D_{q\mu} - m_0) \end{bmatrix} \qquad (18.35q)$$

and

$$\Psi_C(x) = \begin{bmatrix} \psi_{Cu}(x) \\ \psi_{Cd}(x) \end{bmatrix} \qquad (18.36q)$$

---

[247] The global SU(3) symmetry of complexons makes their identification with quarks reasonable. However, the complexon theory that we develop and use for quark dynamics in the Standard Model is not required. Our Standard Model could use Dirac fermion dynamics for the up-type quarks and tachyon dynamics for down-type quarks. We choose to use complexon dynamics for quarks because they have an internal SU(3)-like structure suggestive of color SU(3). More importantly, their spin dynamics is different and this difference may resolve the differences between theory and experiment for the deep inelastic parton spin-dependent structure functions. Nevertheless, quarks could be similar to leptons in this regard and form a doublet of a Dirac fermion and an ordinary tachyon. Whether quarks are complexons or not is an experimental question!

[248] If we wish to obtain the Standard Model without complexon quarks (which is the conventional Standard Model) then $D_{q\mu} = \partial/\partial x^\mu$. If we wish to obtain the Standard Model with complexon quarks (which is a new form of the Standard Model) then $D_{q\mu}$ is given by eq. 17.59.

The upper 4-component field is a u-type field and the lower 4-component field is a d-type tachyonic field. The generalized equation eq. 18.37q is covariant under 8 × 8 L(C) transformations similar to eqs. 18.38 – 18.41. The generalized quark fermion equation eq. 18.35q is also covariant under conventional Lorentz transformations in the 8 × 8 representation.

The free quark sector lagrangian density that corresponds to the 8-dimensional fermion equation eq. 18.37q is

$$\mathcal{L}_{\text{freeQuark}} = \overline{\Psi}_C(x)\not{d}_q(x)\Psi_C(x) \tag{18.55}$$

where

$$\overline{\Psi}_C(x) = \Psi_C^{\dagger}\Gamma_C^{\,0} \tag{18.56}$$

with

$$\Psi_C^{\dagger} = (\Psi_C^{\dagger})|_{\mathbf{x_i} = -\mathbf{x_i}} \tag{18.57a}$$

for complexon quarks, and

$$\Psi_C^{\dagger} = \text{the hermitean conjugate of } \Psi_C \tag{18.57b}$$

in the case of non-complexon quarks. The † in the parentheses on the right side of eq. 18.57a indicates hermitean conjugation. Also

$$\Gamma_C^{\,0} = \begin{bmatrix} \gamma^0 & 0 \\ 0 & i\gamma^0\gamma^5 \end{bmatrix} \tag{18.58}$$

C. The free quark field lagrangian is explicitly

$$\mathcal{L}_{\text{freeQuark}} = \Psi_C^{\dagger} \begin{bmatrix} \gamma^0(i\gamma^{\mu}D_{q\mu} - m_0) & 0 \\ 0 & \gamma^0\gamma^5 i(\gamma^{\mu}D_{q\mu} - m_0) \end{bmatrix} \Psi_C \tag{18.59}$$

Focussing on the derivative term we see that it can be put in the form

$$\Psi_C{}^{\dagger}\gamma^0 \begin{bmatrix} I_4 & 0 \\ 0 & \gamma^5 \end{bmatrix} i\gamma^{\mu}D_{q\mu}\Psi_C = \Psi_C{}^{\dagger}\gamma^0[C^{+}I + C^{-}\sigma_3]i\gamma^{\mu}D_{q\mu}\Psi_C \tag{18.60}$$

where I_4 is a 4×4 identity matrix, where I, and the Pauli matrix σ_3, are 2×2 matrices, and C^+ and C^- are defined in eq. 17.69.

Expression 18.60 can be expressed in terms of left-handed and right-handed fields as

$$\Psi_{CL}{}^{\dagger}\gamma^0 i\gamma^{\mu}D_{q\mu}\Psi_{CL} + \Psi_{CR}{}^{\dagger}\gamma^0 i\gamma^{\mu}D_{q\mu}\sigma_3\Psi_{CR} \tag{18.61}$$

D. At this point we are in a position to introduce couplings to gauge fields. In view of the doublet nature of the fields $\Psi_{CL}(x)$ and $\Psi_{CR}(x)$ it would again appear, at first glance, that the symmetry group of the gauge fields would be $SU(2)_L \otimes SU(2)_R$. However the right-handed tachyonic field has the wrong sign in the lagrangian, as has been noted in the earlier discussion of the free tachyon lagrangian and anti-commutator. Consequently, **if quarks were *not* confined,** the right-handed quark tachyon field could not have trilinear or higher order couplings. **If free tachyon quarks existed, and had interactions, then they would rapidly degrade to lower and lower energy by the emission of particles.**

However, because of quark confinement in bound states with discrete energy levels, a bound tachyon quark can only emit particles if a lower energy bound state exists. As a result right-handed tachyon quarks can have interactions, such as the electromagnetic interaction, because quark confinement "tames" their propensity to emit particles due to the "wrong sign" in the lagrangian.

Again there is an analogy to Gupta-Bleuler QED quantization. In Gupta-Bleuler quantization physical states are required to have equal numbers of time-like and longitudinal photons thus canceling their physical effects. Similarly, right-handed tachyon quarks are required to be bound to other quarks by quark confinement to avoid continuous emission of

particles.[249] Since interactions are allowed for right-handed tachyon quarks the Higgs mechanism (or the Dimensional Mass Mechanism (sections 21.3 and 21.4)) can be used to change their mass.

The doublet nature of the left-handed sector implies at least local SU(2) symmetries. The appearance of σ_3 in the right-handed term in expression 18.61 explicitly breaks SU(2) invariance if the left-handed covariant derivative (eq. 18.62 below) were substituted for $D_{q\mu}$ in the right-handed term. Thus the right-handed fields can only have a U(1) local gauge field interaction, and are SU(2) singlets. We thus obtain a left-handed covariant derivative implementing local SU(2)⊗U(1) covariance:

$$D_{qL\mu} = D_{q\mu} + \tfrac{1}{2} i g_2 \boldsymbol{\sigma}\cdot\mathbf{W}_\mu + i g_1 B_\mu/6 \qquad (18.62)$$

and a right-handed covariant derivative[250]

$$D_{qR\mu} = D_{q\mu}\sigma_3 + i g_1 B_\mu |Q| \qquad (18.63)$$

where |Q| is the absolute value of the charge operator (with u eigenvalue 2/3 and d eigenvalue 1/3). The absolute value is used in order to compensate for the minus sign in front of the right-handed tachyon (d quark) term. As a result expression 18.61 becomes

$$\Psi_{CL}{}^{\dagger}(x)\gamma^0 i\gamma^\mu D_{qL\mu}\Psi_{CL}(x) + \Psi_{CR}{}^{\dagger}(x)\gamma^0 i\gamma^\mu D_{qR\mu}\Psi_{CR}(x) \qquad (18.64)$$

Thus the quark sector of the lagrangian[251] (modulo possible mass/Higgs/Dimensional Mass Mechanism (sections 21.3 and 21.4), and Strong interaction, terms) is

---

[249] Therefore quark confinement is required in order to have a properly formulated quark sector. Another interaction – the strong interaction – is required for quark confinement. Presently there is only one accepted mechanism for quark confinement – through a non-abelian gauge coupling. (Higher derivative theories with quark confinement are in disfavor.) An additional non-abelian symmetry must be introduced for quarks. As discussed later, SU(3) appears to be a natural choice.

[250] The quark SU(2) coupling constant is, by gauge invariance, required to have the same value as the leptonic SU(2) coupling constant. The U(1) coupling constants are not required to be the same in both sectors and, in fact, are different. The coupling constants here are defined by $e = g_1\cos\theta_W = g_2\sin\theta_W$.

$$\mathcal{L}_{\text{quark}1} = \Psi_{CL}{}^{\dagger}\gamma^{0}i\gamma^{\mu}D_{qL\mu}\Psi_{CL} + \Psi_{CR}{}^{\dagger}\gamma^{0}i\gamma^{\mu}D_{qR\mu}\Psi_{CR} \tag{18.65}$$

$$= \Psi_{CL}{}^{\dagger}\gamma^{0}i\gamma^{\mu}D_{qL\mu}\Psi_{CL} + \overline{\psi}_{CuR}i\gamma^{\mu}D_{qR\mu}\psi_{CuR} - \overline{\psi}_{CdR}i\gamma^{\mu}D_{qR\mu}\psi_{CdR}$$

$$= \Psi_{CL}{}^{\dagger}\gamma^{0}i\gamma^{\mu}D_{qL\mu}\Psi_{CL} + \overline{\psi}_{CuR}i\gamma^{\mu}(D_{q\mu} + \tfrac{2}{3}ig_1B_\mu)\psi_{CuR} -$$
$$- \overline{\psi}_{CdR}i\gamma^{\mu}(D_{q\mu} + \tfrac{1}{3}ig_1B_\mu)\psi_{CdR}$$

where we *provisionally* identify the tachyon as a d-type quark and the Dirac-type particle as a u-type quark. Our quark sector lagrangian is now the usual Standard Model quark sector lagrangian (modulo possible mass/Higgs/Dimensional Mass Mechanism (sections 21.3 and 21.4), and strong interaction, terms) except that d-type quarks are tachyons.

E. The SU(2) gauge field sector has the usual Yang-Mills lagrangian terms, and the B field is a U(1) abelian gauge field.

F. Spontaneous symmetry breaking of gauge symmetry, and of L(C) covariance, via the Higgs mechanism or the Dimensional Mass Mechanism (sections 21.3 and 21.4) can be implemented in such a way as to give the quarks, and massive vector bosons, their "known" masses. Since spontaneous symmetry breaking breaks L(C) covariance of the dynamical equations of motion to Lorentz covariance it is a moot point whether the Higgs sector (if there is one) is manifestly L(C) covariant or not.

Thus the initial L(C) covariance of the doublet quark equations of motion results in the "unusual" features of the ElectroWeak quark sector of the Standard Model.

---

[251] Note that the gauge fields do not have a tachyon equivalent since they are initially massless prior to spontaneous symmetry breaking.

18.9 Geometrical Basis of Parity Violation and ElectroWeak Theory

Ockham's Razor[252] (principle 2 of section 12.1) embodies the precept that of all the proposed solutions to a problem the simplest is preferred. The essence of the derivation of the form of the ElectroWeak sector of the Standard Model is 1) in the extension of the Lorentz group (which has real parameters) to the complex Lorentz group L(C); and 2) in the assumption that the four fermion species follow from L(C) boosts of a spin ½ particle at rest to a particle with a real energy (and thus a real time coordinate). The rest consists of canonical quantum field theory considerations and Yang-Mills gauge fields. No other approach to the derivation/understanding of ElectroWeak theory has such a simple basis. And thus this approach is to be preferred not only because of Ockham's Razor but also because its naturalness connotes its truth.

[252] William of Ockham – Law of Parsimony – "Pluralitas non est ponenda sine necessitate" or "Plurality should not be posited without necessity." First stated by Durand De Saint-Pourçain (1270-1334 AD).

19. Color SU(3)

19.1 Two Possible Approaches to Color SU(3)

There are two approaches to obtaining the Strong interaction and Color SU(3) symmetry, which is believed to be the symmetry of the Strong interaction:

1. Assume up-type and down-type quarks are in $\underline{3}$ representations of Color SU(3). This assumption sheds no light on a deeper origin of the Strong interaction and Color SU(3). It simply restates the features of the Strong interaction sector of the Standard Model. Thus our understanding is not deepened by this assumption. The postulate corresponding to this assumption is:

Postulate 19.1a. Quarks are in the $\underline{3}$ representation of Color SU(3). The SU(3) symmetry is gauged with local Yang-Mills SU(3) fields called gluons that constitute the Strong interaction of the quark sector. Quarks are minimally coupled to the gluons in a gauge covariant fashion.

2. In the preceding chapter the ElectroWeak sector of the Standard Model (modulo generations and their mixing) was shown to essentially follow from the L(C) covariance of the doublet equations of motion for leptons and quarks, minimal coupling to gauge fields and stability requirements for tachyons. Thus we have a geometrical basis for the form of the ElectroWeak sector. We would like to establish a similar geometrical basis for the Strong interaction and Color SU(3). In this regard we note that the generators of GL(4) were shown (section 16.5) to include the generators of the Lorentz group and the generators of U(3) as well as a U(1) generator. Since GL(4) has complex parameters, and has a complex 4-dimensional representation which we take to be an extension of our real coordinate space, GL(4) has L(C) as a subgroup. We used L(C) to extract the form of the ElectroWeak sector in the preceding chapter.

We now suggest that the subset of SU(3) generators of GL(4) is the source of Color SU(3) when we limit the parameter set of the subgroup defined with these generators to real values. If we now extend the parameters to be real functions of the space-time coordinates (i.e. local SU(3) transformations), then we obtain a geometrical basis for color SU(3). A key factor in this interpretation is the global covariance of complexon equations of motion under global SU(3). We establish a parallel between the geometrical formulation of the ElectroWeak sector and the geometrical formulation of Color SU(3) and the Strong interaction.(see Fig. 19.1.) This line of development leads to the following alternative postulate. We then analyze the issues that are raised by this approach to Color SU(3).

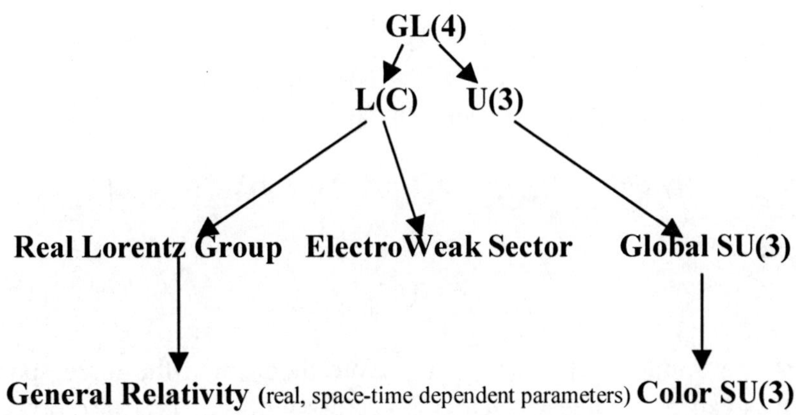

Figure 19.1. Generation of the ElectroWeak sector and color SU(3) from L(C). This diagram is symbolic. For example General Relativity actually originates in 16-dimensional space-time which then breaks down to the real 4-dimensional space-time we experience.

Postulate 19.1b.Quarks are complexons.

We now proceed to derive local Color SU(3).

19.2 Complexon Basis of Color SU(3) and the Strong Interactions

In section 17.6 we showed that the equations of motion of free Dirac-like, complexon, up-type quarks are covariant under the SU(3) subgroup of GL(4). The free, tachyon, complexon, down-type quark equations of motion are also easily seen to be covariant under this SU(3) subgroup. In this section we will show this covariance is the "source" of the local Color SU(3) symmetry of quarks, and then we will introduce the Strong interaction via minimal coupling to SU(3) Yang-Mills gluons in gauge covariant derivatives.

In section 17.6 we considered a global SU(3) rotation of the derivative vector:

$$\mathbf{D_c}^* = \nabla_\mathbf{c} = \nabla_\mathbf{r} + i\nabla_\mathbf{i}$$

namely,

$$D_c{}^{*a} = D_c{}'^{*b}U^{ab*}$$

where U^{ab} is a global SU(3) rotation. We saw

$$[i\gamma^0\partial/\partial t + i\mathbf{D_c}^*\cdot\gamma - m] = [i\gamma^0\partial/\partial t + i\mathbf{D_c}'^*\cdot\gamma' - m] \qquad (17.107)$$

where

$$\gamma^a = U^{ab}\gamma'^b \qquad (17.108)$$

due to the unitarity of U. ($U^\dagger = U^{-1}$) We then noted the three spatial γ-matrices can be treated as SU(3) 3-vectors with eq. 17.108 representing the result of an SU(3) rotation V of the γ-matrices (on the spinor indices)

$$V\gamma^a V^{-1} = U^{ab}\gamma'^b \qquad (17.109)$$

where V is a 4×4 reducible representation[253] of SU(3), namely, $\underline{3} + \underline{1}$. Thus we see that the free complexon equations of motion (eq. 17.49) are covariant under SU(3) transformations:

$$V[i\gamma^0\partial/\partial t + i\gamma\cdot\mathbf{D_c}^* - m]V^{-1}V\psi_C(t, \mathbf{x_r}, \mathbf{x_i}) = 0$$

---

[253] Eqs. 16.50, and 16.53 furnish a reducible 4×4 matrix representation described later. The generators of this reducible representation are ${}_4F_i = F_i + \Omega_{+00} - F_9$ with F_i being the Gell-Mann SU(3) generators for i = 1, 2, …, 8. See eqs. 17.122 and 17.123.

or

$$[i\gamma^{0\prime}\partial/\partial t' + i\mathbf{D_c}'^{*}\cdot\boldsymbol{\gamma}' - m]V\psi_C(t, \mathbf{x_r}, \mathbf{x_i}) = 0 \qquad (17.110)$$

with the SU(3) transformed wave function $\psi_C'(t, \mathbf{x}')$ being

$$\psi_C'(t', \mathbf{x}') = V\psi_C(t, \mathbf{x}) = V\psi_C(t', U\mathbf{x}') \qquad (17.111)$$

(Note $\gamma^{0\prime} = V\gamma^0V^{-1}$ and t' = t.) Thus the complexon Dirac equation is covariant under global SU(3). Note also the relation of the SU(3) vectors $\mathbf{x} = U\mathbf{x}'$ can be expressed in component form by:

$$x^a = U^{ab}x'^b \qquad (19.1)$$

The subsidiary condition, eq. 17.50, is also covariant under an SU(3) rotation:

$$\nabla_{\mathbf{r}}'^{*}\cdot\nabla_{\mathbf{i}}'\psi_C'(t', \mathbf{x}') = \nabla_{\mathbf{r}}\cdot\nabla_{\mathbf{i}}V\psi_C(t, \mathbf{x}) = V\nabla_{\mathbf{r}}^{*}\cdot\nabla_{\mathbf{i}}\,\psi_C(t, \mathbf{x}) = 0 \qquad (17.112)$$

We now introduce a complexon field with a global SU(3) index a which takes values from 1 to 3 making the field a member of the 3 representation of global SU(3):

$$\psi_C^{a}(t, \mathbf{x}) \qquad (19.2)$$

Due to the SU(3) index the transformation property of $\psi_C^{a}(t, \mathbf{x})$ changes from eq. 17.111 to

$$\psi_C''^{a}(t, \mathbf{x}') = U^{ab}V\psi_C^{b}(t, \mathbf{x}) = U^{ab}V\psi_C^{b}(t, U\mathbf{x}') \qquad (19.3)$$

where U^{ab} is an SU(3) rotation of 3 representation "vectors" such as ψ_C^{b} and **x.** V is the corresponding rotation of the spinor indices of $\psi_C^{b}(t, \mathbf{x})$.

Note that the SU(3) rotation of the field factorizes into an SU(3) rotation of the three fields ψ_C^{b} by U^{ab} and an SU(3) rotation of the four spinor components of each individual field ψ_C^{b} by V.

This factorization enables us to consider a global SU(3) rotation of the $\psi_C{}^b$ fields while holding the coordinates fixed:

$$\psi_C{}'^a(t, \mathbf{x}) = U^{ab}\psi_C{}^b(t, \mathbf{x}) \tag{19.4}$$

The equations of motion are covariant under this global transformation

$$0 = U^{ab}[i\gamma^0\partial/\partial t + i\boldsymbol{\gamma}\cdot\mathbf{D_c}^* - m]\psi_C{}^b(t, \mathbf{x_r}, \mathbf{x_i})$$

$$= [i\gamma^0\partial/\partial t + i\boldsymbol{\gamma}\cdot\mathbf{D_c}^* - m]\psi'_C{}^a(t, \mathbf{x_r}, \mathbf{x_i}) \tag{19.5}$$

We now note the form of eq. 19.4 is the same as that of a *local* Yang-Mills rotation:

$$\psi_C{}'^a(t, \mathbf{x}) = \Theta^{ab}(t, \mathbf{x})\psi_C{}^b(t, \mathbf{x}) \tag{19.6}$$

where $\mathbf{x} = \mathbf{x_r} + i\mathbf{x_i}$. Therefore if we introduce a local SU(3) Yang-Mills field $A_{C\nu}(t, \mathbf{x_r}, \mathbf{x_i})$ and define a covariant derivative we can generalize eq. 19.5 to the case of local, color SU(3) if we do <u>not</u> perform the spinor rotation V.[254]

$$\mathcal{D}_\nu = D_\nu - igA_{C\nu} \tag{19.7}$$

where

$$A_{C\nu} = A_C{}^a{}_\nu t^a \tag{19.8}$$

and where D_ν is given by eq. 17.59. The SU(3) 3×3 matrix generators satisfy

$$[t^a, t^b] = if^{abc}t^c \tag{19.9}$$

We can represent $\Theta_{ab}(x)$ in the form:

---

[254] This approach enables us to avoid the dilemmas associated with mixing coordinate and internal symmetries as described by Coleman, S., Phys. Rev. **138** B1262 (1965) and others in the case of SU(6) in the 1960's. Note that the spinor rotation V is expressed in terms of numerical matrices while, in the second quantized formulation, the U^{ab} rotation is expressed in terms of second quantized fields as well as numeric matrices. Thus the factorization is reflected in the form of the transformation.

$$\Theta_{ab}(x) = [\exp(-i\varphi_c(x)t^c)]_{ab} \tag{19.10}$$

where $\varphi_c(x)$ is a local parameter dependent on $x = (x^0, \mathbf{x} = \mathbf{x_r} + i\mathbf{x_i})$, and t^c is an SU(3) generator.

Applying a gauge transformation to the gauge covariant derivative of a complexon fermion field $\mathfrak{D}_v\psi_C(x)$:

$$\Theta\mathfrak{D}_v\psi_C(x) = \Theta D_v\psi_C(x) - ig\Theta A_{Cv}\Theta^{-1}\Theta\psi_C(x) \tag{19.11}$$

$$= D_v\psi_C'(x) - igA_C'{}_v\psi_C'(x) = (\mathfrak{D}_v\psi_C(x))'$$

where

$$\psi_C'(x) = \Theta(x)\psi_C(x) \tag{19.12}$$

we find

$$A_C'{}_v = (-i/g)(D_v\Theta(x))\Theta^{-1}(x) + \Theta(x)A_{Cv}(x)\Theta^{-1}(x) \tag{19.13}$$

The reader will note that the form of eqs. 19.6 – 19.12 is identical to those associated with a conventional non-abelian gauge interaction with the replacement:

$$\partial/\partial x^v \rightarrow D_v \tag{19.14}$$

with D_v given by eq. 17.59. Note that $\varphi_c(x)$, the local parameter in eq. 19.10 is dependent in general, on time, and the real and imaginary parts of the complex spatial 3-vector.

Introducing the SU(3) gauge covariant derivative transforms eq. 19.5 to

$$0 = [i\gamma^v\mathfrak{D}_v - m]\psi_C{}^a(t, \mathbf{x_r}, \mathbf{x_i}) \tag{19.15}$$

The preceding argumentation supports the following posulates (assuming Postulate 19.1b):

Postulate 19.2. Quarks are in a $\underline{3}$ representation of the global SU(3) subgroup of GL(4). Their transformation law under SU(3) transformations is eq. 19.6 wherein the space-time coordinates are not transformed.

Postulate 19.3. The covariance of the quark equations of motion under global SU(3) transformations becomes covariance under local color SU(3) transformations when the equations of motion are generalized to gauge covariant derivatives. We assume the equations of motion are so generalized. The interaction terms so introduced constitute the Strong interaction.

We note the case of tachyon complexon quarks differs only in small details from the above discussion of Dirac-type complexon quarks.

19.3 The One Generation Standard Standard Model

If we assume postulate Postulate 19.1a so that quarks are not complexons then we obtain the conventional form of the Standard Model modulo the values of coupling constants, the introduction of masses (Higgs or Dimensional Mass Mechanisms (sections 21.3 and 21.4)), the introduction of generations and the mixing of the generations.

The lagrangian density that we have derived for the conventional Standard Model[255] (with one generation at this point in the derivation) is

$$\mathcal{L}_{SSM} = \Psi_L{}^{\dagger}\gamma^0 i\gamma^{\mu}D_{L\mu}\Psi_L - \Psi_R{}^{\dagger}\gamma^0 i\gamma^{\mu}D_{R\mu}\Psi_R +$$

$$+ \Psi_{qL}{}^{\dagger}\gamma^0 i\gamma^{\mu}D_{qL\mu}\Psi_{qL} + \Psi_{qR}{}^{\dagger}\gamma^0 i\gamma^{\mu}D_{qR\mu}\Psi_{qR} -$$

$$- \mathcal{L}_{BareMasses} + \mathcal{L}_{Gauge} + \mathcal{L}_{Higgs} \tag{19.16}$$

where

$$D_{L\mu} = \partial/\partial x^{\mu} + \tfrac{1}{2}ig_2\boldsymbol{\sigma}\cdot\mathbf{W}_{\mu} - ig_1B_{\mu}/2 \tag{19.17a}$$

$$D_{R\mu} = \sigma_3\partial/\partial x^{\mu} + ig_1B_{\mu}|Q| \tag{19.17b}$$

$$D_{qL\mu} = D_{\mu} + \tfrac{1}{2}ig_2\boldsymbol{\sigma}\cdot\mathbf{W}_{\mu} + ig_1B_{\mu}/6 \tag{19.18}$$

$$D_{qR\mu} = \sigma_3 D_{\mu} + ig_1B_{\mu}|Q| \tag{19.19}$$

where D_v is given by

---

[255] We call this theory the one generation Standard Standard Model.

$$D_\nu = \partial/\partial x^\nu - igA_\nu \tag{19.20}$$

where $\Psi_q^\dagger$ is the hermitean conjugate of the quark doublet with $\Psi_{qL}^\dagger$ and $\Psi_{qR}^\dagger$ being its left-handed and right-handed parts respectively, where A_ν are local SU(3) Yang-Mills fields, and where

$$\mathcal{L}_{Gauge} = \mathcal{L}_{GaugeEW} + \mathcal{L}_{GaugeColor} \tag{19.21}$$

with

$$\mathcal{L}_{GaugeEW} = -\tfrac{1}{4}\, F_W{}^{a\mu\nu} F_W{}^{a}{}_{\mu\nu} - \tfrac{1}{4}\, F_B{}^{\mu\nu} F_{B\mu\nu} + \mathcal{L}_{EW}{}^{ghost} \tag{19.22}$$

and

$$\mathcal{L}_{GaugeColor} = -\tfrac{1}{4}\, F_{Color}{}^{a\mu\nu} F_{Color}{}^{a}{}_{\mu\nu} + \mathcal{L}_{Color}{}^{ghost} \tag{19.23}$$

$\mathcal{L}_{BareMasses}$ contains the fermion bare mass terms. The ElectroWeak gauge bosons $W_\mu{}^a$ and B_μ field tensors are:

$$F_W{}^{a}{}_{\mu\nu} = \partial W^a{}_\mu/\partial x^\nu - \partial W^a{}_\nu/\partial x^\mu + g_2 f^{abc} W^b{}_\mu W^c{}_\nu \tag{19.24}$$

$$F_{B\mu\nu} = \partial B_\mu/\partial x^\nu - \partial B_\nu/\partial x^\mu \tag{19.25}$$

$\mathcal{L}_{EW}{}^{ghost}$ are the Faddeev-Popov ghost terms for the ElectroWeak $W_\mu{}^a$ gauge bosons. The color gauge field tensors are

$$F_{Color}{}^{a}{}_{\mu\nu} = \partial A^a{}_\mu/\partial x^\nu - \partial A^a{}_\nu/\partial x^\mu + g_2 f_{su(3)}{}^{abc} A^b{}_\mu A^c{}_\nu \tag{19.26}$$

$\mathcal{L}_{Color}{}^{ghost}$ is the color SU(3) ghost terms defined using the Faddeev-Popov mechanism. The Higgs sector(?) $\mathcal{L}_{Higgs}$ or the Dimensional Mass Mechanism (sections 21.3 and 21.4) generates fermion masses and ElectroWeak vector boson masses.

The effective action for the path integral formulation is

$$I_{SSM} = \int dx^0 d^3x\ \mathcal{L}_{SSM} \tag{19.27}$$

19.4 The One Generation Complexon Standard Model

If we assume Postulates 19.1b, 19.2, and 19.3 then we obtain the (one generation) Complexon Standard Model which differs from the "Standard Standard Model" by having complexon quarks and gluons. Which of these Standard Models is the correct one chosen by Nature? That is a question that only experiment can answer. However the naturalness of color SU(3) within the complexon formulation suggests that the Complexon Standard Model is the correct Standard Model.

In this section we show the results of the derivation of previous sections for the Complexon Standard Model: its lagrangian density and action.

$$\mathcal{L}_{CSM} = \Psi_L^{\dagger}\gamma^0 i\gamma^\mu D_{L\mu}\Psi_L - \Psi_R^{\dagger}\gamma^0 i\gamma^\mu D_{R\mu}\Psi_R + $$

$$+ \Psi_{CL}^{\dagger}\gamma^0 i\gamma^\mu \mathfrak{D}_{qL\mu}\Psi_{CL} + \Psi_{CR}^{\dagger}\gamma^0 i\gamma^\mu \mathfrak{D}_{qR\mu}\Psi_{CR} -$$

$$- \mathcal{L}_{BareMasses} + \mathcal{L}_{Gauge} + \mathcal{L}_{Mass} \tag{19.28}$$

using eq. 18.65 where $D_{L\mu}$ and $D_{R\mu}$ are given by eq. 19.17, and where the left-handed complexon quark covariant derivative[256] is

$$\mathfrak{D}_{qL\mu} = \mathfrak{D}_\mu + \tfrac{1}{2} i g_2 \boldsymbol{\sigma}\cdot\mathbf{W}_\mu + i g_1 B_\mu/6 \tag{19.29}$$

and the right-handed complexon quark covariant derivative is

$$\mathfrak{D}_{qR\mu} = \sigma_3\mathfrak{D}_\mu + i g_1 B_\mu|Q| \tag{19.30}$$

where $\mathfrak{D}_\mu$ is given by eq. 19.7. The other covariant derivatives are

$$D_{qR\mu} = D_{q\mu}\sigma_3 + i g_1 B_\mu|Q| \tag{19.31}$$

---

[256] The quark SU(2) coupling constant is, by gauge invariance, required to have the same value as the leptonic SU(2) coupling constant. The U(1) coupling constants are not required to be the same in both sectors and, in fact, are different. The coupling constants here are defined by $e = g_1\cos\theta_W = g_2\sin\theta_W$.

where $\Psi_C{}^\dagger$ is given by eq. 18.57a, and $\Psi_{CL}{}^\dagger$ and $\Psi_{CR}{}^\dagger$ are its left-handed and right-handed parts respectively. $\mathcal{L}_{BareMasses}$ contains the fermion bare mass terms. Also,

$$\mathcal{L}_{Gauge} = \mathcal{L}_{GaugeEW} + \mathcal{L}_{GaugeC} \tag{19.32}$$

with

$$\mathcal{L}_{GaugeEW} = -\tfrac{1}{4}\, F_W{}^{a\mu\nu}F_W{}^a{}_{\mu\nu} - \tfrac{1}{4}\, F_B{}^{\mu\nu}F_{B\mu\nu} + \mathcal{L}_{EW}{}^{ghost} \tag{19.22}$$

and

$$\mathcal{L}_{GaugeC} = \mathcal{L}_{CCG} + \text{Faddeev-Popov terms} \tag{19.33}$$

See eqs. 19A.17, 19A.18, 19A.26 and 19A.28 appearing later for the Faddeev-Popov terms. The ElectroWeak gauge bosons $W_\mu{}^a$ and B_μ field tensors are:

$$F_W{}^a{}_{\mu\nu} = \partial W^a{}_\mu/\partial x^\nu - \partial W^a{}_\nu/\partial x^\mu + g_2 f^{abc} W^b{}_\mu W^c{}_\nu \tag{19.24}$$

$$F_{B\mu\nu} = \partial B_\mu/\partial x^\nu - \partial B_\nu/\partial x^\mu \tag{19.25}$$

$\mathcal{L}_{EW}{}^{ghost}$ contains the Faddeev-Popov ghost terms for the ElectroWeak $W_\mu{}^a$ gauge bosons. The complexon color gluon lagrangian term $\mathcal{L}_{CCG}$ is defined by

$$\mathcal{L}_{CCG} = -\tfrac{1}{4}\, F_{CC}{}^{a\mu\nu}(x) F_{CC}{}^a{}_{\mu\nu}(x) \tag{19.34}$$

where

$$F_{CC}{}^a{}_{\mu\nu} = D_\nu A_C{}^a{}_\mu - D_\mu A_C{}^a{}_\nu + g f_{su(3)}{}^{abc} A_C{}^b{}_\mu A_C{}^c{}_\nu \tag{19.35}$$

where $A_C{}^a{}_\nu$ is the color gluon complexon gauge field, and D_μ is defined by eq. 17.59, g is the color coupling constant, and the $f_{su(3)}{}^{abc}$ are the SU(3) structure constants.

In addition $\mathcal{L}_C{}^{ghost}$ contains the color SU(3) Faddeev-Popov ghost terms defined in appendix 19A for the complexon Lorentz gauge. $\mathcal{L}_{CC}{}^{ghost}$ contains the complexon color SU(3) constraint (eq. 19A.8) ghost terms also defined through the Faddeev-Popov mechanism. The mass sector terms $\mathcal{L}_{Mass}$ are Higgs terms or the Dimensional Mass Mechanism (sections 21.3 and 21.4) terms. They generates masses and generation mixing.

The coordinates are

$$x = (x^0, \mathbf{x_r} + i\mathbf{x_i}) \tag{19.36}$$

and the action is therefore

$$I_{CSM} = \int dx^0 d^3x_r d^3x_i \, \mathcal{L}_{CSM} \tag{19.37}$$

The lagrangian is supplemented with the following condition on all complexon fields $\Phi_{...}$:

$$\nabla_r \cdot \nabla_i \Phi_{...} = 0 \tag{19.38}$$

Non-complexon fields $\Omega_{...}$ in the left-handed formulation under consideration satisfy the subsidiary condition:

$$[\nabla_r \cdot \nabla_i - (\nabla_r^2 \nabla_i^2)^{1/2}]\Omega_{...} = 0 \tag{19.39}$$

which guarantees a complexon's real momentum is parallel to its imaginary momentum.

19.4.1 Renormalization and Divergence Issues

The theory derived from $\mathcal{L}_{CSM}$ when calculated perturbatively in conventional quantum field theory has divergences that are not renormalizable due to its 7-dimensional nature. This issue does not appear to be curable by known renormalization methods.[257] However if the theory is reformulated in the two-tier formalism (next section) developed by Blaha[258] then a finite theory results. This is also true for the three or four generation Complexon Standard Model derived in the next chapter.

---

[257] The known methods pioneered by t'Hooft and Veltmann do renormalize the "Standard Standard Model. "

[258] Blaha (2005a) and Blaha (2003).

19.5 Two-Tier Formulation of the One Generation Complexon Standard Model

The general formulation of Two-Tier quantum field theory, and its application to the Standard Model, were presented in Blaha (2003) and Blaha (2005a). In this section we will develop a Two-Tier formulation of the one generation Complexon Standard Model described in the previous section. This formulation goes beyond our earlier books by using complex spatial coordinates which necessitate a slightly more complicated Two-Tier formulation.

The basic ansatz of the Two-Tier formalism is to replace every appearance of a coordinate x with a variable that is in part a quantum field:

$$x^{\mu} \rightarrow X^{\mu} = (y^0, \mathbf{y} + \mathbf{Y}(y^0, \mathbf{y})) \qquad (19.40)$$

where $\mathbf{Y}(y^0, \mathbf{y})$ is the spatial part of a free massless vector field with features that are identical to the free QED field.

Then one finds that the momentum space free field Feynman propagators G(k) of all particles acquires a Gaussian factor exp(h(k)):

$$G(k) \rightarrow G(k) \exp(h(k)) \qquad (19.41)$$

so that all perturbation theory diagrams are finite. The result is a finite perturbative result for all calculations to any order in perturbation theory. Blaha (2005a) shows that two-tier theories are finite, Poincare covariant, and unitary. (See Blaha (2005a) for a complete discussion.)

19.5.1 Simple Two-Tier Formalism

In this subsection we will describe the basic two-tier formalism. Taking the lagrangian described in Blaha (2005A):[259]

$$\mathscr{L}(y) = \mathscr{L}_F(X_{\mu}(y))J + \mathscr{L}_C(X^{\mu}(y), \partial X^{\mu}(y)/\partial y^{\nu}, y) \qquad (19.42)$$

where

$$X_{\mu}(y) = y_{\mu} + i\, Y_{\mu}(y)/M_c^2 \qquad (19.43)$$

[259] Eq. 7.1.

with M_c being a large mass scale, $Y_\mu(y)$ a vector quantum field, and where J is the absolute value of the Jacobian of the transformation from X to y coordinates:

$$J = |\partial(X)/\partial(y)| \tag{19.44}$$

The lagragian term $\mathscr{L}_C$ is

$$\mathscr{L}_C = +\tfrac{1}{4}\, M_c^4 F^{\mu\nu}F_{\mu\nu} \tag{19.45}$$

with

$$\begin{aligned} F_{\mu\nu} &= \partial X_\mu/\partial y^\nu - \partial X_\nu/\partial y^\mu \\ &\equiv i\,(\partial Y_\mu/\partial y^\nu - \partial Y_\nu/\partial y^\mu)/M_c^2 \end{aligned} \tag{19.46}$$

The sign in eq. 19.45 is not negative – superficially contrary to the conventional electromagnetic Lagrangian. The reason for this difference is that the quantum field part of X^μ is imaginary. Thus $\mathscr{L}_C$ ends up having the correct sign after taking account of the factor of i in the field strength $F_{\mu\nu}$.

Defining

$$F_{Y\mu\nu} = (\partial Y_\mu/\partial y^\nu - \partial Y_\nu/\partial y^\mu) \tag{19.47}$$

we see the Lagrangian assumes the form of the conventional electromagnetic Lagrangian:

$$\mathscr{L}_C = -\tfrac{1}{4}\, F_Y^{\mu\nu}F_{Y\mu\nu} \tag{19.48}$$

The action of this theory has the form

$$I = \int d^4y\, \mathscr{L}(y) \tag{19.49}$$

The further development of the theory is described in Blaha (2005a).

19.5.2 Two-Tier L(C)-based Standard Model Theory

In the present case of the Complexon Standard Model we will need two variables X_r^μ and X_i^μ since we have complex spatial 3-coordinates. We define them similarly to the previous case:

$$X_{r_\mu}(y_r) = y_{r_\mu} + i\, Y_{r_\mu}(y_r)/M_c^2 \tag{19.50}$$

$$X_{i_\mu}(y_i) = y_{i_\mu} + i\, Y_{i_\mu}(y_i)/M_c^2 \tag{19.51}$$

where we choose the same mass scale for both the "real" and "imaginary" variables. The two-tier, single generation, version of the Complexon Standard Model given in eq. 19.37 then has an action of the form

$$I_{CSMtt} = \int dy^0 d^3y_r d^3y_i \,(\mathscr{L}_{CSM}(X_r^\mu(y_r), \mathbf{X}_i^k(y_i))J_2)|_{y_i^0 = 0,\ Y_r^0 = Y_i^0 = 0} +$$

$$+ \int dy_r^0 d^3y_r\, \mathscr{L}_C(X_r^\mu(y_r), \partial X_r^\mu(y_r)/\partial y_r^\nu, y_r) +$$

$$+ \int dy_i^0 d^3y_i \mathscr{L}_C(X_i^\mu(y_i), \partial X_i^\mu(y_i)/\partial y_i^\nu, y_i) \tag{19.52}$$

where the replacements

$$x^\mu \equiv x_r^\mu \rightarrow X_r^\mu(y_r) \tag{19.53}$$

$$x_i^k \rightarrow X_i^k(y_i) \tag{19.54}$$

for $\mu = 0, 1, 2, 3$ and $k = 1, 2, 3$ are made in $\mathscr{L}_{CSM}$ in eq. 19.37 followed by defining $y_r^0 = y^0$ and making an L(C) transformation to a frame where $y_i^0 = 0$. J_2 is the absolute value of the Jacobian of the transformation from (X_r, X_i) to (y_r, y_i) coordinates:

$$J_2 = |\partial(X_r, X_i)/\partial(y_r, y_i)| \tag{19.55}$$

We also choose gauges where $Y_r^0 = Y_i^0 = 0$. These transformations and gauge choices are discussed in detail in Blaha (2005a). The lagragian terms $\mathscr{L}_C(X_r^\mu(y_r), \partial X_r^\mu(y_r)/\partial y_r^\nu, y_r)$ and $\mathscr{L}_C(X_i^\mu(y_i), \partial X_i^\mu(y_i)/\partial y_i^\nu, y_i)$ have the same form:

$$\mathscr{L}_C = +\tfrac{1}{4} M_c^4 F^{\mu\nu} F_{\mu\nu} \tag{19.56}$$

with

$$F_{\mu\nu} = \partial X_\mu/\partial y^\nu - \partial X_\nu/\partial y^\mu \quad (19.57)$$

$$\equiv i\,(\partial Y_\mu/\partial y^\nu - \partial Y_\nu/\partial y^\mu)/M_c^2 \quad (19.58)$$

or defining

$$F_{Y_{\mu\nu}} = (\partial Y_\mu/\partial y^\nu - \partial Y_\nu/\partial y^\mu) \quad (19.59)$$

we see each lagrangian assumes the form of the conventional electromagnetic Lagrangian:

$$\mathscr{L}_C = -\tfrac{1}{4}\, F_Y{}^{\mu\nu} F_{Y_{\mu\nu}} \quad (19.60)$$

The lagrangian is supplemented with the following condition on all complexon fields $\Phi_{\ldots}$:

$$(\partial/\partial X_r{}^k(y_r))\,(\partial/\partial X_i{}^k(y_i))\Phi\ldots = 0 \quad (19.61)$$

summed over k = 1, 2, 3. Non-complexon fields Ω… in our left-handed formulation satisfy the subsidiary condition:

$$\{(\partial/\partial X_r{}^k(y_r))(\partial/\partial X_i{}^k(y_i)) - [(\partial/\partial X_r{}^k(y_r))^2(\partial/\partial X_i{}^m(y_i))^2]^{1/2}\}\Omega\ldots = 0 \quad (19.62)$$

summed over k = 1, 2, 3 and over m = 1, 2, 3 separately in each of the two terms.

19.5.3 Two-Tier Feynman Propagators for Real Space-time

The momentum space, free field, Feynman propagators G…(k) of all particles and ghosts in two-tier QFT acquire a Gaussian factor exp(h(k)):

$$G\ldots(k) \rightarrow G\ldots(k)\,\exp(h(k)) \quad (19.63)$$

so that all perturbation theory diagrams are finite. The result is finite perturbative results in all calculations to any order in perturbation theory. Blaha (2005a) shows that two-tier theories are finite, Poincaré covariant, and unitary.

An example of the two-tier effect on propagators is the simpler case of the two-tier photon propagator. The two-tier photon propagator[260] in two-tier QED is:

$$iD_F^{TT}(y_1 - y_2)_{\mu\nu} = -i \int \frac{d^4p\, e^{-ip\cdot z}\, g_{\mu\nu} R(\mathbf{p}, z)}{(2\pi)^4 (p^2 + i\varepsilon)} \quad (19.64)$$

(since the imaginary parts can be taken to be zero: $y_{1i}^{\mu} - y_{2i}^{\mu} = 0$) where

$$z^{\mu} = y_{1r}^{\mu} - y_{2r}^{\mu} \quad (19.65)$$

$$R(\mathbf{p}, \mathbf{z}) = \exp[-p^i p^j \Delta_{Tij}(z)/M_c^4] \quad (19.66)$$
$$= \exp\{ -\mathbf{p}^2[A(v) + B(v)\cos^2\theta] / [4\pi^2 M_c^4 |\mathbf{z}|^2]\} \quad (19.67)$$

with i, j = 1, 2, 3, and with $\Delta_{Tij}(z)$ the commutator of the positive frequency part $Y^+_k(y)$ and the negative frequency part $Y^-_k(y)$ of $Y_k(y)$:

$$\Delta_{Tij}(z) = [Y^+_j(y_{1r}), Y^-_k(y_{2r})] = \int d^3k\, e^{ik\cdot(y_{1r} - y_{2r})} (\delta_{jk} - k_j k_k/\mathbf{k}^2)/[(2\pi)^3 2\omega_k] \quad (19.68)$$

and

$$v = |z^0|/|\mathbf{z}|$$

$$A(v) = (1 - v^2)^{-1} + .5v \ln[(v-1)/(v+1)] \quad (19.69)$$
$$B(v) = v^2(1 - v^2)^{-1} - 1.5v \ln[(v-1)/(v+1)] \quad (19.70)$$
$$\mathbf{p}\cdot\mathbf{z} = |\mathbf{p}|\, |\mathbf{z}| \cos\theta \quad (19.71)$$

with $|\mathbf{p}|$ denoting the length of a spatial vector **p**, $|\mathbf{z}|$ denoting the length of a spatial vector **z**, and with $|z^0|$ being the absolute value of z^0.

The gaussian factors R(**p**, z) which appear in all two-tier propagators damp the large momentum behavior of all perturbation theory integrals producing a completely finite perturbation theory and yet give the usual results of perturbation theory at energies that are small compared to the mass scale M_c.

---

[260] Blaha (2005a).

19.5.4 Complexon Feynman Propagators

In the case of complexons the two-tier Feynman propagator differs from the non-complexon case by having an integration over imaginary spatial 3-momenta, a derivative of a delta function embodying the orthogonality of the real and imaginary 3-momenta, and two factors of R(**p**, z): one factor being R($\mathbf{p}_r$, z_r) and the other factor being R($\mathbf{p}_i$, z_i) (where the time components $z_r^0 = z^0$ and $z_i^0 = 0$ since there is only one real time coordinate[261]). Thus we obtain large momentum convergence for both real and imaginary 3-momentum integrations.

For a <u>scalar complexon</u> particle in <u>conventional</u> quantum field theory we find the Feynman propagator:

$$i\Delta_{CTF}(x - y) = \theta(x^+ - y^+)<0|\phi_{CT}(x)\ \phi_{CT}(y)|0> + \\ + \theta(y^+ - x^+)<0|\phi_{CT}(y)\phi_{CT}(x)|0>$$

$$= i\int d^4p_r d^3p_i \frac{(2\pi)^{-7}\delta'(\mathbf{p_r}\cdot\mathbf{p_i}/m^2)e^{-ip^+(x^- - y^-) - ip^-(x^+ - y^+) + ip_\perp\cdot(x_\perp - y_\perp) - ip_i\cdot(x_i - y_i)}}{(p^2 + m^2 + i\varepsilon)} \tag{19.72}$$

In the case of <u>Two-Tier quantum field a scalar complexon</u> has the the Feynman propagator

$$i\Delta_{CTFtt}(x - y) = i\int d^4p_r d^3p_i (2\pi)^{-7}\delta'(\mathbf{p_r}\cdot\mathbf{p_i}/m^2)\ R(\mathbf{p}_r, z_r)R(\mathbf{p}_i, z_i)\cdot \\ \cdot e^{-ip^+(x^- - y^-) - ip^-(x^+ - y^+) + ip_\perp\cdot(x_\perp - y_\perp) - ip_i\cdot(x_i - y_i)}/(p^2 - m^2 + i\varepsilon) \tag{19.73}$$

where the time components $z_r^0 = z^0$ and $z_i^0 = 0$ since there is only one time coordinate and where $p^2. = p^{0\,2} - p_r^{\,2} + p_i^{\,2}$. Note both the real and imaginary integrations have gaussian damping at high momenta.

Propagators for other types of particles are similarly modified in the two-tier formalism (See Blaha 2005a).

---

[261] We can arrange for $z_i^0 = 0$ by making a L_C transformation to an inertial frame where z is real.

19.6 Reason for Parity Violating Form of the Standard Model

The issue of parity violation[262] that was discovered 50 years ago is now "explained" by the (broken) L(C) covariance of all forms of the Standard Model.

[262] T. D. Lee and C. N. Yang, Phys. Rev. **104**, 254 (1956).

Appendix 19A. Complexon Gauge Field Dynamics

19A.1 Color SU(3)

Our derivation yields the general form of the ElectroWeak quark sector of the Complexon Standard Model with a dynamics modified by the presence of complex spatial 3-momenta. This difference is a positive for the theory. In the discussion in chapter 19 we showed that "normal" and tachyonic complexons have an SU(3) symmetry.

Complexon quarks have a complex spatial 3-momentum. There are Dirac-type complexons and tachyonic complexons. If we think of quark gluon interactions, and in particular, perturbative diagrams with quark-gluon loops such as the simplest quark self-energy diagram, then imaginary momentum should "flow" around a loop as well as real momentum. Thus we see that gluons should be massless spin 1 complexon SU(3) gauge fields if quarks are complexons for the sake of consistency. More importantly, this is required because the color gluon fields appearing in the gauge covariant derivatives in the fermion field equations are necessarily functions of time and complex spatial 3-coordinates. The color gluon lagrangian density is

$$\mathcal{L}_{CCG} = -\tfrac{1}{4} F_{CC}{}^{a\mu\nu}(x) F_{CC}{}^{a}{}_{\mu\nu}(x) \qquad (19A.1)$$

where

$$F_C{}^a{}_{\mu\nu} = D_\nu A_C{}^a{}_\mu - D_\mu A_C{}^a{}_\nu + g f^{abc} A_C{}^b{}_\mu A_C{}^c{}_\nu \qquad (19A.2)$$

where D_μ is defined by eq. 17.59, g is the color coupling constant, and the f^{abc} are the SU(3) structure constants.

The theory of the strong color interaction must be developed within a path integral framework that takes account of Faddeev-Popov gauge fixing. Before doing so we must consider the effect of the complex spatial 3-vector on the non-abelian gauge formalism.

19A.2 Pure Complexon Gauge Groups

The commutator of of covariant derivatives is

$$F_{C\mu\nu} = F_{C\ \mu\nu}^{\ a} t^a = (i/g)[\mathcal{D}_\mu, \mathcal{D}_\nu] \tag{19A.3}$$

where $\mathcal{D}_\mu$ is given by eq. 19.7. $F_{CC\mu\nu}$ is covariant under gauge transformations:

$$F_{CC'\mu\nu} = \Theta F_{CC\mu\nu}\Theta^{-1} \tag{19A.4}$$

The strong interaction part of the action has terms of the form

$$I = \int d^7x[\mathcal{L}_{CCG} + \bar{\Psi}_q i\gamma^\mu \mathcal{D}_\mu \Psi_q + \ldots] \tag{19A.5}$$

where the "extra" three integrations are over the imaginary spatial coordinates, and Ψ_q represents a generic Dirac quark complexon field. The gauge field part of the action must be supplemented with a constraint similar to those seen earlier for complexons, which ensures the reality of the lagrangian term $\mathcal{L}_{CCG}$ (and thus the corresponding Hamiltonian terms, and ultimately the unitarity of the theory.) The constraint simply specifies the imaginary part of $\mathcal{L}_{CCG}$ is zero:

$$\begin{aligned}[\partial A_C^{a\mu}/\partial x_{ik}]\mathrm{Re}\, F_{C\ \mu k}^{\ a} = [\partial A_C^{a\mu}/\partial x_{ik}](\partial A_{C\ \mu}^{\ a}/\partial x_r^{\ k} - \partial A_{C\ k}^{\ a}/\partial x_r^{\ \mu} + \\ + g f^{abc} A_{C\ \mu}^{\ b} A_{C\ k}^{\ c}) \\ = 0\end{aligned} \tag{19A.6}$$

where r indicates the real part of the coordinate $\mathbf{x} = \mathbf{x}_r + i\mathbf{x}_i$, and $\partial/\partial x_i^{\ k}$ is the derivative with respect to the kth component of the imaginary coordinate 3-vector $\mathbf{x}_i$. This restriction is not a choice of gauge but rather a restriction on the dependence of the $A_{C\nu}$ field on the real and imaginary spatial 3-vectors.

If we assume that we can integrate by parts in the imaginary coordinates then we can re-express the constraint as

$$A_C^{a\mu}\,\mathrm{Re}\,\partial F_{C\ \mu k}^{\ a}/\partial x_i^{\ k} = 0 \tag{19A.7}$$

which is explicitly

$$A_C{}^{a\mu}[\partial^2 A_C{}^a{}_\mu/\partial x_r{}^k \partial x_i{}^k - \partial^2 A_C{}^a{}_k/\partial x_r{}^\mu \partial x_i{}^k + gf^{abc}\partial(A_C{}^b{}_\mu A_C{}^c{}_k)/\partial x_i{}^k] = 0 \quad (19A.8)$$

where $x_r^0 = x^0$. This restriction can be implemented within the framework of the path integral formalism by using the Faddeev-Popov Method with the introduction of ghosts in a manner similar to the use of the Faddeev-Popov Method to implement gauge conditions. The form of the restriction stated in eqs. 19A.7 and 19A.8 leads to second order derivative Faddeev-Popov ghost terms in the path integral while eq. 19A.6 would lead to a fourth order derivative ghost terms in the path integral with attendant unitarity (and possibly other) issues.

19A.3 Pure Gauge Complexon Path Integral Formulation and Faddeev-Popov Method

The path integral formalism for complexon, non-abelian, pure, Yang-Mills fields differs significantly from the conventional gauge field path integral formalism. The path integral for a complexon gauge field can be written symbolically as:

$$Z(J^\mu)=N\int DA_C\Delta_{FP}(A_C)\delta(F(A_C))\Delta_C(A_C)\delta(F_C(A_C))\exp\{i\int d^7y[\mathscr{L}+ + J^\mu(y)A_{C\mu}(y)]\} \quad (19A.8a)$$

where $\delta(F(A_C))$ specifies the gauge, $\Delta_{FP}(A_C)$ is its Faddeev-Popov determinant; and $\delta(F_C(A_C))$ specifies the complexon condition (eq. 19A.8) with $\Delta_C(A_C)$ the Faddeev-Popov determinant for the complexon condition. In both cases the Faddeev-Popov determinant can be calculated in the standard way.[263]

First we consider the gauge fixing delta function. Note that it can be written as a delta function in the gauge times a determinant:

$$\delta(F(A_C{}^\omega)) = \delta(\omega - \omega_0)|\det \delta F(A_{C\mu}{}^\omega(x))/\delta\omega(x)|^{-1}|_{F(A_C)=0} \quad (19A.9)$$

where ω_0 is a reference gauge, where

[263] See for example Huang (1992).

$$A_{C\ \mu}^{a\ \omega}(x) = A_{C\ \mu}^{a}(x) - g^{-1}D_\mu\omega^a + f^{abc}\,\omega^b(x)A_{C\ \mu}^{c}(x) \qquad (19A.10)$$
$$= A_{C\ \mu}^{a}(x) + \delta A_{C\ \mu}^{a\ \omega}(x)$$

and

$$\text{Re } F_{C\ \mu k}^{a\ \omega} = \text{Re } F_{C\ \mu k}^{a} + f^{abc}\,\omega^b(x)\text{Re } F_{C\ \mu k}^{c} \qquad (19A.11)$$
$$= \text{Re } F_{C\ \mu k}^{a} + \delta(\text{Re } F_{C\ \mu k}^{a\ \omega})$$

under an infinitesimal gauge transformation, and where

$$\Delta_{FP}(A_C) = |\det \delta F(A_{C\mu}^{\ \omega}(x))/\delta\omega(x)||_{F(A_C) = 0,\ \omega = 0} \qquad (19A.12)$$

We will choose the complexon Lorentz gauge to evaluate the Faddeev-Popov determinant:

$$F^a(A_C) = D_\mu A_C^{a\mu}(x) = 0 \qquad (19A.13)$$

We find

$$F^a(A_{C\mu}^{\ \omega}(x)) = D^\mu(A_{C\ \mu}^{a}(x) - g^{-1}D_\mu\omega^a(x) + f^{abc}\,\omega^b(x)A_{C\ \mu}^{c}(x))$$

$$= -g^{-1}D^\mu D_\mu\omega^a(x) + f^{abc}A_{C\ \mu}^{c}(x)D^\mu\omega^b(x) \qquad (19A.14)$$

Thus

$$\delta F^a(A_{C\mu}^{\ \omega}(x))/\delta\omega^b(x) = -g^{-1}\,\delta^{ab}D^\mu D_\mu \ + f^{abc}A_C^{c\mu}(x)D_\mu \qquad (19A.15)$$

and

$$\Delta_{FP}(A_C) = |\det\,(g^{-1}\delta^{ab}D^\mu D_\mu \ - f^{abc}A_C^{c\mu}(x)D_\mu)| \qquad (19A.16)$$

where | … | represent absolute value.

We can rewrite the Faddeev-Popov determinant as a path integral over anti-commuting c-number fields with a ghost Lagrangian:

$$\Delta_{FP}(A_C) = \int D\chi^* D\chi \exp[\, i\int d^7x\, \mathscr{L}_C^{\ ghost}(x)] \qquad (19A.17)$$

where

$$\mathscr{L}_C^{\ ghost}(x) = \chi^{a*}(x)[\delta^{\,ab}D^\mu D_\mu \ - gf^{abc}A_C^{c\mu}(x)D_\mu]\chi^b(x) \qquad (19A.18)$$

19A.3.1 Faddeev-Popov Application to the Complexon Condition

The complexon condition can be also implemented within the path integral formalism using the Faddeev-Popov Mechanism. Using the identity

$$1 = \int DA_C \Delta_C(A_C)\delta(F_C(A_C)) \tag{19A.19}$$

we see that an infinitesimal gauge transformation yields eqs. 19A.10 and 19A.11. This enables us to relate $\Delta_C(A)$ to the determinant

$$\delta(F_C(A_C{}^{\omega})) = |\det \delta F_C(A_{C\mu}{}^{\omega}(x))/\delta\omega(x)|^{-1}|_{F_C(A_C)=0,\ \omega=0} \tag{19A.20}$$

and

$$\Delta_C(A_C) = |\det \delta F_C(A_{C\mu}{}^{\omega}(x))/\delta\omega(x)||_{F_C(A_C)=0,\ \omega=0} \tag{19A.21}$$

From eq. 19A.8 we see

$$F_C(A_{C\mu}(x)) = A_C{}^{a\mu}[\partial^2 A_C{}^a{}_\mu/\partial x_r{}^k \partial x_i{}^k - \partial^2 A_C{}^a{}_k/\partial x_r{}^\mu \partial x_i{}^k + \\ + g f^{abc}\partial(A_C{}^b{}_\mu A_C{}^c{}_k)/\partial x_i{}^k] \tag{19A.22}$$

with

$$F_C{}^a(A_{C\mu}) = 0 \tag{19A.23}$$

Inserting eq. 19A.10 and 19A.11 we find

$$[\delta F_C(A_{C\mu}{}^{\omega}(x))/\delta\omega^a(x)]|_{F_C(A_C)=0,\ \omega=0} = \delta[\delta A_C{}^{b\mu\omega}\, \mathrm{Re}\ \partial F_C{}^b{}_{\mu k}/\partial x_{ik} + \\ + A_C{}^{b\mu}\, \partial\delta(\mathrm{Re}\ F_C{}^b{}_{\mu k}{}^{\omega})/\partial x_{ik}\ ]/\delta\omega^a(x)|_{\omega=0}$$

$$= -g^{-1}(\mathrm{Re}\ \partial F_C{}^a{}_{\mu k}/\partial x_{ik})D^\mu - f^{abc}A_C{}^{b\mu}(\mathrm{Re}\ F_C{}^c{}_{\mu k})\partial/\partial x_{ik} \tag{19A.24}$$

Thus

$$\Delta_C(A_C) = |\det (g^{-1}(\mathrm{Re}\ \partial F_C{}^a{}_{\mu k}/\partial x_{ik})D^\mu + f^{abc}A_C{}^{b\mu}(\mathrm{Re}\ F_C{}^c{}_{\mu k})\partial/\partial x_{ik}| \tag{19A.25}$$

where | … | represent absolute value.

We can rewrite this Faddeev-Popov determinant as a path integral over anti-commuting c-number fields with a ghost Lagrangian:

$$\Delta_C(A_C) = \lim_{r \to \infty} \int D\chi_C{}^* D\chi_C \exp[ir^{-2}\int d^7x\, \mathscr{L}_{CC}{}^{ghost}(x)] \qquad (19A.26)$$

where r is a constant that is taken to the limit ∞, and where

$$\mathscr{L}_{CC}{}^{ghost}(x) = \chi_C{}^*(x)\{D^\mu D_\mu + r^2 t^a[(\text{Re}\, \partial F_C{}^a{}_{\mu k}/\partial x_{ik})D^\mu + \\ + g f^{abc} A_C{}^{b\mu}(\text{Re}\, F_C{}^c{}_{\mu k})\partial/\partial x_{ik}]\}\chi_C(x) \qquad (19A.27)$$

where t^a is a 3 × 3 matrix of the 3 representation of color SU(3) and $\chi_C(x)$ is a three row field in the 3 representation. *The introduction of $D^\mu D_\mu$ is based on consistency with the complexon formalism. It is needed to establish a perturbative expansion of the path integral. Its effect vanishes in the limit r → ∞ reducing ghost loops of this type to point interactions.* The reader will note that second order and third order derivative terms appear in the interaction in $\mathscr{L}_C{}^{ghost}(x^\mu)$ and raise the issue of non-renormalizable divergences. If one uses the two-tier approach to quantum field theory developed by Blaha (2005a) then all potential divergences disappear. *The two-tier formulation of the pure, complexon, Yang-Mills theory that we are discussing is finite.*

The complete pure complexon, Yang-Mills path integral is

$$Z(J^\mu) = N\int DA_C D\chi^* D\chi D\chi_C{}^* D\chi_C \Delta_{FP}(A_C)\delta(F(A_C))\Delta_C(A_C)\delta(F_C(A_C))\cdot \\ \cdot\exp\{i\int d^7y\, [\mathscr{L}_{CCG} + J^\mu A_{C\mu}]\} \qquad (19A.28)$$

Eq. 19A.28 appears previously in eq. 19.33 in the definition of the complexon color gauge lagrangian terms.

19A.4 Complexon Quark-Gluon Perturbation Theory

In this section we will give an impression of complexon strong interaction theory by considering a simple diagram. Fig. 19A.1 is a self-energy diagram for a quark in which a gluon is emitted and absorbed. Complexon particles have complex spatial momentum. From the form of the propagators and wave functions previously considered it is clear that the real part of the 3-momentum and the imaginary part of the 3-

momentum will be separately conserved at each vertex. The general form of the self-energy integral corresponding to Fig. 19A.1 is

$$I_{ab}(p) = \int d^7q\, P_{ab}(q, p)\exp(\text{Gaussian})/Q(q, p) \qquad (19A.29)$$

where a and b are SU(3) color indices, $P_{ab}(q, p)$ is a polynomial in q and p together with Dirac gamma matrix factors and color SU(3) matrix factors, and Q(q, p) is the product of a quark and a gluon propagator denominator factor. The Gaussian exponential factor appears if we use a two-tier formulation[264] of the complexon quark and gluon theory. This exponential factor guarantees the convergence of the integral resulting in a finite result.

Thus while perturbation theory is only partly useful in calculating strong interaction phenomena due to the "largeness" of the coupling constant in general, we see complexon perturbation theory is well defined, finite, and sensible if one uses two-tier quantum field theory.

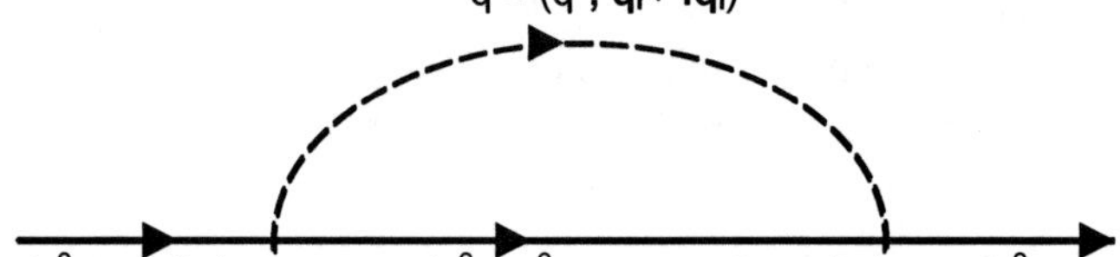

Figure 19A.1. Quark self-energy diagram in which a quark emits and subsequently absorbs a gluon. The quark momentum is $p = (p^0, \mathbf{p}_r + i\mathbf{p}_i)$ and the gluon momentum is $q = (q^0, \mathbf{q}_r + i\mathbf{q}_i)$ with conservation of energy, and separate conservation of real spatial momentum and imaginary spatial momentum at each vertex.

The integral corresponding to Fig. 19A.1 has the form:

$$I_{1ab}(p) = \int dq^0 d^3q_r d^3q_i\, P_{ab}(q, p)\exp(\text{Gaussian})/Q(q, p) \qquad (19A.30)$$

and includes integrations over both real and imaginary 3-momenta.

---

[264] See Blaha (2005a).

19A.5 Complexon Quark ElectroWeak Interactions

In the Complexon Standard Model quarks and gluons are complexons with complex spatial 3-momenta. On the other hand, the ElectroWeak bosons: $W^{\pm}{}_{\mu}$, Z_{μ} and A_{μ} are not complexons and so can be viewed as having totally real 3-momenta (or complex 3-momenta with parallel real and imaginary parts). In this section we address the issue of quark ElectroWeak perturbation theory. *The general perturbation theory rule for this situation is that real and imaginary momenta are separately conserved at each vertex.*

Fig. 19A.2 shows a quark self-energy diagram in which a quark emits a photon and subsequently reabsorbs it. Since ElectroWeak bosons can be viewed as having real energies and 3-momenta, and since real and imaginary momenta are separately conserved at each vertex by the above stated rule, the photon momentum is real while the quark spatial 3-momentum is complex.

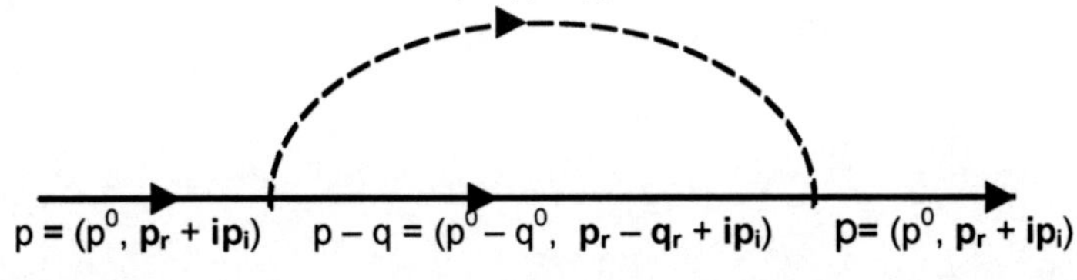

Figure 19A.2. Quark self-energy diagram in which a quark emits and subsequently absorbs a photon. The quark momentum is $p = (p^0,\ \mathbf{p}_r + i\mathbf{p}_i)$ and the photon momentum is $q = (q^0,\ \mathbf{q}_r)$ with conservation of energy, and separate conservation of real spatial momentum and imaginary spatial momentum at each vertex.

The integral corresponding to Fig. 19A.2 has the form:

$$I_{2ab}(p) = \int dq^0 d^3q_r d^3q_i \delta^3(\mathbf{q}_i) P_{ab}(q, p)\exp(\text{Gaussian})/Q(q, p)$$
$$= \int dq^0 d^3q_r P_{ab}(q, p)\exp(\text{Gaussian})/Q(q, p) \qquad (19A.31)$$

In cases where there are complexon quark loops or gluon loops the imaginary quark or gluon 3-momenta affects the result. For example, Fig. 19A.3 shows a complexon quark loop contribution to the photon self-energy in which the imaginary quark 3-momenta integration contributes.

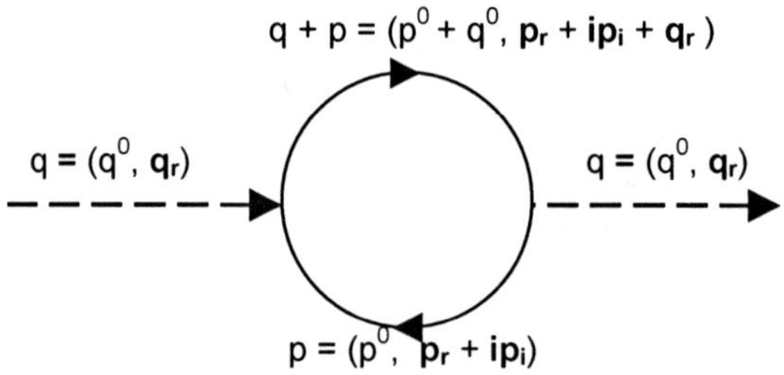

Figure 19A.3. Photon self-energy diagram with a quark loop. The photon momentum is $q = (q^0, \mathbf{q_r})$ and quark loop momentum is $p = (p^0, \mathbf{p_r} + i\mathbf{p_i})$.

Thus complexon imaginary 3-momenta can appear in many parts of the Complexon Standard Model.

20. Origin of the Generations

20.1 Introduction

In the preceding chapters we have seen how most of the features of the Standard Model originate in the L(C) group: Parity Violation, the four species of fermions, the ElectroWeak sector, and color SU(3).

The three known generations of each species appear to have a different origin. Unlike the ElectroWeak vector bosons, whose masses and mixing angle (the Weinberg angle) are fairly well known, the masses of the fermions are only well known for the charged leptons, and the masses and mixing angles are only fairly well-known for the quark species. Neutrino masses are known to be small and the mixing angles are approximately known.[265]

This experimental situation opens the door to many theoretical approaches. In this chapter we will postulate that the generations are the result of an SU(3) structure of the fermions that results from the breaking of another GL(4) to SU(3).[266] Thus we have a GL(4) group at the base of the developments in the preceding chapters and another GL(4) (denoted GL'(4)) at the base of the generations and their mixing. One can extend this thought process (as we do) to conjecture that the fundamental symmetry of the universe is GL(16) which is broken in what we can view as a series of stages to the world we know. This breakdown hierachy is depicted in Fig. 20.1.

Thus the sixteen generators of GL(4) consist of a set of 6 generators that form the L(C) generators; a set of nine generators that are the generators of the U(3) group; and one U(1) generator.

The L(C) generators do not commute with the U(3) generators. So we will postulate two GL(4) groups: a GL(4) which ultimately leads

---

[265] E. Ma, arXiv:hep-ph/0612013 (2006) and references therein; P. F. Harrison, D. H. Perkins, and W. G. Scott, Phys. Lett. **B530**, 167 (2002).

[266] Another possibility that we will explore is four generation Standard Models based on the SU(4) subgroup of GL(4).

to the space-time of our experience; and a GL(4) which leads to the three (or four) fermion generations.

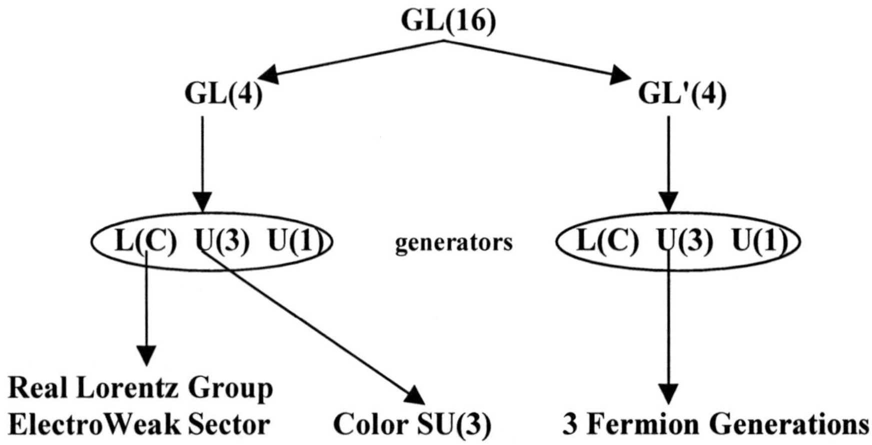

Figure 20.1. Group hierarchy of extended "space-time" leading to the Standard Model. Nature takes L(C) from one GL(4) subgroup to define the space-time structure, and U(3) from GL'(4) to define 3 generations and their mixing. (Note: the ' serves only to differentiate between the GL(4)'s.)

Thus the Standard Model is built on L(C) and U(3) from GL(4) and U(3)[267] *from GL'(4). The generators of each GL(4) commute with the generators of the other GL(4) thus avoiding a lethal combination of non-commuting space-time and internal symmetries. The consanguinity and compactness of this symmetry approach is remarkable.*[268]

20.2 The Fermion Generations

One generation Standard Models were developed in earlier chapters that captured the form of the conventional Standard Model. In

[267] Or U(4).

[268] Seeing that an L_C set of generators is extracted from GL(4) and a U(3) set of generators is extracted from GL'(4) to build the Standard Model, one is led to speculate that perhaps a "twin" universe exists based on U(3) extracted from GL(4) and L_C extracted from GL'(4). This twin universe concept is not related to the many universes theory of Everitt and others. Since the twin universes, if they exist, have different space-times, there would be no known connection between them – not even gravity.

this section we will state postulates that generalize the one generation Standard Model to a three generation Standard Model.

Postulate 20.1. The four complex dimensions of GL'(4) symmetry are reduced to a point. However the fundamental $\underline{3}$ representation of its U(3) subgroup generates the three generations of fermions for each fermion species.

The mixing of the three known generations forces the ElectroWeak and color SU(3) interactions to be the same for the three generations.

20.2.1 Generation Field index

Based on postulate Postulate 20.1 we now add a generation index n to lepton and quark fields:

Lepton Fields

$$\psi^n(t, \mathbf{x})$$

Quark Fields

$$\psi_C{}^{an}(t, \mathbf{x})$$

and introduce a lagrangian mass term term for each fermion species with generation mixing with the form:

$$\bar{\psi}(x)M\psi(x)$$

prior to separation into left-handed and right-handed fields.

20.3 Speculation: Fourth Generation = Dark Matter?

The set of generators of GL'(4) contains the subgroup U(3) consisting of a block diagonal 3×3 representation of U(3) within 4×4 matrices (eqs. 16.50 and 16.53), and a U(1) generator represented by the matrix $\Omega_{+00} - F_9$ with only one non-zero entry, namely, 1 in the fourth row and column.

We now consider whether this fourth dimension is the source of Dark Matter and Dark Energy. Suppose a one generation Standard Model as described in previous chapters exists for this fourth generation with no mixing with the first three generations.[269] Then we could have different ElectroWeak interactions and a different color SU(3) strong interaction from the known three generation ElectroWeak interactions and Strong Interactions. The only interaction between the first three generations and the fourth generation would be gravity. This scenario corresponds to the known facts of Dark Matter. And it would be reasonable to further assume that Dark Energy consists primarily of Dark Photons – photons of the fourth generation ElectroWeak theory.

Thus the group generated by the U(3) and U(1) generators of section 16.5 provide a possible solution of the problems of Dark Matter and Dark Energy within a framework analogous to the known three generation Standard Model.

---

[269] Some recent cosmological data on Dark Matter suggest that Dark Matter can decay or interact very weakly to produce particles of "normal" matter. Whether these results stand the test of time remains to be seen.

21. A Possible Origin for Particle Masses—The Dimensional Mass Mechanism

21.1 The Origin of Mass

The origin of mass is both a physics question and a metaphysical question. On the metaphysical question we defer consideration until later. On the physical question, the Higgs Mechanism is a solution that is widely accepted and meets the needs for mass terms in the Standard Model. However, the Higgs particles that are required by this mechanism have not been found experimentally. As new experimental results have appeared in recent years the experimental lower bound on the masses of Higgs particles has increased significantly. Theoretical physicists have responded by "tweaking" the predicted masses to higher and higher values. While this process can be iterated more times, there is a point where the existence of Higgs particles will lose credibility and an alternate mechanism for generating particle masses will be required.

In this chapter we will consider the possibility of a different mechanism for generating the known fermion particle masses. This mechanism has several advantages over the Higgs Mechanism:

1. Economy – no extra particles beyond the known Standard Model particles are required.
2. It fits very nicely within the GL'(4) explanation of the origin of the fermion generations as presented in the previous chapter.
3. It is well motivated by the method by which the curvature constant, usually denoted k, with the dimensions of $[\text{mass}]^2$ arises in the derivation of the Robertson-Walker metric of General Relativity.

Thus this new approach to the origin of mass has both an experimental and theoretic advantage over the Higgs Mechanism at the present time.

21.2 The Origin of k (with Dimension [mass]^2) in the Robertson-Walker Metric

The Robertson-Walker metric:

$$d\tau^2 = dt^2 - R^2(t)[dr^2/(1 - kr^2) + r^2(d\theta^2 + \sin^2\theta\, d\varphi^2)] \qquad (21.1)$$

contains the constant k which is a measure of the cuvature of the universe. This constant has the dimensions of [mass]^2 and arises in the solution of the Einstein equations of General Relativity. As Weinberg (1972) and Blaha (2004) show[270] the constant k arises as a dimensionful constant in the separation of the Einstein equations. In particular Blaha (2004) obtained the solution for a generalization of the generalized Robertson-Walker metric equation

$$\dot{a}^2 - 8\pi G\rho a^2/3 - (\check{r}b^2)^{-1}\, \partial(\check{r}\, b'/b)/\partial\check{r} = 0 \qquad (21.2)$$

where a (the scale factor) and b are certain functions of time and a radial coordinate ř, by noting that the first two terms in eq. 21.2 are solely functions of t while the third term is solely a function of ř. Thus we can separate the differential equation in the usual way to obtain:

$$\dot{a}^2 - 8\pi G\rho a^2/3 = -k \qquad (21.3)$$

and

$$k + (\check{r}b^2)^{-1}\, \partial(\check{r}\, b'/b)/\partial\check{r} = 0 \qquad (21.4)$$

where k turns out to be the curvature parameter k in eq. 21.1 after some manipulations.

Thus we find that, in perhaps the most fundamental of theories, that a dimensionful parameter arises as a separation constant. This is the key to an alternate mechanism to obtaining fermion particle masses within the framework of the GL(4)⊗GL'(4) derivation of The Standard Model.

---

[270] Weinberg (1972) p. 343 and Blaha (2004) p. 174.

21.3 Origin of the Fermion Masses and Mass Mixing Matrices

Having seen that a parameter k with the dimensions of mass squared naturally appears in certain separable solutions of the Einstein equations (in particular Robertson-Walker solutions), we now consider the possibility that the masses of fermions appear as separation constants in the GL(16) → GL(4)⊗GL'(4) based derivation of the Standard Model presented in prior chapters. We will call this method of giving particles masses the *Dimensional Mass Mechanism*.[271]

In chapter 20 we saw how either a Standard Model with three generations or four generations of fermions could be derived by assuming that each species of fermions were 3-vectors under the SU(3) subgroup of GL'(4) or 4-vectors under the SU(4) subgroup of GL'(4).

In either case the dynamical equation of each fermion species (as shown in chapter 17) takes the form

$$(\gamma^{\mu}\mathcal{D}_{\mu}(x) - m)\psi(x) = 0 \tag{21.5}$$

where $\mathcal{D}_{\mu}(x)$ is a differential operator in the complex coordinates x^{ν} of GL(4) before breakdown to the real Lorentz group. Since the full group GL(16) is broken down to GL(4)⊗GL'(4), eq. 21.5 should be extended to the case of GL(4)⊗GL'(4). If we denote the GL(4) 4-vector coordinates as x^{ν} and the GL'(4) 4-vector coordinates as y^{μ} then each species equation becomes a 3 generation or 4 generation equation of the form

$$(\gamma^{\nu}\mathcal{D}_{\nu}(x)M_{ab} + \gamma\,'^{\mu}\mathcal{D}'_{\mu}(y)\delta_{ab} - m_0\delta_{ab})\psi_b(x, y) = 0 \tag{21.6}$$

where the $\gamma\,'^{\mu}$ matrices are the Dirac matrices but with the spinor indices of GL'(4) while the γ^{μ} matrices are the Dirac matrices with spinor indices of GL(4). $\mathcal{D}'_{\mu}(y)$ is the GL'(4) differential operator, m_0 is the bare mass, and a and b are generational indices. (We suppress all other indices for the sake of clarity.)

M_{ab} is a non-singular generation mixing matrix. This mixing matrix reflects the nature of the breakdown of the original symmetry

---

[271] We use the phrase "dimensional mass" since the origin of masses in this approach follows from the separation of the part of the dynamical equations dependent on the coordinates (dimensions) of GL'(4).

from GL(16) to GL(4)⊗GL'(4).[272] In the absence of detailed information about the cause and nature of this breakdown mechanism we will not speculate on its form and merely assume it appears in the broken theory.

Eq. 21.6 is a manifestly separable equation. So we can express the wave function as[273]

$$\psi_b(x, y) = \psi_{1b}(x)\psi_{2b}(y) \tag{21.7}$$

where the repeated subscript b on the right side of the equation does not indicate a summation over b. If we introduce a triplet or quadruplet of separation constants $\mathcal{M}_a$, as the case may be, then we find that eq. 21.6 is solved using the solutions of the separated equations:[274]

$$(\gamma^\nu \mathcal{D}_\nu(x)M_{ab} + \mathcal{M}_a\, \delta_{ab})\psi_{1b}(x) = 0 \tag{21.8}$$

$$[\gamma\, '^\mu \mathcal{D}'_\mu(y)\delta_{ab} - (\mathcal{M}_a + m_0)\delta_{ab}]\psi_{2b}(y) = 0 \tag{21.9}$$

Eq. 21.8 exhibits the generation mixing of the charged ElectroWeak currents and the separation constants $\mathcal{M}_a$ are seen to be the physical masses of the fermions of each species. Thus there are four sets of *physical* fermion masses denoted $\mathcal{M}_a$ (one set for each species), and four mixing matrices that combine to give the physical mixing matrices of the charged ElectroWeak currents.[275]

Since we assume the coordinate space for GL'(4) is compressed to a point, then eq. 21.9 becomes a relation between the field and its derivatives in that limit.

We note that the separation constants $\mathcal{M}_a$ must have the dimensions of [Mass] by the nature of eq. 21.6. Thus we obtain particle

---

[272] In some sense it shows that the visible fermion parts are twisted in "generation space" with respect to their invisible GL'(4) parts.

[273] The reader who is concerned about treating the wave function classically rather than as a second quantized field should remember that the manipulations that we are performing can be redone in a path integral formalism in which the manipulation of fields as classical fields is perfectly acceptable.

[274] The appearance of m_0 in eq. 21.6 rather than eq. 21.7 is arbitrary. It could have been otherwise with no change in results by simply shifting the $\mathcal{M}_a$ values by m_0.

[275] The neutral ElectroWeak currents do not have generational mixing.

masses as separation constants just as the constant k with the dimension of $[\text{Mass}]^2$ appears in the solution for the Robertson-Walker metric.

Therefore we have a simple mechanism—the Dimensional Mass Mechanism—for fermion mass generation that does not require the introduction of hitherto unseen scalar fields.[276]

Once we have obtained equations of the form of eq. 21.8 for each of the fermion species then we can see that the leptonic ElectroWeak charge changing lepton current has the form:

$$J_{\text{leptonic}}{}^{\alpha} \sim [\bar{e}\ \ \bar{\mu}\ \ \bar{\tau}\,]_a \gamma^{\alpha}\, M^{e\dagger}{}_{ab} M^{\nu}{}_{bc} \begin{bmatrix} \nu_e \\ \nu_\mu \\ \nu_\tau \end{bmatrix}_c \tag{21.10}$$

in the 3 generation case[277] with other indices and labeled not shown. Thus the leptonic generation mixing matrix is

$$M_{\text{leptonic}} = M^{e\dagger} M^{\nu} \tag{21.11}$$

where M^e is the charged lepton species mixing matrix and M^ν is the neutral lepton species mixing matrix with † indicating complex conjugate. The quark ElectroWeak charge changing current has the form:

$$J_{\text{quark}}{}^{\alpha} \sim [\bar{d}\ \ \bar{s}\ \ \bar{b}\,]_a \gamma^{\alpha}\, M^{d\dagger}{}_{ab} M^{u}{}_{bc} \begin{bmatrix} u \\ c \\ t \end{bmatrix}_c \tag{21.12}$$

in the three generation case with other indices and labeled not shown. where M^u is the up-type quark species mixing matrix and M^d is the down-type quark species mixing matrix. Thus the quark generation mixing matrix is

---

[276] The philosophically inclined reader may balk at thrusting the origin of the known physical fermion masses onto an unseen point-like universe, which is, in the present state of physical knowledge, unknown. However this assumption is in reality a challenge to examine the point-like universe in more detail—something which is beyond the scope of the present derivation.

[277] The four generation case is a direct extension of the three generation case.

$$M_{quark} = M^{d\dagger}M^{u} \tag{21.13}$$

The generalization of eqs. 21.10 and 21.12 to four generations is direct.

Thus we find that the generalization of our four Dirac-like fermion equations to GL(4)⊗GL'(4) in eq. 21.6 naturally leads to fermion generations, differing fermion masses, and fermion generation mixing in general.

Postulate 21.1a. The masses, and mixing angles, of fermions result from the Higgs mechanism.

Postulate 21.1b. The masses, and mixing angles, of fermions appear as separation constants in fermion dynamical equations in the breakdown GL(16) → GL(4)⊗GL'(4).

21.4 Origin of the Massive Vector Boson Masses

The Higgs Mechanism was needed to give masses to ElectroWeak vector bosons in the t'Hooft-Veltmann renormalization program for ElectroWeak Theory. Since experiment has been unable to discover Higgs scalar bosons up to this point in time and may well not find them at future accelerators we may need to develop other mechanisms for mass generation for massive vector bosons.

We could choose to use massive vector bosons ab initio and obtain a *fully renormalizable* (indeed finite) ElectroWeak Theory by using the Two-Tier quantization procedure (Blaha (2003) and (2005a)). A massive vector boson lagrangian would then have the form:

$$\mathcal{L}_{Y\text{-}M} = -\tfrac{1}{4}\, F^{a\mu\nu}(x)F^{a}{}_{\mu\nu}(x) + \tfrac{1}{2}\, m^2 A^{a\mu}A^{a}{}_{\mu} \tag{21.14}$$

where

$$F^{a}{}_{\mu\nu}(x) = D_\nu(x)A^{a}{}_{\mu}(x) - D_\mu(x)A^{a}{}_{\nu}(x) + gf^{abc}A^{b}{}_{\mu}(x)A^{c}{}_{\nu}(x) \tag{21.15}$$

The form of the differential operator D_ν depends on whether the Yang-Mills field is "normal" or complexon. The constants f^{abc} are the algebra's structure constants.

Since the structure of the ElectroWeak sector that we derived in earlier chapters comes exclusively from GL(4) (and not GL'(4)) it is not reasonable to generate vector boson masses from the interplay of GL(4) and GL'(4) as we did in the case of the fermion masses in section 21.3. Thus explicit mass terms such as in eq. 21.14 are the mechanism in the GL(4)⊗GL'(4) derivation at our current level of understanding if Higgs particles are not found experimentally.

Two alternate postulates are possible:

Postulate 21.2a. The masses of vector bosons arise from the Higgs Mechanism.

Postulate 21.2b. The masses of vector bosons result from a non-Higgsian mechanism.

22. The "Standard" Standard Model

22.1 Postulates

If we assume Postulate P20a so that quarks are not complexons then we derive the conventional form of the Standard Model which we call the "Standard" Standard Model to distinguish it from the Complexon Standard Model that is described in the following chapter. The postulates of the Standard Standard Model are:

Postulate 15.1. Any observer can define a set of time and space coordinates called a coordinate system in which the observer is at rest. One can define a transformation that relates the coordinate systems of two observers traveling at a constant velocity with respect to each other.

Postulate 15.2. The speed of light, c, is the same in all coordinate systems.

Postulate 15.3. The invariant interval or distance $d\tau$ is defined by

$$d\tau^2 = g_{\mu\nu}dx^\mu dx^\nu \qquad (15.2)$$

It is invariant under a change of coordinate systems. The 16 quantities $g_{\mu\nu}$ are known as the metric tensor. The four quantities dx^μ are infinitesimal displacements in space and time.

Postulate 16.1. The universe has sixteen complex dimensions and is described by a curved, 16-dimensional complex extension of 4-dimensional General Relativity. Complex, 16-dimensional General Relativity is invariant under complex coordinate transformations of the form:

$$X^\mu = f^\mu(X'^0, X'^1, \ldots, X'^{15}) \qquad (16.1)$$

where X'^{μ} denotes an alternate complex coordinate system that is related to X^{μ} by continuous, holomorphic functions f^{μ}.

Postulate 16.2. We define an invariant interval with a metric $g_{\mu\nu}$ where

$$d\tau^2 = g_{\mu\nu}dX^{\mu}dX^{\nu} \tag{16.2}$$

Postulate 16.3. We assume a 16-dimensional Equivalence Principle: At any point one can transform to the coordinates of a locally flat space-time where

$$g'^{\mu\nu} = \eta_{16}{}^{\alpha\beta}\, \partial X'^{\mu}/\partial X^{\alpha}\ \partial X'^{\nu}/\partial X^{\beta} \tag{16.9a}$$

with

$$\eta_{16\mu\nu} = \text{diag}(1, -1, -1, -1, -1, 1, 1, 1, -1, 1, 1, 1, -1, 1, 1, 1) \equiv \eta_{16}{}^{\mu\nu} \tag{16.12}$$

Postulate 16.4. We assume that GL(16) is simply broken to GL(4)⊗GL'(4) through some mechanism and this group is the basis of the form of the Standard Model.

Postulate 16.5. After breakdown, a complex 16-dimensional vector can be represented as the outer product of two complex 4-dimensional vectors

$$x^{\mu} \rightarrow y^{i}y'^{j} \tag{16.30}$$

where μ ranges from 0 to 15, and i, j range from 0 through 3. The 4-dimensional metrics are assumed to be the outer product of the Lorentz metric

$$\eta_{16\mu\nu} \rightarrow \text{diag}(1, -1, -1, -1)\otimes\text{diag}(1, -1, -1, -1) \tag{16.31}$$

$$= (\eta_4)\otimes(\eta'_4)$$

using a matrix notation.

Postulate 16.6. The GL(4) group has a familiar coordinate space representation except that it is a complex space-time with a complex Lorentz group denoted L(C) whose elements Λ_C satisfy $\Lambda_C{}^T G_4 \Lambda_C = G_4$. L(C) is in turn broken to the real Lorentz group L of space-time that we physically observe.

Postulate 16.7. The GL'(4) group has a space-time that is reduced to a point. However it has a U(3) subgroup that gives the 3 known fermion generations, OR it may have 4 generations based on GL'(4)'s U(4) subgroup.

Postulate 17.1. The dynamical equations of the allowed types of free fermions can be generated by L(C) boosts from the Dirac equation for a fermion at rest subject to the requirement that there be only one real time variable in each resulting dynamical differential equation.

Postulate 17.2. Each free fermion field has the simplest possible lagrangian from which the canonically conjugate momentum, the energy-momentum tensor, the equal-time (or equal light-front) anti-commutation relations, and other physically important quantities can be derived.

Postulate 17.3. The Standard Model is a quantum field theory with a lagrangian and a path integral formulation. The mathematics of these constructs is assumed to be correct.

Postulate 17.4. The free field approximation to the Standard Model is a canonical quantum field theory but not necessarily quantized on light-like surfaces.

Postulate 18.1. A Dirac field and a tachyon field form a doublet whose dynamical equation is covariant under the 8×8 L(C) transformation eq. 18.5.

Postulate 18.2. A lepton doublet in any generation consists of a charged Dirac field and a neutral tachyon field (the neutrino). By convention the neutrino field is the topmost component in a lepton doublet. The equations of motion of the lepton doublet are eqs. 18.4 with $\Psi(x) = \Psi_{1_e}(x)$ (eq. 18.25).

Postulate 18.3. A quark doublet in any generation consists of an up-type quark Dirac field and a down-type quark tachyon field. By convention the up-type quark field is the topmost component in a quark doublet. (The quark fields may be "normal" or complexon – an experimental issue.)

Postulate 18.4. The form of the equations of motion of the unbroken leptonic sector of the Standard Model are determined by covariance under the L(C) group, and local gauge symmetries. Electron number is not gauged.

Postulate 18.5. In each generation leptonic particles consist of spin ½ Dirac particles (electron-like) with charge -1 and tachyons (neutrinos) of charge zero as well as their anti-particles.[278]

Postulate 18.6. A neutrino is a tachyon with a non-zero bare mass that may be changed by symmetry breaking effects.

Postulate 18.7. The interacting extensions of doublet equations of motion have interactions consisting of local Yang-Mills fields. Gauge fields are massless "before" spontaneous symmetry breaking and are thus conventional local Yang-Mills gauge fields without a tachyon equivalent.

Postulate 18.8. L(C) group covariance, and gauge symmetries of the dynamical equations of motion, are spontaneously broken through the appearance of additional fermion and vector boson mass terms generated by a mechanism such as the Higgs mechanism or the Dimensional Mass Mechanism (sections 21.3 and 21.4).

Postulate 18.9. The equations of motion of free quark doublets in the unbroken form of the ElectroWeak sector of the Standard Model are

---

[278] If neutrinos are Majorana particles then the derivation must be modified.

determined by covariance under the complex Lorentz group L(C), and local gauge symmetries. Baryon number is not gauged.

Postulate 19.1a. Quarks are in the 3 representation of Color SU(3). The SU(3) symmetry is gauged with local Yang-Mills SU(3) fields called gluons that constitute the Strong interaction of the quark sector. Quarks are minimally coupled to the gluons in a gauge covariant fashion.

Postulate 20.1. The four complex dimensions of GL'(4) symmetry are reduced to a point. However the fundamental 3 representation of its U(3) subgroup generates the three generations of fermions for each fermion species.

Postulate 21.1a. The masses, and mixing angles, of fermions result from the Higgs mechanism.

Postulate 21.2a. The masses of vector bosons arise from the Higgs Mechanism.

The number of postulates, and their complexity, far exceeds Euclid's formulation of geometry. However, the subject is also far more complex.

The author's view is that a smaller, more compact set of postulates will emerge from the set upon which this derivation is based. While the ability to derive the Standard Model is in itself a worthwhile pursuit because it informs us of the cohesion of the theory, it is also important in considering possible generalizations of the theory. Many theorists have proposed a plethora of generalizations (and variants) of the Standard Model. Extending the postulates of the basic Standard Model gives us a way of understanding the content of the proposed extension in relation to the basic Standard Model.

In addition, it is clear that the basic Standard Model has a strong geometric foundation in GL(16) or GL(4)⊗GL(4). Can one develop a similarly strong geometric basis for a proposed extension? If not then the probability of the validity of a proposed extension is not as high as it would be if were geometrically based in a way amiable with the geometric basis of the Standard Model.

Perhaps the most important reason to establish the Standard Model on the basis of fundamental postulates is that we thereby obtain a deeper level of understanding of the physics of the Standard Model. Working at this deeper level is in itself an advance in our understanding of the fundamental basis of physical reality – a deeper layer in the onion of fundamental physics.

22.2 The Standard Standard Model Lagrangian

The lagrangian density that we have derived for the conventional Standard Model with three generations is

$$\begin{aligned}\mathcal{L}_{SSM} = \Psi_L^{a\dagger}\gamma^0 i\gamma^\mu D_{L\mu}\Psi_L^{\ a} - \Psi_R^{a\dagger}\gamma^0 i\gamma^\mu D_{R\mu}\Psi_R^{\ a} + \\ + \Psi_{qL}^{a\dagger}\gamma^0 i\gamma^\mu D_{qL\mu}\Psi_{qL}^{\ a} + \Psi_{qR}^{a\dagger}\gamma^0 i\gamma^\mu D_{qR\mu}\Psi_{qR}^{\ a} - \\ - \mathcal{L}_{BareMasses} + \mathcal{L}_{Gauge} + \mathcal{L}_{Mass}\end{aligned} \tag{22.1}$$

where a is the generation index, and where

$$D_{L\mu} = \partial/\partial x^\mu + \tfrac{1}{2} i g_2 \boldsymbol{\sigma}\cdot\mathbf{W}_\mu - i g_1 B_\mu/2 \tag{22.2}$$

$$D_{R\mu} = \sigma_3\partial/\partial x^\mu + i g_1 B_\mu|Q| \tag{22.3}$$

$$D_{qL\mu} = D_\mu + \tfrac{1}{2} i g_2 \boldsymbol{\sigma}\cdot\mathbf{W}_\mu + i g_1 B_\mu/6 \tag{22.4}$$

$$D_{qR\mu} = \sigma_3 D_\mu + i g_1 B_\mu|Q| \tag{22.5}$$

where D_ν is given by

$$D_\nu = \partial/\partial x^\nu - i g A_\nu \tag{22.6}$$

where $\Psi_q^{\ \dagger}$ is the hermitean conjugate of the quark doublet with $\Psi_{qL}^{\ \dagger}$ and $\Psi_{qR}^{\ \dagger}$ being its left-handed and right-handed parts respectively, where D_ν is the color covariant derivative, and where

$$\mathcal{L}_{Gauge} = \mathcal{L}_{GaugeEW} + \mathcal{L}_{GaugeColor} \tag{22.7}$$

with

$$\mathcal{L}_{GaugeEW} = -\tfrac{1}{4}\, F_W{}^{a\mu\nu} F_W{}^{a}{}_{\mu\nu} - \tfrac{1}{4}\, F_B{}^{\mu\nu} F_{B\mu\nu} + \mathcal{L}_{EW}{}^{ghost} \quad (22.8)$$

and

$$\mathcal{L}_{GaugeColor} = -\tfrac{1}{4}\, F_{Color}{}^{a\mu\nu} F_{Color}{}^{a}{}_{\mu\nu} + \mathcal{L}_{Color}{}^{ghost} \quad (22.9)$$

$\mathcal{L}_{BareMasses}$ contains the fermion bare mass terms. The ElectroWeak gauge bosons $W_\mu{}^a$ and B_μ field tensors are:

$$F_W{}^{a}{}_{\mu\nu} = \partial W^a{}_\mu/\partial x^\nu - \partial W^a{}_\nu/\partial x^\mu + g_2 f^{abc} W^b{}_\mu W^c{}_\nu \quad (22.10)$$

$$F_{B\mu\nu} = \partial B_\mu/\partial x^\nu - \partial B_\nu/\partial x^\mu \quad (22.11)$$

$\mathcal{L}_{EW}{}^{ghost}$ are the Faddeev-Popov ghost terms for the ElectroWeak $W_\mu{}^a$ gauge bosons. The color gauge field tensors are

$$F_{Color}{}^{a}{}_{\mu\nu} = \partial A^a{}_\mu/\partial x^\nu - \partial A^a{}_\nu/\partial x^\mu + g_2 f_{su(3)}{}^{abc} A^b{}_\mu A^c{}_\nu \quad (22.12)$$

$\mathcal{L}_{Color}{}^{ghost}$ is the color SU(3) ghost terms defined using the Faddeev-Popov mechanism. The mass sector $\mathcal{L}_{Mass}$ (using the Higgs Mechanism or the Dimensional Mass Mechanism (sections 21.3 and 21.4)) gives masses to the ElectroWeak vector bosons and the fermions as well as specifying generational mixing.

The effective action for the path integral formulation is

$$I_{SSM} = \int dx^0 d^3x\, \mathcal{L}_{SSM} \quad (22.13)$$

Thus we have derived the *form* of the conventional Standard Model (modulo the mass terms which may be generated by the Higgs Mechanism or Dimensional Mass Generation). Chapter 22 describes the missing pieces in this derivation: coupling constant values, generation mixing, and the mass spectrum that are not derived.

It is a somewhat amazing fact that we can *precisely* derive the *form* of the Standard Model with the *constants* appearing in the theory *not determined* and thus, in a sense, *free*.

22.3 Renormalization

Since neutrinos and d-type quarks are tachyons in this theory the t'Hooft-Veltmann renormalization program should be reconsidered for this case if one wishes to use conventional renormalization.

However as pointed out in Blaha (2005a) and earlier books (as well as section 19.5) one can use the two-tier formalism to obtain a finite, renormalized theory in the present case. Consequently renormalization is not an issue.

23. The Complexon Standard Model

23.1 Postulates

If we assume postulate Postulate 19.1b is correct so that quarks and gluons *are* complexons then we obtain the Complexon Standard Model The postulates of the Complexon Standard Model are:

Postulate 15.1. Any observer can define a set of time and space coordinates called a coordinate system in which the observer is at rest. One can define a transformation that relates the coordinate systems of two observers traveling at a constant velocity with respect to each other.

Postulate 15.2. The speed of light, c, is the same in all coordinate systems.

Postulate 15.3. The invariant interval or distance $d\tau$ is defined by

$$d\tau^2 = g_{\mu\nu}dx^\mu dx^\nu \qquad (15.2)$$

It is invariant under a change of coordinate systems. The 16 quantities $g_{\mu\nu}$ are known as the metric tensor. The four quantities dx^μ are infinitesimal displacements in space and time.

Postulate 16.1. The universe has sixteen complex dimensions and is described by a curved, 16-dimensional complex extension of 4-dimensional General Relativity. Complex, 16-dimensional General Relativity is invariant under complex coordinate transformations of the form:

$$X^\mu = f^\mu(X'^0, X'^1, \ldots, X'^{15}) \qquad (16.1)$$

where X'^{μ} denotes an alternate complex coordinate system that is related to X^{μ} by continuous, holomorphic functions f^{μ}.

Postulate 16.2. We define an invariant interval with a metric $g_{\mu\nu}$ where

$$d\tau^2 = g_{\mu\nu}dX^{\mu}dX^{\nu} \tag{16.2}$$

Postulate 16.3. We assume a 16-dimensional Equivalence Principle: At any point one can transform to the coordinates of a locally flat space-time where

$$g'^{\mu\nu} = \eta_{16}{}^{\alpha\beta}\, \partial X'^{\mu}/\partial X^{\alpha}\; \partial X'^{\nu}/\partial X^{\beta} \tag{16.9a}$$

with

$$\eta_{16\mu\nu} = \text{diag}(1, -1, -1, -1, -1, 1, 1, 1, -1, 1, 1, 1, -1, 1, 1, 1) \equiv \eta_{16}{}^{\mu\nu} \tag{16.12}$$

Postulate 16.4. We assume that GL(16) is simply broken to GL(4)⊗GL'(4) through some mechanism and this group is the basis of the form of the Standard Model.

Postulate 16.5. After breakdown, a complex 16-dimensional vector can be represented as the outer product of two complex 4-dimensional vectors

$$x^{\mu} \rightarrow y^{i}y'^{j} \tag{16.30}$$

where μ ranges from 0 to 15, and i, j range from 0 through 3. The 4-dimensional metrics are assumed to be the outer product of the Lorentz metric

$$\eta_{16\mu\nu} \rightarrow \text{diag}(1, -1, -1, -1)\otimes\text{diag}(1, -1, -1, -1) = (\eta_4)\otimes(\eta'_4) \tag{16.31}$$

using a matrix notation.

Postulate 16.6. The GL(4) group has a familiar coordinate space representation except that it is a complex space-time with a complex Lorentz group denoted L(C) whose elements Λ_C satisfy $\Lambda_C^T G_4 \Lambda_C = G_4$. L(C) is in turn broken to the real Lorentz group L of space-time that we physically observe.

Postulate 16.7. The GL'(4) group has a space-time that is reduced to a point. However it has a U(3) subgroup that gives the 3 known fermion generations, OR it may have 4 generations based on GL'(4)'s U(4) subgroup.

Postulate 17.1. The dynamical equations of the allowed types of free fermions can be generated by L(C) boosts from the Dirac equation for a fermion at rest subject to the requirement that there be only one real time variable in each resulting dynamical differential equation.

Postulate 17.2. Each free fermion field has the simplest possible lagrangian from which the canonically conjugate momentum, the energy-momentum tensor, the equal-time (or equal light-front) anti-commutation relations, and other physically important quantities can be derived.

Postulate 17.3. The Standard Model is a quantum field theory with a lagrangian and a path integral formulation. The mathematics of these constructs is assumed to be correct.

Postulate 17.4. The free field approximation to the Standard Model is a canonical quantum field theory but not necessarily quantized on light-like surfaces.

Postulate 18.1. A Dirac field and a tachyon field form a doublet whose dynamical equation is covariant under the 8×8 L(C) transformation eq. 18.5.

Postulate 18.2. A lepton doublet in any generation consists of a charged Dirac field and a neutral tachyon field (the neutrino). By convention the neutrino field is the topmost component in a lepton doublet. The equations of motion of the lepton doublet are eqs. 18.4 with $\Psi(x) = \Psi_{1_e}(x)$ (eq. 18.25).

Postulate 18.3. A quark doublet in any generation consists of an up-type quark Dirac field and a down-type quark tachyon field. By convention the up-type quark field is the topmost component in a quark doublet. (The quark fields may be "normal" or complexon – an experimental issue.)

Postulate 18.4. The form of the equations of motion of the unbroken leptonic sector of the Standard Model are determined by covariance under the L(C) group, and local gauge symmetries. Electron number is not gauged.

Postulate 18.5. In each generation leptonic particles consist of spin ½ Dirac particles (electron-like) with charge -1 and tachyons (neutrinos) of charge zero as well as their anti-particles.[279]

Postulate 18.6. A neutrino is a tachyon with a non-zero bare mass that may be changed by symmetry breaking effects.

Postulate 18.7. The interacting extensions of doublet equations of motion have interactions consisting of local Yang-Mills fields. Gauge fields are massless "before" spontaneous symmetry breaking and are thus conventional local Yang-Mills gauge fields without a tachyon equivalent.

Postulate 18.8. L(C) group covariance, and gauge symmetries of the dynamical equations of motion, are spontaneously broken through the appearance of additional fermion and vector boson mass terms generated by a mechanism such as the Higgs mechanism or the Dimensional Mass Mechanism (sections 21.3 and 21.4).

Postulate 18.9. The equations of motion of free quark doublets in the unbroken form of the ElectroWeak sector of the Standard Model are

[279] If neutrinos are Majorana particles then the derivation must be modified.

determined by covariance under the complex Lorentz group L(C), and local gauge symmetries. Baryon number is not gauged.

Postulate 19.1b. Quarks are complexons.

Postulate 19.2. Quarks are in a 3 representation of the global SU(3) subgroup of GL(4). Their transformation law under SU(3) transformations is eq. 19.6 wherein the space-time coordinates are not transformed.

Postulate 19.3. The covariance of the quark equations of motion under global SU(3) transformations becomes covariance under local color SU(3) transformations when the equations of motion are generalized to gauge covariant derivatives. We assume the equations of motion are so generalized. The interaction terms so introduced constitute the Strong interaction.

Postulate 20.1. The four complex dimensions of GL'(4) symmetry are reduced to a point. However the fundamental 3 representation of its U(3) subgroup generates the three generations of fermions for each fermion species.

Alternative 1:
Postulate 21.1a. The masses, and mixing angles, of fermions result from the Higgs mechanism.

Postulate 21.2a. The masses of vector bosons arise from the Higgs Mechanism.

Alternative 2:
Postulate 21.1b. The masses, and mixing angles, of fermions appear as separation constants in fermion dynamical equations in the breakdown GL(16) → GL(4)⊗GL'(4).

Postulate 21.2b. The masses of vector bosons result from a non-Higgsian mechanism.

If the LHC accelerator (at CERN) does not find Higgs particles then alternative 2 becomes more probable in this author's view. One cannot chase rainbows forever.

23.2 The Complexon Standard Model Lagrangian

If we assume Postulates 19.1b, 19.2, and 19.3 then we obtain the Complexon Standard Model which differs from the "Standard Standard Model" by having complexon quarks and gluons. Which of these Standard Models is the correct one chosen by Nature? That is a question that only experiment can answer – perhaps deep inelastic electron-nucleon spin experiments.

In this section we show the results of the derivation of the three generation Complexon Standard Model: its lagrangian density and action.

$$\mathcal{L}_{CSM} = \Psi_L^{a\dagger}\gamma^0 i\gamma^\mu D_{L\mu}\Psi_L^{\ a} - \Psi_R^{a\dagger}\gamma^0 i\gamma^\mu D_{R\mu}\Psi_R^{\ a} +$$
$$+ \Psi_{CL}^{a\dagger}\gamma^0 i\gamma^\mu \mathfrak{D}_{qL\mu}\Psi_{CL}^{\ a} + \Psi_{CR}^{a\dagger}\gamma^0 i\gamma^\mu \mathfrak{D}_{qR\mu}\Psi_{CR}^{\ a} -$$
$$- \mathcal{L}_{BareMasses} + \mathcal{L}_{Gauge} + \mathcal{L}_{Mass} \tag{23.1}$$

where a is the generation index, using eq. 18.65 where $D_{L\mu}$ and $D_{R\mu}$ are given by eq. 19.17, and where the left-handed complexon quark covariant derivative

$$\mathfrak{D}_{qL\mu} = \mathfrak{D}_\mu + \tfrac{1}{2} i g_2 \boldsymbol{\sigma}\cdot\mathbf{W}_\mu + i g_1 B_\mu/6 \tag{23.2}$$

and the right-handed complexon quark covariant derivative is

$$\mathfrak{D}_{qR\mu} = \sigma_3\mathfrak{D}_\mu + i g_1 B_\mu|Q| \tag{23.3}$$

where $\mathfrak{D}_\mu$ is given by eq. 19.7.

$$D_{qR\mu} = D_{q\mu}\sigma_3 + i g_1 B_\mu|Q| \tag{23.4}$$

and where $\Psi_C^\dagger$ is given by eq. 18.57a, and $\Psi_{CL}^\dagger$ and $\Psi_{CR}^\dagger$ are its left-handed and right-handed parts respectively. $\mathcal{L}_{BareMasses}$ contains the fermion bare mass terms. Also,

$$\mathcal{L}_{Gauge} = \mathcal{L}_{GaugeEW} + \mathcal{L}_{GaugeC} \tag{23.5}$$

with

$$\mathcal{L}_{GaugeEW} = -\tfrac{1}{4} F_W{}^{a\mu\nu} F_W{}^a{}_{\mu\nu} - \tfrac{1}{4} F_B{}^{\mu\nu} F_{B\mu\nu} + \mathcal{L}_{EW}{}^{ghost} \tag{23.6}$$

and

$$\mathcal{L}_{GaugeC} = \mathcal{L}_{CCG} + \mathcal{L}_C{}^{ghost} + \mathcal{L}_{CC}{}^{ghost} \tag{23.7}$$

The ElectroWeak gauge bosons $W_\mu{}^a$ and B_μ field tensors are:

$$F_W{}^a{}_{\mu\nu} = \partial W^a{}_\mu/\partial x^\nu - \partial W^a{}_\nu/\partial x^\mu + g_2 f^{abc} W^b{}_\mu W^c{}_\nu \tag{23.8}$$

$$F_{B\mu\nu} = \partial B_\mu/\partial x^\nu - \partial B_\nu/\partial x^\mu \tag{23.9}$$

$\mathcal{L}_{EW}{}^{ghost}$ is the Faddeev-Popov ghost terms for the ElectroWeak $W_\mu{}^a$ gauge bosons. The complexon color gluon lagrangian $\mathcal{L}_{CCG}$ is defined by

$$\mathcal{L}_{CCG} = -\tfrac{1}{4} F_{CC}{}^{a\mu\nu}(x) F_{CC}{}^a{}_{\mu\nu}(x) \tag{23.10}$$

where

$$F_{CC}{}^a{}_{\mu\nu} = D_\nu A_C{}^a{}_\mu - D_\mu A_C{}^a{}_\nu + g f_{su(3)}{}^{abc} A_C{}^b{}_\mu A_C{}^c{}_\nu \tag{23.11}$$

where $A_C{}^a{}_\nu$ is the color gluon gauge field, and D_μ is defined by eq. 17.59, g is the color coupling constant, and the $f_{su(3)}{}^{abc}$ are the SU(3) structure constants.

In addition $\mathcal{L}_C{}^{ghost}$ is the color SU(3) Faddeev-Popov ghost terms defined in appendix 19A for the complexon Lorentz gauge and $\mathcal{L}_{CC}{}^{ghost}$ is the complexon color SU(3) constraint ghost terms defined through the Faddeev-Popov mechanism. The mass sector $\mathcal{L}_{Mass}$ based on the Higgs Mechanism or the Dimensional Mass Mechanism (sections 21.3 and

21.4) creates the fermion and ElectroWeak vector boson masses and generation mixing.

The coordinates are

$$x = (x^0, \mathbf{x_r} + i\mathbf{x_i}) \tag{23.12}$$

and the action is therefore

$$I_{CSM} = \int dx^0 d^3x_r d^3x_i \, \mathcal{L}_{CSM} \tag{23.13}$$

The lagrangian is supplemented with the following condition on all complexon fields $\Phi_{\ldots}$:

$$\nabla_r \cdot \nabla_i \Phi\ldots = 0 \tag{23.14}$$

Non-complexon fields $\Omega\ldots$ in the left-handed formulation under consideration satisfy the subsidiary condition:

$$[\nabla_r \cdot \nabla_i - (\nabla_r^2 \nabla_i^2)^{1/2}]\Omega\ldots = 0 \tag{23.15}$$

which guarantees a complexon's real momentum is parallel to its imaginary momentum.

23.3 Renormalization

Since quarks are complexons, since neutrinos and d-type quarks are tachyons, and since we have effectively a 7-dimensional lagrangian theory, the t'Hooft-Veltmann renormalization program is not applicable to the Complexon Standard Model.

However as pointed out in Blaha (2005a) and earlier books (as well as section 19.5 in some detail) one can use a Two-Tier formalism to obtain a finite, renormalized Complexon Standard Model thus eliminating renormalization as an issue.

24. Resolved and Unresolved Issues

24.1 Resolved Issues in the Standard Standard Model and the Complexon Standard Model

We have seen that the form of both types of the Standard Model is ultimately based on the "broken" geometry of 16-dimensional space-time. Thus we can derive the form of the Electroweak sector, parity violation, the Strong interaction – color SU(3), and the existence of fermion generations.

24.2 Unresolved Issues in the Standard Standard Model and the Complexon Standard Model

However there are several major issues that are not resolved by these derivations. The unresolved issues can be divided into conceptual issues and experimental issues. These issues overlap to some extent.

The conceptual issues center around

1. The breakdown mechanism(s) presumably from GL(16) to the Lorentz group in several conceptual stages, and to the generations with a residue of SU(3) (or SU(4) if there are four generations).

2. The values of coupling constants, mixing parameters, and particle masses. These values may be an artifact of the breakdown mechanism or of consistency conditions as discussed previously.

The significant experimental issues that fix the structure of the Standard Model are:

1. The nature of quarks: Are they complexons?
2. Are neutrinos and down-type quarks tachyons?
3. Is there a fourth generation of fermions, and does it mix with the three known generations.

4. Are there other subtle effects such as CP violation that give further insight into realistic possible extensions of the Standard Model.

Experimental issues are further addressed in chapter 25.

24.3 The Breakdown Mechanism

We have been constantly referring to the breakdown(s) of symmetries ultimately from GL(16). While a mechanism such as the Higgs Mechanism may account for these breakdowns, at least in part, we will not speculate as to the breakdown mechanism in the absence of decisive experimental evidence. The fact that the Higgs Mechanism offers a framework for breakdowns with the Standard Model is encouraging. However, the "proof of the pudding" is the experimental discovery of Higgs particles, which at the moment, is lacking.

The investigation of other possible methods of symmetry breaking such as the Dimensional Mass Mechanism (sections 21.3 and 21.4) should be pursued. It would be somewhat surprising if the Higgs Mechanism were the unique, one and only, non-trivial, symmetry breaking mechanism.

24.4 Determination of the Values of the Coupling Constants and Mass Matrices

While the derivation has determined the form of the possible Standard Model lagrangians the values of the Electroweak coupling constants and Weinberg angle, the strong interaction coupling constant, (and the gravitational coupling constant upon unification with quantum gravity), the masses of ElectroWeak vector bosons, and the mass matrices (and mixing angles) of the four fermion species remain to be determined.

There are two general methods to pin down the values of constants within the framework of Quantum Field Theory: 1) imposition of a symmetry group structure that relate the values of constants; and 2) the use of consistency conditions that set the values of some or all of the theory's constants. There are many model theories illustrating approach 1.

The only major 4-dimensional quantum field theory illustrating approach 2 is the Johnson-Baker-Willey model[280] of massless Quantum Electrodynamics which used the requirement of cancellation of all divergences as a consistency condition on QED. Johnson, Baker, and Willey showed that the consistency condition would be met if a certain function (called the eigenvalue condition) had a zero[281] at the physical value of the fine structure constant α, which is approximately 1/137.

This author[282] developed an approximate solution of the Schwinger-Dyson equations to calculate the eigenvalue function for α.

The Johnson-Baker-Willey model has been made obsolete by the success of ElectroWeak theory and thus does not itself appear to be of interest any longer. Nevertheless it illustrates the possibility of using consistency conditions to determine the value of constants appearing in a quantum field theory.

---

[280] M. Baker and K. Johnson, Phys. Rev. **D8**, 1110 (1973) and references therein.

[281] Adler later showed the zero, if it exists, must be an essential zero.

[282] S. Blaha, Phys. Rev. **D9**, 2246 (1974). Although this solution summed an infinite number of diagrams contributions – more than any other calculation in QED known to this author, and although the solution agreed with the exact calculated values to 4th order in α, an essential singularity was not found (although a complex analytic structure was found). The first zero was at α = 1. For this reason the Johnson-Baker-Willey program for creating a finite QED seems to have failed.

25. Experimental Tests

25.1 Experimental Tests to Nail Down the Form of the Standard Model

Two possible forms of the Standard Model have been derived based on two possible sets of postulates. The postulates embody several experimental tests. In this chapter we state the tests that would prove or disprove various postulates.

25.2 Tachyonic Neutrinos and Down-type Quarks

In both Standard Models, the Standard Standard Model and the Complexon Standard Model, fermions are assumed to appear in doublets. In both Standard Models each doublet is assumed to have a tachyon member. A major experimental test of this derivation is to determine whether neutrinos are tachyons and down-type quarks are tachyons. In the case of neutrinos the known smallness of the neutrino masses makes an experimental test difficult. In the case of down-type quarks, quark confinement makes the determination of whether they are tachyons indirect.

One possible area of experimental tests are parton models of nucleons in electron-nucleon scattering. Another experimental test area would be accelerator experiments that create quark-gluon plasmas such as Au-Au scattering experiments.

25.3 Quarks and Gluons: Complexons or "Normal"

In our study of the implications of our postulates we have found two natural forms of the Standard Model: the Standard Standard Model and the Complexon Standard Model. The difference between these theories is in the 3-momenta of quarks and gluons. Are their 3-momenta real? If so, the Standard Standard Model is the correct model. Or are their 3-momenta complex? If so, then the Complexon Standard Model is the correct model of elementary particles.

How can we detect the difference between complexon and "normal" quarks and gluons? Color confinement makes this determination difficult. And since the Strong Interaction does not have good, accurate, perturbative calculation predictions like QED the determination of the nature of quarks and gluons is that much more difficult. At the moment it appears that fits to emerging quark-gluon plasma data from Au-Au (and similar) experiments may offer a means of distinguishing complexon from non-complexon particles.

In addition the difficulties in the analysis of the spin-dependent structure functions of electron-nucleon scattering may be explainable by complexon quark constituents since complexons have a richer spin structure than normal spin ½ particles.

25.4 A 4th Generation of Fermions?

Section 20.3 raised the possibility of a fourth generation of fermions possibly with its own one generation Standard Model. If a fourth generation of fermions exist the overall pattern of masses of the three known generations suggest that the masses of the charged fourth generation fermions would be extremely large.

There are three possible scenarios for a fourth fermion generation:

1. The fourth generation mixes with the three known generations. In this case they would constitute an extension of the set of particles of the Standard Model and could possibly be found at high energies.
2. The fourth generation does not mix with the three known generations. In this case the fourth generation could constitute a separate one generation Standard Model, which would then would be a candidate for the origin of Dark Matter and Dark Energy. Experimentally fourth generation particles could not be created in accelerators from known particle interactions except possibly to an undetectable extent from gravitons.
3. The fourth generation does not mix with the three known generations but has the same interactions as the known three generations. In this case they would have to be very massive to have escaped detection as part of unusual baryons created in accelerator experiments.

Possibility 1 is currently the most favored based on preliminary experimental data. It appears Dark Matter does decay into normal matter.

25.5 Higgs Mechanism or Dimensional Mass Mechanism

If Higgs particles are not found, then the "God-particle" concept is dead and an alternate mechanism for mass generation will be required. The alternate mechanism can be as simple as inserting mass terms in the Standard Model lagrangian and thus deferring the solution of the mystery of mass to future. Or, in the case of fermion masses, the Dimensional Mass Mechanism described in section 21.3, which neatly combines the concepts of fermion generations, fermon masses, and mixing matrices between the generations. It appears to satisfactorily resolve the issue for fermions. In the case of ElectroWeak vector bosons, which do not appear in generations, but are merely four particles, the direct insertion of mass terms in the Standard Model lagrangian is completely acceptable if one uses Two-Tier quantization.[283] A fully renormalizable theory results.

The primary experimental question in this case is then do Higgs particles exist? Hopefully the question will be resolved by the new LHC (Linear Hadron Collider) that is just beginning operation (June, 2008).

If Higgs particles are not found then the Dimensional Mass Mechanism presents itself as an alternative. If there are four generations of fermions then, in this author's opinion, there would be strong presumptive evidence for GL'(4) and the Dimensional Mass Mechanism. A more decisive test is not apparent at the present time.

---

[283] The Two-Tier formalism has the added advantage of producing a fully renormalizable theory of quantum gravity. Thus a unified theory of the Standard Model and Quantum Gravity can be realized in a straightforward manner.

26. Summary of the Eight Possible Variations of the Derived Standard Model Due to Experimental Uncertainties

Due to experimental uncertainties with respect to the features of the Standard Model there are eight possible variations of the derived "basic" Standard Model. In addition there are certain known detailed features of the "full" Standard Model such as CP violation that remain to be understood. Lastly, there are undoubtedly additional, currently unknown, features of the Standard Model that will become apparent as experiments are done at much higher energies in the future.

The eight variants of the "basic" Standard Model that we have explored that are derivable from reasonable variations of our set of postulates are summarized in the table of Fig. 26.1.

Standard Model	Number of Generations	Masses: Higgs (H) Dimensional (D)	Complexon Quarks/Gluons
StdMod 1	4	D	Yes
StdMod 2	4	H	Yes
StdMod 3	4	D	No
StdMod 4	4	H	No
StdMod 5	3	D	Yes
StdMod 6	3	H	Yes
StdMod 7	3	D	No
StdMod 8	3	H	No

Figure 26.1. A Table of the possible variants of the "basic" Standard Model due to experimental uncertainties.

StdMod 1 is the Complexon Standard Model and StdMod 8 is the Standard Standard Model.

27. The Layer Below the Standard Model Lagrangian

At this point it is clear that we have uncovered a deeper level of particle physics that exactly implies the known form of the Standard Model with all its unusual features and "peculiarities." We have phrased the description of this deeper level in terms of sets of postulates. Undoubtedly the correct set of postulates remains to be specified due to experimental uncertainties detailed in the preceding chapters that are consistent with eight variations of the Standard Model.

But we have good reason to believe that the key symmetry and space that lies at the base of the Standard Model is GL(16) with sixteen complex dimensions. We further see that this symmetry is broken through an as yet unknown mechanism to GL(4)⊗GL'(4). The fundamental representation of these groups are 4-dimensional complex spaces.

The 4-dimensional, complex representation of the group that we denote as GL(4) is the source of the four species of fermions (two of which are tachyonic), parity violation, ElectroWeak doublets, the form of the ElectroWeak sector of the Standard Model, and the Strong interactions of quarks. This complex, 4-dimensional representation is broken to the 4-dimensional real space-time that we experience. Again the breakdown mechanism remains to be determined.

The 4-dimensional, complex representation of the group that we denote as GL'(4) is the source of the three or four generations of fermions, and, if Higgs particles are not found, the source of the fermion masses and mixing matrices of each of the four fermion species. This space collapses to a point through an unknown mechanism.

The thought process that evolved with this series of books has the form:

Lagrangian → postulate set → refined postulate set → ?...

The success of our approach in deriving the *exact* form of the Standard Model suggests that we are on the right track to a deeper level of physics. The major questions that currently confront this approach are: the source of the breakdown mechanisms, and the reason for values of the various constants in the theory: coupling constants, masses, and mixing matrices. It is somewhat amazing that we can derive the *form* of the Standard Model and yet have the various constants and masses completely "free." This situation opens the door to speculation on universes with a Standard Model of the same form but differing fundamental constants.

Another open question is whether more physics will be found at higher energies. The fact that our derivation follows so directly from geometry suggests that what we have found will remain true. What remains to be found will extend these results. The Standard Model now has a firm foundation in theory as well as experiment.

28. Beyond the Standard Model

28.1 Approaches to a Deeper Level of Reality

While we have shown that Operator Logic furnishes a basis for the development of the Standard Model of Particles, the additional assumptions – what we call the Knowledge Base – remain to be determined. It may be that, like Euclidean geometry, we can go no further then a set of postulates/axioms (assumptions). The number of explicit axioms needed to obtain the Standard Model in Blaha (2008) was twenty-three.[284] There were also a number of implicit assumptions in the derivation in Blaha (2008) just as there are a number of implicit assumptions in Euclid's'geometry which are only evident when figures are drawn in the process of proving theorems.

Irrespective of the question of the number and content of the axioms required to prove the Standard Model of Elementary Particles and thus establish the basis of Reality as we know it, the question that immediately arises is the source/reason for these particular axioms.

The possible sources for the axioms leading to the Standard Model are:

1. There is no reason. They are just the basis of Reality.
2. They follow from an unknown deeper unifying principle(s).
3. They follow from a deeper known theory such as Superstring Theory.
4. They follow from an unknown mechanism that establishes order in the form of the axioms from chaos. A mechanism or explanation for the persistence of such order over billions of years must also be found.
5. They follow because they are the only totally consistent set of physical axioms.

[284] The number of postulates in the Complexon Standard Model in this book is 29.

6. They are the result of chance and one of many possible sets of axioms.
7. They follow because they are required for life, as we know it, to exist. (The Anthropic Principle) This case is clearly a subcase of cases 1 and 6.

While case 1 is possible it is intellectually unsatisfying and so we will not pursue it. We will discuss case 2 in the following subsections. We will not discuss case 3 and refer the reader to the physics literature.

Case 4 seems unlikely to the author because of the secondary requirement that order persists indefinitely.

Case 5 brings us to Gödel's Consistency Theorem, which roughly states that a set of axioms cannot be proved to be consistent within the mathematical-deductive system that they define. Thus case 5 is unprovable.

Case 6 introduces an infinity of possible choices for the set of axioms for physics. The broadness of the set of choices makes this an unattractive possibility.

Case 7 is certainly true for humanity but recent studies have shown there is an extremely wide range of environments that can support life, and possibly, intelligent life. Thus, unless one posits religious reasons, it is not particularly clear that there is a compelling case for the Anthropic Principle. Case 7 then degenerates to case 6 – mere chance. And mere chance is not susceptible to intelligent discussion unless one can show that there is a "probability distribution" for sets of axioms and the set of axioms that governs our universe is amongst the most likely sets of axioms. Then case 7 becomes a subcase of case 2.

Therefore we find that case 2 is the most attractive case to study although case 3 (discussed elsewhere) is also of interest.

28.2 Extremum Rule for the Lagrangian of Physical Reality?

In Blaha (2005b) we developed a classification scheme for the "space" of all possible lagrangians and divided the space into subsets based on a definition of a Gödel number for lagrangians:[285]

[285] Blaha (2005b) pp. 151-154.

Definition: Suppose the Lagrangian L is an expression (string) of symbols, β_1, β_2, β_3, β_4, ... β_n (including field operator symbols, mathematical symbols, integral signs, the number of space dimensions, the number of time dimensions, space-time indices, and any other quantities that appear directly or indirectly in the Lagrangian); and v_1, v_2, v_3, v_4, ... v_n is the sequence of tokens corresponding to these symbols, then the Gödel number of a Lagrangian L is the integer

$$gn(L) = \text{Min} \prod_{m=1}^{n} Pr(m)^{v_m}$$

where $Pr(m)^{v_m}$ is the mth prime number raised to the power v_m with the 1st prime number taken to be 2; and where Min indicates the minimum is taken over all possible permutations of the order of symbols in the lagrangian that maintain its well-formed nature and all possible permutations of the assignment of tokens to the symbols in the lagrangian. The token numbers of the n symbols consist of a set of odd numbers ranging from 3 + p to 3 + p + 2(n – 1) where p is an even number greater than or equal to zero. (We set p = 0 for the sake of convenience.) *The set of token numbers consists of all odd integers* ≥ 3. A *well-formed lagrangian* has its symbols ordered in such a way that their order conforms to the syntax rules of the operators in the lagrangian. If L is an empty string (containing no symbols) then gn(L) = 1.

Since any number has a unique decomposition as a product of powers of prime numbers the following corollaries hold.

21. Corollary: If L_1 and L_2 are expressions such that $gn(L_1) = gn(L_2)$, then $L_1 = L_2$ (the Lagrangians are the same Lagrangian.)

22. Corollary: If the Gödel numbers of two Lagrangians are the same then they generate the same physical theory.

If we take an arbitrary lagrangian and generate all possible lagrangians from it through substitutions, definitions of new variables (and field operators if the lagrangian is for a field theory), and other Rules of Inference that are used to transform lagrangians to a new form, then the set of lagrangians so constructed constitutes a set of physically equivalent theories. (Certain additional changes are made in the case of theories with path integral formulations.) As stated, we denote these sets of physically equivalent theories as $LPset_i$ for i = 1, 2, ... These sets are subsets of the set of all lagrangians Lset. The use of physically equivalent lagrangians is exemplified by a Higgs particle lagrangian, which is transformed to a different form to show the effects of its constant ground state value.

Those subsets of Lset, $LPset_i$, that contain classical (deterministic) physical theories such as theories in classical mechanics require no special treatment. Goldstein (1965) describes various forms of transformations: Legendre transformations and other transformations between equivalent physical theories.

In the case of subsets of Lset, $LPset_i$, that contain quantum theories, changes of variables (fields) require an additional modification in the path integral formulation of the theories. The path integral integrand must have a Jacobian factor that reflects the change of variables in the lagrangian in order for the original lagrangian theory and the transformed lagrangian theory to have the same physical implications. For example if the path integral for a certain lagrangian in a certain subset of lagrangian theories is

$$Z(J) = N \int D\phi \exp\{i\int d^4y\, [\mathscr{L}(\phi) + J(y)\phi(y)]\}$$

where N is a normalization factor. Then a change of field variable from ϕ to ψ

$$\phi = \phi(\psi)$$

in the lagrangian necessitates the introduction of a Jacobian factor $\mathcal{J}(\psi)$ in the path integral integrand (except for linear changes of variable)

$$Z(J) = N \int D\psi\, \mathcal{J}(\psi)\exp\{\, i\int d^4y\, [\mathscr{L}(\psi) + J(y)\phi(\psi(y))]\}$$

See Huang (1965) or Weinberg (1998) for examples in the case of Faddeev-Popov gauge fixing.

Thus we have shown that it is possible to define subsets of theories for deterministic and quantum lagrangians where each lagrangian in the subset embodies the same physical theory.

We can define a Gödel number for a subset $LPset_k$ as:

23. Definition: The Gödel number of a subset $LPset_k$ is the minimum of the Gödel numbers of the members of the subset.

Due to the unique decomposition of a whole number into its prime factors we have the lemma:

24. Lemma: If the Gödel numbers of two subsets of Lpset are the same then they are the same subset.

In the next part we will define a criteria for selecting the "simplest" lagrangian in a given subset. This definition will be of some importance since potential Theories of Everything are judged by their simplicity as well as their physical implications.

The criteria of simplicity and elegance, which are much talked about in discussions of the Theory of Everything, are subjective, ambiguous, and anthropomorphic. The assignment of Gödel numbers to lagrangians enables us to create a mathematical definition of simplicity and, to some extent, of elegance since these criteria are usually subjectively related. A lagrangian's Gödel number is a measure of the number of symbols and the number of times each symbol appears in the lagrangian. Therefore lagrangians in fewer dimensions, with fewer species of particles or fields, with lower spin entities, and with fewer terms will have a smaller Gödel number. Based on this observation we define the simplicity of a lagrangian as:

25. Definition: For any two lagrangians L_1 and L_2 if $gn(L_1) < gn(L_2)$ then L_1 is simpler than L_2.

which implies the following lemma:

26. Lemma: The Gödel number of the subset $LPset_{TE}$ that contains the Theory of Everything is the Gödel number of the simplest lagrangian in the subset.

The preceding quote gives us a method of finding the Gödel number of the set of equivalent lagrangians that define the "Theory of Everything". However an extremum method for determining this set of lagrangians is lacking. Clearly it is not defined by the least Gödel number. And, in fact, there does not appear to be an obvious connection between Gödel numbers and lagrangians that would lead to an extremum method.

Leibniz, who invented calculus including extending extremal (minimax) methods,[286] suggested perhaps the most concrete concept for an extremum rule to determine the fundamental laws of nature: *the laws of nature are of maximal simplicity yet capable of producing a universe of maximal complexity.*[287]

---

[286] Nova Methodus pro Maximis et Minimis ("New Method for the Greatest and the Least") 1684.

[287] In Leibniz's words, "at the same time the simplest in hypotheses [i.e., its laws] and the richest in phenomena." Quoted in Rescher (1967) p.19.

The union of these two features was one of the motivations that led Leibniz to describe this universe as "the best of all possible worlds." The mathematical framework needed to express this elegant, but currently unquantifiable, extremum principle, or any extremum principle that leads to the laws of space, time and matter, remains to be developed. One can hope that someday we will reach that level of understanding if finding the "right" lagrangian with this approach is the correct approach. It seems difficult to see how this approach can determine the features of space-time and of internal symmetries and interactions.

If we do, then we will perhaps be able to create "universes of the mind" in which we can explore the implications of other designs for a universe. And perhaps we may be able in that latter age to experimentally create bubbles of those other possibilities for a universe in the laboratory.

Whether there may be other sister universes that realize these other designs is also a question of importance. Today physicists speculate about such universes but if the history of physics has taught us anything, it has taught us that very few speculations about future physical theory turn out to be correct.

28.3 Axiomatic Approach to the Knowledge Base of Reality

Blaha (2008), and his earlier books, developed an axiomatic approach that led *exactly* to the established form of the lagrangian of the Standard Model of Particles. In this approach an axiomatic method was used and the established parts of the Standard Model derived.

The success of this approach leads us to consider physics to ultimately be the result of a set of fundamental axioms that lead to the Standard Model lagrangian. It is possible to define Gödel numbers for a set of axioms in a manner similar to that of the preceding subsection for lagrangians. Blaha (2005b) defined Gödel numbers for sets of axioms:[288]

> 27. Definition: If A is a non-Lagrangian, axiomatic theory, and a_1, a_2, a_3, … , a_n are the *sequences* of symbols of the n (different) axioms of A, then the Gödel number of the theory can be defined to be

[288] Blaha (2005b) pp. 154-156.

$$gn(A) = \text{Min} \prod_{m=1}^{n} Pr(m)^{gn(a_m)}$$

where $Pr(m)^{gn(a_m)}$ is the mth prime number raised to the power $gn(a_m)$, where $gn(a_m)$ is the Gödel number of the axiom a_m and where Min indicates the minimum is taken over all possible permutations of the order of the axioms, all possible orderings of the symbols in the axioms that result in well-formed axioms, and all possible permutations of the assignment of tokens to the symbols in the production rules.

Thus if there are n axioms rules that use m symbols, then there at least n!m! Gödel numbers that correspond to the production rules (assuming the token numbers consist of the set of odd numbers ranging from 3 + k to 3 + k + 2(m – 1) where k is an even number greater than or equal to zero. (We set k = 0 for the sake of convenience.) A unique Gödel number for A is obtained by taking the prescribed minimum.

28. Theorem: If the Gödel numbers of two axiomatic theories are the same then they generate the same physical theory.

15.6.2 Subsets of NLset Containing Physically Equivalent Theories

Given a set of primitives and a set of axioms that define a physics theory, we can create a mathematical-deductive system that describes the physical theory in a manner similar to that used to define a mathematical-deductive system for a mathematics theory such as the theory of whole numbers.

Suppose we now take the primitive terms p_1, p_2, ... of the theory A and define a new set of primitive terms p_i' with

$$p_k = p_k(p')$$

and substitute the new primitives in the axioms creating a new theory A'. The new theory is physically equivalent to the original theory A but its mathematical-deductive system will look different.

Now consider the set of physical theories generated by all possible redefinitions of the primitive terms as above. This set of physical theories will be a subset of the set of all non-Lagrangian theories NLset. All theories in this subset are physically equivalent.

A moment's consideration leads to the realization that NLset is composed of subsets of equivalent axiomatic theories which we will denote $NLset_i$ for i = 1, 2, ... The elements of each subset have different Gödel numbers in general. Therefore we will define the Gödel number of a subset by:

29. Definition: The Gödel number of a subset $NLset_k$ is the minimum of the Gödel numbers of the members of the subset.

Due to the unique decomposition of a whole number into its prime factors we have the lemma:

30. Lemma: If the Gödel numbers of two NLset subsets are the same then they are the same subset.

In the next section we will define a criteria for selecting the "simplest" axiomatic theory in a given subset.

15.6.3 Simplicity Criteria for Subsets of NLset Containing Physically Equivalent Theories

A Gödel number is a measure of the number of symbols and the number of times each symbol appears in the set of axioms. Therefore sets of axioms for physics theories in fewer dimensions, with fewer species of particles or fields, with lower spin entities, and with fewer terms will have a smaller Gödel number. Based on this observation we define the simplicity of an axiomatic theory by:

31. Definition: For any two axiomatic theories A_1 and A_2 if $gn(A_1) < gn(A_2)$ then A_1 is simpler than A_2.

which implies the following lemma:

32. Lemma: The Gödel number of the subset $NLset_{TE}$ that contains the Theory of Everything, if it is axiomatic in nature, is the Gödel number of the simplest theory in the subset.

15.7. Escaping the Trap of Anthropomorphism

The form of Theories of Everything, and of physical theories in general, normally has anthropomorphic trappings: the choice of symbols, the form of the expressions, the characterization of the symmetries, and so on.

If we use the Gödel-number-based approach to defining lagrangian and non-lagrangian theories using the definitions of simplicity that we have developed we arrive at a non-anthropomorphic characterization of physical theories.

Thus we can hope that we can arrive at a universal characterization of the Theory of Everything that would pass this test: if we meet an intelligent alien scientist we could compare our theories of everything based on Gödel numbers! If the theories are the same

> their Gödel numbers should be equal. (We assume that they define tokens in the same way.)

As the passage shows one can define Gödel numbers for sets of axioms as well as for lagrangians. Unfortunately the same issue arises: there is no principle(s) that can be used to select the set of axioms actually used by Nature. Blaha's axioms are quite simple, relatively speaking, and lead directly to the Standard Model of Particles. It seems the best approach to progress in this area is to refine, and to reduce the number of, this set of axioms as well as to develop physical motivations for the axioms. If progress is made in this area then perhaps the Gödel number classification scheme may prove of interest.

29. Relativistic Quantum Metaphysics

"Beauty is truth, truth beauty;" —that is all
Ye know on earth, and all ye need to know.
***Poems**, Ode on a Grecian Urn – John Keats*

In section 13.6 we saw that a simple Operator Logic statement could be transformed into the Dirac equation upon the introduction of space-time and the use of a Lorentz boost. Later in section 15.11 and chapter 17 we saw that the use of complex Lorentz boosts could generate the four species of fermions[289] from an Operator Logic statement. From this beginning we were able to construct the form of The Standard Model and explicate its seemingly bizarre features.

Upon one fact physicists and mathematicians agree —logic is the only certain foundation for science and mathematics. Without logic neither is possible. On this solid core of truth we have constructed a fundamental physics of Reality based on a set of postulates that are contingent —not necessary; but consistent and simple compared to the universe that they lead to. Thus in seeking the basis of truth we have found beauty in The Standard Model of Elementary Particles. The key factor in this development was the extension of simple logic to Operator Logic and Quantum Operator Logic, and thence to fermion field equations.

29.1 Metaphysical Truths in The Standard Model

The metaphysical concepts and principles that have been developed in the course of the past two millenia have been based on the analysis of everyday phenomena. In the past hundred years we have witnessed an explosion of our knowledge of the physical universe. This new understanding of physical phenomena requires a new metaphysics of physical Reality based on our understanding of the implications of the

[289] Charged leptons, neutral leptons, up-type quarks, and down-type quarks.

Standard Model of Elementary Particles. Sections 12.6 – 12.9 outlined general aspects of the new view of physical Reality that a new metaphysics must come to grips with. The reader may find rereading those sections of interest.

29.2 "What Are" Questions

Concerning the nature of things we still are —and most likely will continue to be —at the level of Lucretius. For the nature of fundamental things are by their very nature unknowable. So the nature of time, space, matter, mass, energy, quantum numbers, and so on are unstable except by analogy, metaphor or vague words that in the end leave the question open. To some extent we can clarify our understanding of the most fundamental entities and properties of our universe. But an absolute understanding of the nature of any of them is not possible since it would be words about words. It is in the nature of things and, in a sense, a principle that

The nature of things must ultimately be expressible only in terms of undefined terms. Some of these undefined terms may have several possible meanings.

So with that all-encompassing limitation in mind we turn to a discussion of the entities and properties of ultimate physical Reality as we now know them at the beginning of the 21st century.

29.3 What We Have Learned of Metaphysical Import

Based on our current knowledge of infinitesimal, fundamental physical Reality we can specify the essential ingredients of a new metaphysics which we call Relativistic Quantum Metaphysics.

The ingredients combine to form a coherent view of fundamental physical Reality that differs significantly from the phenomena, and the metaphysics, of everyday experience. In short, the ingredients consist of space, time, particles of various kinds, the vacuum, and the forces between the particles.

29.3.1 Substance and Properties

The particles forming the "substances" are actually made of one substance that takes different forms. The vacuum can be viewed as that substance as well. Thus we find one substance at the most fundamental level of Reality.[290] All the varied substances encountered in everyday phenomena are combinations and aggregations of the fundamental particles and so essentially composites of the one substance in its various forms. The fermion elementary particles can be grouped in table by generations: Fig. 29.1 below.[291]

Leptons			Quarks	
Generation 1:	e	ν_e	$u_1\ u_2\ u_3$	$d_1\ d_2\ d_3$
Generation 2:	μ	ν_μ	$c_1\ c_2\ c_3$	$s_1\ s_2\ s_3$
Generation 3:	τ	ν_τ	$t_1\ t_2\ t_3$	$b_1\ b_2\ b_3$

Each type of quark has three varieties (colors is the descriptive word used by physicists) represented by subscripts 1, 2, 3. There are also twelve particles of spin 1 that embody the interactions (forces) between the fermions (and each other) – the ElectroWeak interactions (4 including the photon) and the Strong interactions (8 particles called gluons because they "glue" quarks together so that they cannot be individually seen as far as we know). The total number of particles is 36 (not counting the yet unseen graviton – the carrier of the force of gravity). From this small number of entities all of the phenomena of everyday life results. (We neglect discussing Dark Matter and Dark Energy since they are not part of The Standard Model and their features are largely unknown.)

The number of properties of the thirty-six fundamental particles is finite and we can list them:[292]

Mass
Energy
Spatial momentum
Spin
Lepton number
Baryon number

[290] See sections 12.8 for a detailed discussion of substance and being.

[291] The letters are the standard symbols for leptons and quarks. The import of the table is the simplicity of the set of quarks and leptons.

[292] There is some redundncy in this list. For example, electric charge is related to weak hypercharge and the third component of weak isospin.

Color SU(3)
Electric Charge
Weak hypercharge
Weak isospin
The Generation Number

The properties of space and time were discussed in section 12.7. We will discuss them further below. But we note that the properties of space and time were crucial in our derivation of The Standard Model.

Thus the typical metaphysical interest in the variety of properties (attributes, universals, bundles, and so on) at the level of everyday phenomena is a simple matter at the fundamental level that we have reached.

One can ask two metaphysical questions about these properties: What are each of them in Reality? Why are each of them required? The second question is answered in the course of our derivation of the Standard Model in previous chapters. The first question cannot be answered definitively at present. An answer would require a deeper knowledge of Reality then currently available. And there is a possible problem with answers based on a deeper view of Reality – the answers may be couched in abstract terms that may have multiple interpretations or be vague in some fashion.

29.3.2 Time and Space

Time and Space are usually considered separately in metaphysical studies although some studies do bring in their relation through the Special Theory of Relativity.

In the preceding derivation of the Standard Model it became clear that time and space – expecially through their connection through the complex Lorentz group are crucial to understanding the origin and form of The Standard Model. Time and space are inextricably intertwined although in everyday phenomena they appear to be separate.

A stunning example of the interrelation of time and space is the situation in the interior of a spherical black hole. Inside a black hole the time coordinate becomes effectively a spatial coordinate and the radius coordinate becomes effectively a time coordinate. An infalling particle has its distance from the center of the black hole constantly decreasing just as time always increases for us outside a black hole on earth.

And so we see the distinction between time and space, while clear for us in everyday phenomena, collapses in certain cosmological situations.

29.3.3 Causation

In subsection 3.2.3 we saw that causation and change took place through the interchange of particles at the fundamental level of Reality. The everyday experience of causation and change – varied as it is and sometimes seemingly paradoxical – is due to this conceptually simple general mechanism. In the case of matter in bulk many particles are interchanged. For example a magnet deflecting a stream of electrons exchanges many photons with it. The transfer of energy and momentum results in the deflection of the electron stream. We summarized the nature of causation and change in The Standard Model with the statement:

III. At the level of ultimate Reality, Causation and Change take place through interactions (forces) mediated (conveyed) by particles.

in subsection 3.2.3.

29.3.4 Origin of Physical Laws – The Knowledge Base

In section 13.4 we discussed the Knowledge Base of Reality. The Standard Model lagrangians found in the preceding chapters when implemented by a path integral formulation provides the complete Knowledge Base of Reality. The form of the lagrangians was shown to be ultimately constructed from Operator Logic with the addition of space and time and other relevant postulates. Thus we have obtained the Platonic Connection of Ideas to Reality.

30. Generalized Metaphysics

30.1 Goals of Generalized Metaphysics

In the preceding chapters we have considered the metaphysics of our universe which we have called Relativistic Quantum Metaphysics. This metaphysics embodied the deepest known physical information about our universe and established a metaphysical view based on it (particularly in sections 12.6 – 12.9 as well as chapter 29).

We now turn to universes in general and inquire into their physical Realities and their metaphysics. The word universe[293] is a bit vague when it is used in the physics and metaphysics literature. One reads of worm holes between universes, quantum tunneling between universes, and so on. These concepts and gedanken (thought) experiment discussions raise the question – if one can connect "universes" are they really separate universes or are they parts of the same universe with difficult paths between them? We therefore will define a *universe* to be specifically a completely self-contained entity totally unconnected with any other entity in any way whatsoever. Thus beings within a universe cannot communicate/contact/obtain knowledge of or even be aware of other possible universes either physically or conceptually. Beings within a universe can speculate on possible other universes and their features. These other universes are "universes of the mind" – assemblages of thought and logic created in imagination.

The metaphysical studies and conceptualizations of aspects of universes of the mind we will call *Generalized Metaphysics.* In this chapter we will consider some fundamentals of Generalized Metaphysics.

---

[293] Metaphysicians use the word "world" where physicists use the word "universe" generally speaking.

30.2 Logic is Pan-Universal

The various logics that we have considered and developed in the earlier chapters of this book would apply in any universe since they stand independent of physical details and are based on the rules of rational thought. There is one type of exception to this principle. One can conceive of a universe – perhaps empty or perhaps with contents – in which there are no physical laws and all behavior is capricious and chaotic. In this type of universe there are two possibilities: the laws of probability, statistics and chaos theory may apply (the *chaotic* class of universes)[294]; or no laws whatsover apply (the *lawless* class of universes). In the absence of physical laws logic is irrelevant for physical metaphysics.

Logic applies to all other types of universes.

30.3 All Nontrivial Postulates in all Universes are Contingent

Modality concerns the distinction between the necessary and the contingent. Properties and propositions that are necessary are present in all possible worlds. Properties and propositions that are contingent are present in some worlds and absent in other worlds. (See subsection 3.2.4 for more detail.) Modalists phrase their discussions in terms of worlds; we will use the word universes instead. The assumption of multiple worlds is required to meaningfully distinguish between the necessary and the contingent.

However in the case of properties and propositions of physical Reality it is not possible to examine the properties and propositions (laws) of other universes since by definition any other universe is completely separate from our universe and no communication or detection of the properties and features of other universes is possible. Thus the acid test of necessary/contingent of modalists is not possible for aspects of physical Reality.

Some counterexamples showing universes with no necessary properties are:

1. A point universe consisting of one point

---

[294] Universes in which quantum theory holds are not part of this class of universes since quantum theory is a physical theory with physical laws albeit probabilistic.

2. An empty universe containing nothing but having extension
3. A non-quantum universe
4. A non-relativistic universe with a non-Lorentzian space-time transformation group

We conclude that properties and postulates (propositions) are necessarily contingent. An examination of the postulates of The Standard Model listed earlier supports this view. Modality is physically irrelevant.

30.4 Towards a Generalized Metaphysics of Universes

It would seem that a Generalized Metaphysics would include Leibniz's seven principles listed in section 2.3. However the generality of these principles reduces their value to a minimum that almost amounts to triviality as far as contributing to an understanding of ultimate physical Reality.

One might also think of listing all possible properties (attributes) and bundles of properties that might be relevant for classes of universes. This exercise seems to be less than illuminating in grasping the range of ultimate Realities.

Considerations of Modality which superficially seems relevant to a consideration of "many worlds" or universes also seem to be less than relevant in the study of possible ultimate Realities for reasons given in the preceding section.

Thus one is left with the approaches broached in chapter 28. One of the primary difficulties of developing an ultimate Reality[295] is the consistency of postulates required to define an ultimate Reality. Are there deep connections between postulates that make certain combinations of postulates inconsistent?

Another area of study is the range of ultimate Realities subject to certain specified constraints? For example a constraint could be that a set of postulates for an ultimate Reality lead to a set of stable chemical elements, or, more interestingly, a set of chemical elements that would support the evolution of some form of life. Clearly, these questions would require a Generalized Metaphysics that is mathematically deep and not qualitative in nature (as most past metaphysics studies have been.) Also clear, is our current inability to develop the consequences of

---

[295] Presumably different universes will have different ultimate Realities in general.

a set of ultimate postulates. For example, given the postulates that lead to The Standard Model in this book, what mind or eye could see the chemical elements to which they lead, or to the combinations of chemical elements that led to life in this universe?

We conclude we are only at the beginning of our understanding of this universe's ultimate Reality, and not even at the beginning to understanding the ultimate Realities of other possible universes.

31. Logic, Language and the Universe

A reading of this book brings us to the conclusion that the ultimate Reality of our universe is expressible in words in the general sense of both verbal language and mathematics. We also conclude that the stuff of the universe is insubstantial[296] – form imposed on nothingness. So in the end we can only view ultimate Reality as language embodying logic. So it is appropriate to end our study with a brief word on language.

31.1 Logical Equivalence of Languages

All logic is expressed in human or symbolic languages. As a result there is an intimate connection between logic and language. Statements and deductive systems require a sufficiently robust language to express their content. A language usually must have the equivalent of predicates, subjects, connectives and quantifiers.

A question of some interest is the equivalence of languages. When are two languages equally capable of expressing the statements of a universe of discourse? Clearly they must have sets of equivalent terms although a term in one language might be a combination of terms in the other. However in some cases of human languages this type of equivalence is hard to achieve. A classic example is the Greek language in comparison to English. The Greek language has hundreds of words expressing various forms and nuances of love while English, in comparison, has few words for love. In other areas of the world some

---

[296] Evoking Shakespeare's prescient vision of Reality in *The Tempest*:

"The baseless fabric of this vision,
The cloud-capp'd towers, the gorgeous palaces,
The solemn temples, the great globe itself,
Yea, all which it inherit, shall dissolve,
And, like this insubstantial pageant faded,
Leave not a rack behind. We are such stuff
As dreams are made on …"

languages have a plethora of words for one aspect of nature, or another, which have no simple analogue in European languages.

One might think that the equivalence of languages is not of great importance. However, the growth of culture and science is directly tied to the growth in their terminology and the concepts that they embody. An example is the growth in the knowledge of quantum theory in the twentieth century, which introduced a host of new terms in physics. Thus the equivalence of languages reflects, to some degree, the equivalence of the range of universes of discourse (and their intellectual content) that the languages can support.

The requirements for two languages to be equivalent are:

1. Equivalent expressions in the two languages must have the same truth value.

2. Any expression in one language must have an expression with equivalent meaning in the other language.

3. The primitive terms of one language must be equivalent to the primitive terms in the other language, or to combinations of the primitive terms in the other language.

31.2 Operator Languages

Hitherto the languages that have been considered in studies of Logic have been human languages or symbolic languages that we will call c-number languages following the terminology of Quantum Theory. A c-number is a quantity, a number, character, or string of characters, that is not an operator. We have introduced the concept of a q-number language for logic in the form of Operator Logic and Quantum Operator Logic. Primitive terms are represented by operators in a Linear vector space. This q-number language is as valid as c-number languages to express statements.

Q-number languages have the added advantages of

1. Being able to project out undecidable statements and thus resolving the profound issues raised by the Gödel Undecidability Theorem. C-number formulations of logic are

an inadequate framework. Q-number logic is the proper framework for Logic.

2. Enabling the creation of subuniverses of discourses, direct products of universes of discourse, and direct sums of universes of discourse.

3. Furnishing a unifying framework for deterministic logic and quantum probabilistic logic that is guaranteed to be well-formed and consistent since it is based on Quantum Measurement Theory, which, embodying Reality, cannot be inconsistent or incomplete.

4. Providing matrix formulations of Operator Logic and Quantum Operator Logic.

5. Implementing the Platonic concept of Ideas connected mathematically to Reality.

31.3 Quantum Languages, Grammar, Turing Machines, Computers, and Computer Programs

This book develops Operator Logic and Quantum Operator Logic. Blaha (2005b) developed the concepts of Quantum Languages, Quantum Grammars, Quantum Turing Machines, Quantum Computers, and Quantum Computer Programs; and proved Gödel's Undecidability Theorem required the fundamental laws of Nature to be quantum.

These books complement each other by bringing the Quantum concept, which is undoubtedly the deepest knowledge that we have of Reality, to logic and language.

A reading of Blaha (2005b) displays a remarkably similarity in the concepts and mathematics of quantum languages with our present development of Quantum Operator Logic.

Together they give us a coherent weltanschauung of Thought and Reality. The major open question is the determination of the knowledge base of Reality. In this author's view this question will be resolved by an extension of our understanding of the formulation of the nature of space, time, and substance within the framework of Quantum Operator Logic and Quantum Language.

REFERENCES

Bjorken, J. D., Drell, S. D., 1965, *Relativistic Quantum Fields* (McGraw-Hill, New York, 1965).

Blaha, S., 2004, *Quantum Big Bang Cosmology: Complex Space-time General Relativity, Quantum Coordinates™ Dodecahedral Universe, Inflation, and New Spin 0, ½, 1 & 2 Tachyons & Imagyons* (Pingree-Hill Publishing, Auburn, NH, 2004).

Blaha, S., 2005a, Quantum Theory of the Third Kind: A New Type of Divergence-free Quantum Field Theory Supporting a Unified Standard Model of Elementary Particles and Quantum Gravity based on a New Method in the Calculus of Variations (Pingree-Hill Publishing, Auburn, NH, 2005).

Blaha, S., 2005b, *The Metatheory of Physics Theories, and the Theory of Everything as a Quantum Computer Language* (Pingree-Hill Publishing, Auburn, NH, 2005).

Blaha, S., 2005c, *The Equivalence of Elementary Particle Theories and Computer Languages: Quantum Computers, Turing Machines, Standard Model, Superstring Theory, and a Proof that Gödel's Theorem Implies Nature Must Be Quantum* (Pingree-Hill Publishing, Auburn, NH, 2005).

Blaha, S., 2006, *A Derivation of ElectroWeak Theory based on an Extension of Special Relativity; Black Hole Tachyons; & Tachyons of Any Spin.* (Pingree-Hill Publishing, Auburn, NH, 2006).

Blaha, S., 2007b, *The Origin of the Standard Model: The Genesis of Four Quark and Lepton Species, Parity Violation, the ElectroWeak Sector, Color SU(3), Three Visible Generations of Fermions, and One Generation of Dark Matter with Dark Energy* (Pingree-Hill Publishing, Auburn, NH, 2007).

Blaha, S., 2008, *A Complete Derivation of the Form of the Standard Model With a New Method to Generate Particle Masses Second Edition* (Pingree-Hill Publishing, Auburn, NH, 2008)

Blaha, S., 2009, *The Algebra of Thought & Reality: The Mathematical Basis for Plato's Theory of Ideas, and Reality Extended to Include A Priori Observers and Space-Time Second Edition* (Pingree-Hill Publishing, Auburn, NH, 2009).

Blaha, S., 2010a, *Operator Metaphysics: A New Metaphysics Based on a New Operator Logic and a New Quantum Operator Logic that Lead to a Mathematical Basis for Plato's Theory of Ideas and Reality* (Pingree-Hill Publishing, Auburn, NH, 2010).

Braithwaite, R. B., 1960, *Scientific Explanation* (Harper Torchbook, New York, 1960).

Carnap, R., 1956, *Meaning and Necessity* (Univ. Chicago Press, Chicago, 1956).

Carnap, R., (Ed. M. Gardner), 1995, A*n Introduction to the Philosophy of Science* (Dover Publications, New York, 1995).

Curry, H. B., 1976, *Foundations of Mathematical Logic* (Dover Publications, New York, 1976).

Davis, M., 1982, *Computability and Unsolvability* (Dover Publications, New York, 1982).

Dirac, P. A. M., 1931, *Quantum Mechanics* Third Edition (Oxford University Press, Oxford, 1947).

Frege, G., (Ed. M. Beaney), 1997, *The Frege Reader* (Blackwell Publishing, Malden, MA, 1997).

Garson, J. W., 2006, *Modal Logic for Philosophers* (Cambridge University Press, Cambridge, 2006).

Gödel, K., 1992, Tr. Meltzer, B., Introduction by R. B. Braithwaite, *On Formally Undecidable Propositions of Principia Mathematica and Related Systems* (Dover Publications, New York, 1992).

Gottfried, K., 1989, *Quantum Mechanics I Fundamentals* (Addison-Wesley, Reading, MA, 1989).

Hilbert, D. and Ackermann, W. (Tr. L. M. Hammond et al), 1950, *Principles of Mathematical Logic* (Chelsea Publishing Co., New York, 1950).

Kleene, S. C., 1967, *Mathematical Logic* (Dover Publications, New York, 1967).

Konyndyk, K., 1986, *Introductory Modal Logic* (University of Notre Dame Press, Notre Dame, Indiana, 1986).

Lavine, S., 1994, *Understanding the Infinite* (Harvard University Press, Cambridge, MA, 1994).

Loux, M. J., 2006, *Metaphysics: A Contemporary Introduction* (Routledge, New York, 2006).

Lowe, E. J., 2002, *A Survey of Metaphysics* (Oxford University Press, Oxford, 2002).

Mackey, G. W., 1963, Mathematical Foundations of Quantum Mechanics (W. A. Benjamin, New York, 1963).

Messiah, A., 1965, *Quantum Mechanics* Volume I (John Wiley & Sons, New York, 1965).

Potter, M., 2004, *Set Theory and its Philosophy* (Oxford University Press, Oxford, 2004).

Quine, W. van O., 1962, *Mathematical Logic* (Harper Torchbooks, New York, 1962).

Rescher, N., (1967), *The Philosophy of Leibniz* (Prentice-Hall, Englewood Cliffs, NJ, 1967).

Révész, G. E., 1983, *Introduction to Formal Languages* (Dover Publications, New York, 1983).

Smullyan, R. M., 1995, *First-Order Logic* (Dover Publications, New York, 1995).

Tarski, A., 1995, *Introduction to Logic and to the Methodology of Deductive Sciences* (Dover Publications, New York, 1995).

Tiles, M., 1989, *The Philosophy of Set Theory* (Dover Publications, New York, 1989).

Van Inwagen, P., 2009, *Metaphysics* (Westview Press, Boulder, CO, 2009).

Weinberg, S., 1995, *The Quantum Theory of Fields Volume I* (Cambridge University Press, New York, 1995).

Weyl, H., 1950, *Space, Time, Matter* (Dover, New York, 1950).

Weyl, H., (Tr. S. Pollard et al), 1987, *The Continuum* (Dover Publications, New York, 1987).

About the Author

Stephen Blaha is an internationally known physicist with extensive interests in Science, the Arts, and Technology. He received his Ph.D. in Theoretical Physics from Rockefeller University (NY). He has written a highly regarded book on physics, consciousness and philosophy – *Cosmos and Consciousness*, a book on Science and Religion entitled *The Reluctant Prophets*, a book applying physics concepts to the history of civilizations, and books on Java and C++ programming. He developed a mathematical theory of civilizations that is described in *The Life Cycle of Civilizations*. Recently he completed a major new study of Cosmology: *Quantum Big Bang Cosmology: Complex Space-time General Relativity, Quantum Coordinates, Dodecahedral Universe, Inflation, and New Spin 0, ½, 1 & 2 Tachyons & Imagyons.* He has served on the faculties of several major universities. He was an Associate of the Harvard Physics Faculty for twenty years (1983-2003). He was also a Member of the Technical Staff at Bell Laboratories, a member of management at the Boston Globe Newspaper, a Director at Wang Laboratories, President of Blaha Software Inc and Janus Associates Inc. (NH), and 2008 Program Chair of the International Society for the Comparative Study of Civilizations.

Among other achievements he was a co-discoverer of the "r potential" for heavy quark binding developing the first (and still the only demonstrable) non-abelian gauge theory with an "r" potential; first suggested the existence of topological structures in superfluid He-3; first proposed Yang-Mills theories would appear in condensed matter phenomena with non-scalar order parameters; first developed a grammar-based formalism for quantum computers and applied it to elementary particle theories; first developed a new form of quantum field theory without divergences (thus solving a major 60 year old problem that enabled a unified theory of the Standard Model and Quantum Gravity without divergences to be developed); first developed a formulation of complex General Relativity based on analytic continuation from real space-time; first developed a generalized non-homogeneous Robertson-Walker metric that enabled a quantum theory of the Big Bang to be developed without singularities at $t = 0$; first generalized Cauchy's theorem and Gauss' theorem to complex curved multi-dimensional spaces; first developed a physically acceptable theory of faster-than-light particles – tachyons – of any spin; first showed a universe with three complex spatial dimensions has an icosahedral symmetry; first developed the form of the composition of extrema in the Calculus of Variations; first quantitatively suggested that inflationary periods in the history of the universe were not needed; first proved Gödel's Theorem implies Nature must be quantum, first derived the form of the Standard Model, first showed how to resolve logical paradoxes including Gödel's Undecidability Theorem by developing Operator Logic and Quantum operator Logic, first developed a quantitative harmonic oscillator-like model of the life cycle, and interactions, of civilizations, and first developed an axiomatic derivation of the Standard Model.

Blaha was also a pioneer in the development of UNIX for financial and scientific applications, database benchmarking, and networking (1982); in the development of Desktop Publishing (1980's); and developed a hybrid shell programming technique (1982) that was a precursor to the PERL programming language. He received Honorable Mention in the Gravity Research Foundation Essay Competition in 1978, and was nominated for three "Awards for Technical Excellence" in 1987 by PC Magazine for PC software products that he designed and developed.

LaVergne, TN USA
21 February 2010

173724LV00003B/3/P

9 780981 904979